THE ROUGH GUIDE to Evolution

Mark Pallen

www.roughguides.com

Credits

The Rough Guide to Evolution

Editing: Joe Staines
Picture research: Mark Pallen & Joe Staines
Typesetting: Nikhil Agarwal
Diagrams and maps: Ed Wright
Proofreading: Adam Smith
Production: Rebecca Short

Rough Guides Reference

Editors: Peter Buckley, Tracy Hopkins, Sean Mahoney, Matthew Milton, Joe Staines, Ruth Tidball
Director: Andrew Lockett

Cover picture credits

Front cover: Chimp portrait © Renee Lynn/Corbis
Inside front cover: *Archaeopteryx* fossil © James L. Amos/Corbis
Back cover: Darwin portrait © Bettmann/Corbis

Publishing Information

Published January 2009 by
Rough Guides Ltd, 80 Strand, London WC2R 0RL
345 Hudson St, 4th Floor, New York 10014, USA
Email: mail@roughguides.com

Distributed by the Penguin Group:
Penguin Books Ltd, 80 Strand, London WC2R 0RL
Penguin Group (USA), 375 Hudson Street, NY 10014, USA
Penguin Group (Australia), 250 Camberwell Road, Camberwell, Victoria 3124, Australia
Penguin Group (Canada), 90 Eglinton Avenue East, Suite 700, Toronto, ON, M4P 2Y3
Penguin Group (New Zealand), Cnr Rosedale and Airborne Roads, Albany, Auckland, New Zealand

Printed and bound in Singapore by SNP Security Printing Pte Ltd

Typeset in DIN, Myriad and Minion

352 pages; includes index

A catalogue record for this book is available from the British Library

ISBN 13: 978-1-85828-946-5
ISBN 10: 1-85828-946-7

3 5 7 9 8 6 4 2

Contents

Introduction

A hundred and fifty years after the publication of *The Origin of Species*, Charles Darwin and his theory of evolution permeate our society. From cult TV shows like *The Simpsons* or *Heroes* to highbrow novels, such as Ian McEwan's *Saturday*, from Guinness adverts to the British £10 note, Darwin and evolution are everywhere.

Darwin's written work, perhaps uniquely among great scientists, is still widely read and remains accessible to the everyday reader. His heavy-browed avuncular face, complete with bushy beard, is instantly recognizable – no other British scientist has achieved such iconic status. Although Darwin now stands as the undisputed founder of modern biology, his influence is also felt in disciplines as diverse as computer science and cosmology, economics and earth science, philosophy and engineering. To quote Ernst Mayr: "Almost every component in modern man's belief system is somehow affected by Darwinian principles."

And yet, two centuries after his birth, most people still only have a hazy and distorted view of how evolution works. And for many, particularly in America, Darwin is more villain than hero. According to a poll in May 2007, nearly half of all Americans, and two-thirds of Republicans, do not believe in the theory of evolution. Creationism – the belief that species were created in more or less their modern form by a supernatural being – still dominates the agenda in school boardrooms across the US, despite repeated legal rulings banning it from American classrooms.

Why evolution matters

In an age in which astrology is more popular than astronomy, never has there been a more urgent need to defend science and rationalism. The evolutionary biologist Theodosius Dobzhansky once wrote: "Nothing in biology makes sense except in the light of evolution". In the era of the human genome, his words are more cogent than ever: the human genome sequence would be impossible to interpret without comparisons to similar sequences from organisms as diverse as yeast, worms, fruit flies and rodents. A genome biologist who does not accept evolution is as unthinkable as an airline pilot who believes the world is flat!

Evolution helps us to know our enemy, to understand – and fight against – changes in the biological world that threaten human health, wealth, happiness and even our very survival: from the emergence of pesticide-resistant head lice to the menace of multi-drug resistant tuberculosis and hospital "superbugs". Harnessing the creative power of evolution in the laboratory allows us to find new treatments for cancer or evolve new microbes able to clear up our own pollution.

In revealing and explaining the fascinating richness of life and its history, evolution also provides the ultimate account of creation, the greatest story ever told. An understanding of the dazzling diversity of life on Earth, and how life has shaped the very air we breathe and the climate we inherit, is central to our custodianship of the world. Evolution teaches that we are not separate from nature but an integral part of it, and an analysis of the evolutionary record brings us face to face with the terrifying reality of global climate change and mass extinction.

But evolution's most inspiring message comes from the realization that our species evolved in Africa no more than ten thousand generations ago. Just one small band of humans left the mother continent a mere two thousand generations ago to people the rest of the world: a breath-taking fact that reveals how all men are brothers and unites the sons of former slaves and the sons of former slave owners as the scatterlings of Africa.

How this book works

The Rough Guide to Evolution aims to present the educated lay reader with a readable and trustworthy introduction to evolution, while also providing a source of inspiration and information for the hardened Darwin fanatic. From this one wide-ranging volume, the reader will not only learn what evolution is and who Darwin was, but will also be exposed to the full impact of evolutionary ideas on modern thought and popular culture. The first part of the book, **Ideas and Evidence**, introduces the key concepts of evolution and outlines Darwin's life and works, documenting the evidence for evolution and providing a historical overview of contemporary evolutionary biology. The next section, **The Greatest Story Ever Told**, describes the rich history of life on Earth and of the evolution of our own human lineage. The third part, **Impact**, provides a unique survey of the influence of evolutionary thinking on science, culture and religion, and includes an account of creationist opposition to Darwin's big idea. Finally, **Resources** provides the reader with a wealth of opportunities to follow up on the exciting facts and ideas surveyed in this book.

Mark Pallen
Malvern, England, 2008

Acknowledgements

For inspiration, advice and encouragement in starting out on writing this book, I thank Carl Zimmer, Bob Henson and Robert Shepard. I am extremely grateful to all my colleagues at the University of Birmingham who read and commented on early drafts: Tony Beech (Psychology), Ian Fairchild (Geography), Nick Loman (Medical School), David Parker (Theology), Tim Reston (Earth Sciences), Peter Robinson (Theology), Paul Smith (Earth Sciences), Dov Stekel (Biosciences) and Susannah Thorpe (Biosciences). I count myself lucky to work in such a constructive and collaborative academic environment! I should also like to express gratitude to friends and fellow academics from outside my university who have read and commented on parts of the manuscript: Emma D, Lauri Lebo, Danny Levin, Ken Miller, Berni Munt, Ronald Numbers, Fred Spoor and John Van Wyhe. I owe a debt of gratitude to the tremendous efforts of the online community of pro-evolution/anti-creationism enthusiasts, whose work I have consulted in creating this book. In particular, I thank all those who have contributed to the talk.origins archive, especially Doug Theobald for his marshalling of the evidence for macroevolution. I thank Nick Matzke for his enthusiastic eyewitness account of the Dover trial. I am grateful to my colleagues from the "university of the train" for their support and patience: David Blackbourn, John Bryson, Julia and Martin Ellis, Steve Kukureka, Steve Minchin, Edward Peck, Martin Russell and Mike Ward. The opinions expressed herein are my own (as are any errors) and the content does not necessarily reflect the views of any of those who have commented on it.

I am extremely grateful to all my collaborators at Rough Guides: to Andrew Lockett for championing the initial proposal, to Ruth Tidball for helpful feedback on my initial efforts, to Ed Wright for the great diagrams, Nikhil Agarwal for typesetting and, most of all, to Joe Staines for all his hard work in editing my wordy prose and for his patience in tolerating a sometimes irascible and pedantic author. I thank my research group (Lewis, Lori, Nick, Rasha, Rob and Sophie) for putting up with a supervisor preoccupied with book writing and I hope my children (Emma, Charlie, Tasha and Tom) will forgive me for so many dull weekends sacrificed to the writing of this book.

Especial thanks go to my wife, Helen, for the numerous blog and web links she has e-mailed me, for the encouragement and support she has provided throughout the project and for her toleration of my obsessive interest in all things Darwinian which, I might point out, she helped fan when she bought me a Darwin biography the year before we married.

Finally, I am grateful to the writers that first inspired my interest in science and in evolution: Arthur C. Clarke, Isaac Asimov, Carl Sagan, Stephen Jay Gould and Richard Dawkins. I thank my parents for supporting my education. But most of all, I thank my teachers, to whom I dedicate this work: in particular, I thank my biology teachers James Mangham and Mary McPherson, my English teacher David Mulhearne and my history teachers, David Doerr and David Bury, for breathing life into the old maxim: a child is not a vessel to be filled but a torch to be set alight!

About the author

Mark Pallen is Professor of Microbial Genomics at the University of Birmingham, England. Having obtained his medical education from the University of Cambridge and the London Hospital Medical College, Mark completed his specialist training as a medical microbiologist at Bart's Hospital in London. In the mid-1990s, while completing a PhD in molecular bacteriology at Imperial College, London, Mark led a team of students to victory in the national quiz show *University Challenge*. In 1999, he took up a chair in microbiology at Queen's University Belfast, before moving to his current position in Birmingham in 2001.

Part 1
Ideas and evidence

The evolution of evolution

The idea of evolution did not spring fully formed from Darwin's mind but was the culmination of centuries of speculation about the origins of man and the natural world. Such questions fascinated the ancients, but the first substantial theory of evolution did not emerge until the work of Lamarck in the early nineteenth century.

The Ancient World

Although the biblical account of creation is often seen as diametrically opposed to the current view of biological evolution, it contains a hint of evolutionary progression in the idea that simpler life forms arrived on the scene before more complex ones – thus, in Genesis, chapter 1, plants are created before animals, marine organisms before terrestrial. Building on Genesis, **Basil of Caesarea** (329–379 AD) linked biological similarities to common ancestry: "Why do the waters give birth also to birds? Because there is a family link between the creatures that fly and those that swim. In the same way that fish cut the waters, using their fins to carry them forward … so we see birds float in the air by the help of their wings … their common derivation from the waters has made them of one family".

In contrast to divine intervention, some early Greek philosophers envisaged a mechanical, naturalistic view of creation. **Thales** (c.624–546 BC) suggested that water was the source of all things, while **Anaximenes** (c.585–525 BC) viewed air as the source of creation. **Empedocles** (c.490–430 BC), in his book *On Animals*, envisaged an almost comical, higgledy-piggledy creation of animals from random body parts, with only those able to provide for themselves managing to survive and reproduce: "...these things joined together as each might chance, and many other things besides them continually arose. Clumsy creatures with countless hands. Many creatures with faces and breasts looking in different directions were born; some, offspring of oxen with faces of people ... and creatures in whom the nature of women and men was mingled, furnished with sterile parts." These ideas, echoed by **Lucretius** (c.99–55 BC), provide a faint foreshadowing of Darwin's ideas of variation and natural selection. **Democritus** (460–370 BC) viewed the universe as made of atoms and claimed that everything was due to "chance and necessity", a phrase adapted in recent times by French geneticist, Jacques Monod, to describe the driving forces of biological evolution.

Among the classical Greek philosophers, **Plato** (427–347 BC) stands out as the anti-hero of evolution on account of his philosophy of **essentialism** – the view that anything we see in the real world is merely a reflection of an ideal form, characterized by a defining essence. Following Plato, biological species reflect real entities, with defined essences, while biological variation within species is dismissed as merely an annoying irrelevance. **Aristotle** (384–322 BC) has also had an ambivalent impact on evolutionary thought. He is rightly celebrated as the father of biological classification, with a hierarchical approach that culminated, two thousand years later, in the *Systema Naturae* of the Swedish botanist **Linnaeus** (1707–78). However, Aristotle believed that biological species were immutable and that the relationships between biological form and function were explained by **teleology**, i.e. as things designed for, or

Plato, whose philosophy of ideal forms ignored biological variation.

Al-Jahiz: an Islamic Darwin?

Al-Jahiz (Arabic for "the boggle-eyed") is the nickname given to the renowned Islamic scholar, **Abu Uthman Amr Ibn Bahr al-Kinani al-Fuqaimi al-Basri** (781–869). Al-Jahiz was born and died in Basra, but spent his most productive years in Baghdad, during the city's triumphant flowering under the Abbasid Caliphate. He was a key member of the **Mu'tazili** School of Islamic philosophy, (deeply influenced by Classical Greek and Hellenistic philosophy, but now deprecated within Islam). More than a thousand years before Darwin wrote *The Origin of Species*, Al-Jahiz wrote *Kitab al-Hayawan* (*Book of Animals*). Here, it is often claimed, Al-Jahiz outlined ideas that later appeared in Darwin's Theory of Evolution (for example, the struggle for existence). However, these claims have not been backed up by direct quotations from Al-Jahiz's work and remain hard to substantiate, given the lack of a complete English translation of *Kitab al-Hayawan*.

Evolutionary ideas surfaced in some Islamic philosophical texts after Al-Jahiz, for example, in the eclectic Neoplatonic work *The Encyclopaedia of the Brethren of Purity* (late 900s) and in the works of **Alhacen** (ibn al-Haytham; 965–1039) and **ibn Miskawayh** (932–1030). However, although thinkers from the Islamic golden age played a crucial role in reviving and enhancing the European intellectual tradition (as evidenced by Arabic words in English, such as "algebra" and "algorithm"), there is no evidence to suggest that Muslim evolutionary writings had any impact on Darwin's own thinking. A recent myth – that Darwin learnt Arabic from fellow Cambridge scholar Samuel Lee – is not supported by documentary evidence. In fact, the only available evidence suggests that the two men met just once, at a dinner in Cambridge!

directed towards, some final result. On this view, an organism will have eyes because it *wants* to see. The pairing of Plato's essentialism and Aristotle's teleology acted as a conceptual straightjacket, hindering the development of evolutionary thought for over two thousand years.

Aristotle's contemporary, Chinese philosopher **Zhuangzi** (c.370–301 BC) suggested that living organisms, including humans, had the power to adapt to their surroundings and underwent transformations from simpler to more complex forms. Evolutionary ideas next surface over a thousand years later, in the works of **Islamic scholars** (see box, above). A few centuries later in Italy, **Thomas Aquinas** (1225–74) argued that what reason tells us to be true about the world has to be taken seriously and cannot be overturned by appeal to sacred texts. In his view, it was acceptable for Christians to avoid a literalist interpretation of Genesis as "the manner and the order according to which creation took place concerns the faith only incidentally".

The Scientific Revolution and its aftermath

The culture and learning of the Renaissance was founded on the rediscovery of Europe's classical heritage; science was no exception. However, a decisive break with classical tradition occurred in the sixteenth century when Polish astronomer **Nicolaus Copernicus** (1473–1543) proposed that the Earth went around the sun, rather then vice versa. The next generation of astronomers, most notably **Galileo Galilei** (1564–1642), refined the Copernican model and sought empirical support for it. This brought them into conflict with the Catholic Church and led to Galileo being tried by the Inquisition and placed under house arrest. In post-Reformation England, where papal authority held no sway, **Francis Bacon** (1561–1626) formulated the scientific method by insisting that information about the natural world should be collected free of prior assumptions and that reasoning by induction should lead one from specific facts to general laws. In 1687, English scientist **Isaac Newton** (1643–1727) published his *Principia Mathematica*, in which he proposed a universal theory of **gravitation** and three laws of motion. Newton's conception of a universe based upon **natural laws**, amenable to rational understanding, fuelled the flowering of knowledge in the seventeenth and eighteenth centuries now known as the **Enlightenment**.

In the late eighteenth century, the Enlightenment ideals of reason and of mechanistic explanation filtered through to natural history, prompting the first clear articulations of biological evolution. French naturalist **George-Louis Leclerc, Comte de Buffon** (1707–78) surveyed all known life in his monumental *Histoire Naturelle*. Buffon pointed out that the large mammals of the tropical regions of the Old and New Worlds were quite different, so that climate alone could not account for the distribution of animals. He made the radical suggestion that species have dispersed from a centre of creation in the far North, degenerating as they entered the Americas. Buffon even speculated on the possibility of a common ancestry between humans and apes.

Baron Georges Cuvier (1769–1832) is generally credited as the founder of comparative anatomy, greatly extending the framework provided by Aristotle. At the end of the eighteenth century, Cuvier provided strong support for the idea of extinction, showing that fossil species, such as mammoths, mastodons and giant sloths, were clearly distinct from their living relatives. This led him to support the theory of catastrophism, the view that the majority of features of geology and of the history of life could be best explained by catastrophic events that had caused the mass extinctions.

Jean-Baptiste Lamarck (1744–1829), curator of the natural history museum in Paris, proposed the first substantial theory of evolution.

Lamarck lampooned

The biologist Conrad Waddington once lamented that "Lamarck is the only major figure in biology whose name has become, to all intents and purposes, a term of abuse". Quite unfairly, Lamarck is now best remembered in parody for "**Lamarckism**", a combination of two ideas which were widely held during Lamarck's time but which did not originate with him: the law of use and disuse, and the law of heritability of acquired characteristics ("soft inheritance"). According to the first of these laws, organisms, through a mysterious act of will, lose characteristics or organs they no longer require and develop and strengthen those that are useful; according to the second law, these changes are then inherited by the organism's offspring. Thus, the effort made by giraffes to reach leaves at the top of the tree leads to offspring with longer necks, or the effect of exercise on the muscles in a blacksmith's arm leads to sons with stronger, larger muscles. Crucially, many attempts to discredit "Lamarckism" fail to target anything Lamarck actually said: for example, the widely cited work of German biologist August Weismann, who repeatedly chopped off the tails of mice and showed that the offspring still grew them. However, Lamarck never claimed that the effects of injury or mutilation were ever inherited – he believed that acquired characteristics resulted not from the actions of external agents, but from an individual's own internal needs.

For Lamarck, reaching for higher leaves led to longer necks.

Despite the vilification he receives in most modern biology textbooks, Lamarck deserves credit for clearly articulating the idea of organic evolution and attempting to provide explanations for it, something that Darwin himself recognized. In fact, Lamarck has the last laugh as his is the only theory of evolution generally known to the public apart from Darwin's: Lamarckian evolution even merited a mention in an episode of *The Simpsons*!

Crucially, he advocated the principle that biological species, including humans, are descended from other species through a smooth continuous process of evolution (thus placing him at odds with Cuvier's catastrophism). In addition, Lamarck envisaged biological evolution as the result of natural laws, rather than miraculous intervention. However, Lamarck's outlook remained more metaphysical than scientific. He saw evolution as the result of two grand mystical forces: "le pouvoir de la vie" (the power of life), which drove organisms to become ever more complex and "l'influence des circonstances" (the influence of circumstances), an adaptive force powered by the interaction between organisms and their environment. Lamarck's ideas were highly influential and flourished well into the twentieth century.

Erasmus Darwin and the epic poetry of evolution

Erasmus Darwin (1732–1802) was the most exuberant figure of the English Enlightenment: England's finest physician, an inventor, a natural philosopher, a botanist and a poet. A man with an enormous appetite, gastronomic and sexual; so fat, that a semi-circle had to be cut from his writing table to accommodate his immense girth. Twice married, he fathered fourteen children, two of them illegitimate. Despite a stammer, he was a witty socialite and a razor-sharp rhetorician. Among his inventions were a talking machine, a canal lift, a steam car, an artesian well, a copying machine, a steering mechanism, a rocket engine and a horizontal windmill; among his predictions: war planes, the Sydney Harbour Bridge and the existence of pores in plant leaves.

Erasmus Darwin: physician, scientist, poet and evolutionist.

Erasmus Darwin wrote two long poems in heroic couplets, *The Loves of Plants* and *The Economy of Vegetation*, published together as *The Botanic Garden* (1791). In *The Loves of Plants*, he fancifully links sex, plants, evolutionary progress and people in an integrated biology. In *The Economy of Vegetation* he celebrates scientific ingenuity and radical politics: inventors, scientists, American rebels and French Revolutionaries are his heroes of a new age. Darwin's greatest scientific work was *Zoonomia* (1794–96),

Around the same time, evolutionary ideas were also circulating in the English-speaking world, particularly among members of the **Lunar Society**. This discussion club of prominent industrialists and intellectuals, centred on the English city of Birmingham, held meetings that coincided with the full moon, to ensure light for the journey home. **Erasmus Darwin**, grandfather of Charles Darwin and a leading light of the society, elaborated a theory of biological and cosmic evolution in poetry and prose (see box, below). **Benjamin Franklin**, who corresponded with members of the society, also played an important role in the development of evolutionary thinking. His description in 1751 of the principle of population pressure influenced English political economist **Thomas Malthus** (1766–1834), who half a century later published his own *Essay on the Principle*

in which he describes a system of pathology and a theory of biological evolution that included mutability of species and common descent: "all warm-blooded animals have arisen from one living filament, which THE FIRST GREAT CAUSE endowed with animality." Darwin elaborates on his evolutionary and revolutionary ideas in his crowning poetic achievement, *The Temple of Nature*, published posthumously in 1803. In this poem, he articulates his belief in evolution: "Hence without parent by spontaneous birth; Rise the first specks of animated earth… ORGANIC LIFE beneath the shoreless waves; Was born and nurs'd in Ocean's pearly caves."

"Sexual reproduction is the chef d'oeuvre, the masterpiece of nature."

Erasmus Darwin, from *Phytologia* (1800)

Opinions differ on the merits of Darwin's poems; Coleridge found them nauseating. But his writings were influential, shaping the work of the Romantic poets: "Dr Darwin" even appears in the opening line of Mary Shelley's *Frankenstein*. Love it or loathe it, Darwin's heady poetry still conveys a joyous optimism from the time when science was young and progress was assured:

"Shout round the globe, how reproduction strives;
With vanquished Death and Happiness survives;
How Life increasing peoples every clime,
And young renascent Nature conquers Time!"

Erasmus Darwin: A Life of Unequalled Achievement Desmond King-Hele (Giles de la Mare, 1999)

Cosmologia: A Sequence of Epic Poems in Three Parts Erasmus Darwin (Stuart Harris, 2002). The epic poems available in one volume.

of Population. Malthus predicted an apocalyptic scenario, in which the human population would, if unchecked, inevitably and catastrophically outrun food supply: "The power of population is so superior to the power of the earth to produce subsistence for man, that premature death must in some shape or other visit the human race." Malthus's vision of the mathematical inevitability of mass misery had a substantial influence on many later thinkers in economics, politics and biology, including Charles Darwin and Alfred Russel Wallace.

Replacing mythology with geology

A key thread in the progress of evolutionary thought was the discovery of **deep time** and the development of **geomorphology** (the study of landforms and the forces that shape them). Though **Shen Kuo** (1031–95) elaborated a theory of geomorphology in China as early as the eleventh

Siccar Point, Scotland: the variation in the dip between the upper rock layer and the near-vertical lower layer convinced Hutton of the vastness of geological time.

century, significant advances in the Western tradition only began in the late eighteenth century with the work of **James Hutton** (1726–97). Hutton's studies of Scottish rock formations led him to the belief, articulated in his *Theory of the Earth* (1795), that the processes that shape the Earth are in a constant balance, poised between the destructive influence of erosion and the offsetting creative mechanism of uplift. Importantly, Hutton was among the first to suggest that the Earth is much older than the biblical accounts allow. He also brought **gradualism** to geology, arguing that the geological processes we see in action today are the same forces that have gradually, but relentlessly, sculpted the Earth over long past eons, a view now known as **uniformitarianism** (in contrast to Cuvier's catastrophism). This outlook is often summarized by the phrase: "the present is the key to the past". **Sir Charles Lyell** (1797–1875), a contemporary of Charles Darwin, extended and popularized Hutton's uniformitarian views in his multi-volume *Principles of Geology* (1830–33).

"The result, therefore, of this physical inquiry is, that we find no vestige of a beginning, no prospect of an end."

James Hutton

From watches to watchmakers: natural theology

Attempts to explain the appearance of design, purpose or mechanism in nature by recourse to a cosmic designer stretch back to classical times: the Roman politician Cicero even made an analogy between the universe and a sundial as far back as 45 BC surmising that both operate by design rather than chance. In the following two millennia, numerous other authorities, including René Descartes, Robert Boyle, Robert Hooke, William Derham, and even sceptics such as Voltaire and Thomas Paine, proposed similar arguments. Turning the argument on its head, theological thinkers in the Islamic and Christian traditions, such as **Ibn Rushd**, known as Averroes (1126–98), and **Thomas Aquinas** (1225–74), attempted to prove the existence of God through observations of apparent design or purpose in the universe. Such **arguments from design** culminated in the book *Natural Theology*, published by English clergyman **William Paley** (1741–1805). Paley, in persuasive

How we know the Earth is old

In his *Natural Theology*, William Paley contrasted a watch (a designed object) with a stone (a natural object) – with the boring old pebble coming off worse from the comparison! However, as Terry Pratchett, Ian Stewart and Jack Cohen point out in *The Science of Discworld III: Darwin's Watch*, Paley was unfair to rocks, which are really clocks in disguise, clocks that have been measuring time for far longer than any man-made timepiece. Even by Darwin's time, geologists had uncovered the relative dates of rocks through **stratigraphy** (see p.158). However, during the twentieth century **radiometric methods** were devised that revealed the absolute ages of rocks. These techniques measure the ratio of a radioactive version of an element (a radioisotope) to one of its breakdown products.

Radiometric dating relies on the fact that radioactive decay occurs at a predictable pace: the amount of starting material halves over a set period of time, known as the half-life, which is distinctive for each isotope. Different radiometric methods work over different time spans, ranging from thousands to billions of years. These approaches assume that the mineral has remained a closed system since its formation, an assumption that can easily be tested by using multiple independent isotopic dating methods. For example, some of the oldest, well-studied rocks from Greenland and Canada have been dated at 3.65–3.85 billion years old by four different isotopic methods (uranium-lead; lead-lead; rubidium-strontium; samarium-neodymium). In addition, radiometric dates can be crosschecked with stratigraphic information to arrive at a consistent view of the Earth's history.

Several other methods can be used to confirm radiometric dates. The Earth's magnetic field has flipped over numerous times during the planet's history, the magnetic North Pole moving to the geographical South Pole and vice versa.

prose and tight argument, reiterates the **watchmaker argument**: that when we see a complex object like a watch, we assume the existence of a watchmaker, so when we observe the complexity of the universe and the elaborate contrivances of nature, we are compelled by analogy to a belief in a divine watchmaker.

Priming the prepared mind

By the 1830s, the shelves of a well-stocked university library would have contained works by half a dozen English thinkers who had already laid the foundations of what we now call the theory of evolution. Among them: Francis Bacon and Isaac Newton with their scientific method and natural laws; Erasmus Darwin, with his upbeat cosmic progression

As minerals often record the direction of the magnetic field at the time of their formation, this provides an independent source of information as to their age. This is particularly evident on the sea floor, where volcanic activity at a mid-ocean ridge produces new oceanic crust, which then gradually moves away from its place of origin. In such cases, a symmetrical pattern of magnetic stripes on the sea floor provide a historical barcode documenting **geomagnetic reversals** for as far back as 160 million years ago (Ma). Similarly, **volcanic hotspots** in the mantle periodically spawn new islands as the crust moves over them. The best example of this is the Hawaiian-Emperor volcanic chain, where there is perfect correlation between the distance of an island or sunken peak from the active hotspot and the age of its volcanic rock for at least 64 million years.

As one gets closer to the present day, a wide range of **incremental dating techniques** allow not just dating but even detailed reconstruction of past climates and environments. These methods rely on seasonal variations in biological, climatic or sedimentary processes that result in the formation of a chronological barcode. For example, analysis of annual tree rings (**dendrochronology**) provides chronologies going back over ten thousand years; **varves**, annual layers of sediment deposited at the bottom of lakes, also provide us with chronologies reaching back over thirteen thousand years, while ice cores removed from Greenland and Antarctic ice sheets reveal series of annual layers that stretch back hundreds of thousands of years. If one makes the assumption that what counts as evidence in recent historical times (one ring or layer per year) can be extrapolated backwards in time, these incremental dating methods alone are enough to discredit Biblical-literalist claims that the Earth is less than ten thousand years old.

and common origin of life; Thomas Malthus and his arithmetic of the apocalypse; William Paley and his powerful articulation of the problem of design in nature; and, finally, Charles Lyell with his emphasis on deep time and geological gradualism. But it fell to the mind of one man – standing at the boundary between conformity and dissent, primed by the adventure of a lifetime and face-to-face with fresh facts from faraway places – to weave the many disparate threads into the fabric of a grand unified theory of life. Cometh the hour, cometh the man: Charles Robert Darwin.

> **"Every indication of contrivance, every manifestation of design, which existed in the watch, exists in the works of nature…"**
>
> **William Paley, from *Natural Theology***

Natural selection before Darwin

In 1941, botanist and science historian Conway Zirkle collated descriptions of natural selection that pre-date Darwin, some strikingly prescient. French freethinker **Jean-Jacques Rousseau** (1712–78) states "children, bringing with them into the world the excellent constitution of their parents ... thus acquire all that strength and vigour of which the human frame is capable. Nature in this case treats them exactly as Sparta treated the children of her citizens: those of them who came well formed into the world, she renders strong and robust, and destroys all the rest." German philosopher **Johann Herder** (1744–1803) wrote: "Each strives with each ... Why acts Nature thus and why does she thus crowd her creatures one upon another? Because she would produce the greatest number and variety of living beings in the least space, so that one crushes another, and an equilibrium of powers can alone produce peace in the creation". Darwin's grandfather, **Erasmus Darwin**, poetically encapsulated the struggle for existence: "From Hunger's arms the shafts of Death are hurl'd; And one great Slaughter-house the warring world!"

Two early descriptions of natural selection have surfaced since Zirkle's paper. One occurs in a neglected work by geologist **James Hutton**: "in conceiving an indefinite variety among the individuals of that species, we must be assured, that, on the one hand, those which depart most from the best adapted constitution, will be most liable to perish, while, on the other hand, those organized bodies, which most approach to the best constitution for the present circumstances, will be best adapted to continue, in preserving themselves and multiplying the individuals of their race." The other stems from an unlikely source, **William Paley**: "every organized body which we see, are only so many out of the possible varieties ... which the lapse of infinite ages has brought into existence ... millions of other bodily forms and other species having perished, being by the defect of their constitution incapable of preservation, or of continuance by generation."

Darwin freely acknowledged the influence of earlier thinkers, even crediting two with priority in later editions of *The Origin*: **William Wells**, a physician who described natural selection in a paper on human evolution from 1818 and **Patrick Matthew**, a Scottish fruit grower who outlined the idea in his *On Naval Timber and Arboriculture* of 1831. Darwin's notebooks also reveal inspiration from a passage penned in 1809 by English agriculturalist John Sebright: "A severe winter, or a scarcity of food, by destroying the weak and the unhealthy, has had all the good effects of the most skilful selection."

Does this matter? Darwin freely acknowledged where he did and did not borrow ideas from his predecessors. To dismiss Darwin as a serial plagiarist is like damning Shakespeare as a second-rate playwright because he reused existing plotlines. As Darwin's son Frank put it: "In science the credit goes to the man who convinces the world, not to the man to whom the idea first occurs. Not the man who finds a grain of new and precious quality, but to him who sows it, reaps it, grinds it and feeds the world on it."

Darwin's life and works

At first glance, there is little in the early life of Charles Darwin to suggest that, fifty years after his birth, he would produce what philosopher Daniel Dennett has described as "the single best idea anyone ever had". But Darwin's intellectual radicalism was rooted in a family history of dissidence and high achievement.

Miseducation of a clergyman-naturalist

Charles Robert Darwin was born on 12 February, 1809 in **Shrewsbury**, a provincial English market town in the Welsh Marches (the borderland between England and Wales; had he been born nine miles further west, Darwin would have been a Welshman). As the fifth of six children born to **Robert** and **Susannah Darwin**, the young Charles Darwin was heir to a Protestant, Unitarian tradition of dissent that not only rejected the authority of the Pope, but also the divinity of Jesus. Although baptized as a baby into the Anglican communion, he was brought up outside the established Church of England. The boy Darwin was no angel, he made up tall stories and was often caught up in practical jokes; once, he beat a puppy just for the sense of power. He developed a keen interest in natural history and outdoor pursuits, which included long solitary walks.

Schooldays in Shrewsbury

Darwin's formal education began at the age of eight, when he joined a small grammar school run by the local Unitarian minister, **George Case**. His mother died a few months later (probably from peritonitis), leaving Darwin to the care of a sister, Caroline, nine years his senior. In September

The Darwins and the Wedgwoods

Charles Robert Darwin sits at the intersection of two prominent English families, the Darwins and the Wedgwoods. His paternal grandfather was **Erasmus Darwin** (1731–1802), physician-poet and proto-evolutionist. His maternal grandfather was **Josiah Wedgwood** (1730–95), an industrialist potter from Stoke-on-Trent. The first Josiah Wedgwood started a dynasty that included at least five more Josiah Wedgwoods. His great-great-grandson **Josiah Wedgwood IV**, later 1st Baron Wedgwood (1872–1943) was a war hero and radical Labour politician. Josiah Wedgwood V was managing director of the family firm from 1930–68. The first Josiah's daughter **Susannah** married Erasmus' son, physician **Robert**. Robert's shrewd investment of his own income and Darwin-Wedgwood inheritance provided the funds that would later subsidize Charles Darwin's life as a gentleman-naturalist; without the wealth from Wedgwood's pottery, there would have been no Darwinian theory of evolution.

Charles Darwin had four sisters (one of whom, **Caroline**, married Josiah Wedgwood III), and an elder brother, **Erasmus "Ras" Darwin** (1804–81). Charles married his cousin **Emma**, daughter of Josiah Wedgwood II and sister of Josiah III. Charles and Emma had ten children, but only three of them left descendants: **George**, an astronomer; **Frank**, a botanist and his father's biographer; and **Horace**, an engineer. **Leonard**, an army officer and liberal MP, married twice (the second time to a Wedgwood cousin) but had no children. Charles and Emma's grandchildren included **Erasmus Darwin**, killed in World War I and memorialized on the Menin Gate; **Gwen Raverat**, a distinguished artist and engraver; **Sir Charles Darwin**, a physicist who probed the nucleus with Rutherford and helped develop radar; **Bernard Darwin**, a golf writer; **Frances Cornford**, a poet; and **Nora Barlow**, a botanist. Great-grandchildren included two physiologists **Horace Barlow** and **Richard Keynes** (who taught the author); two artists **Christopher Cornford** and **Robin Darwin** (respectively Dean and Principal of the Royal College of Art); Robin's sister the potter **Ursula Mommens**, still active at 100 years old, and **John Cornford**, a poet who died fighting with the International Brigade in the Spanish Civil War. This generation saw a merger of the Darwin and Huxley lineages as a great-

1818, Darwin became a boarder, alongside his older brother Erasmus, at **Shrewsbury School**. In Darwin's day, **Samuel Butler** (grandfather of the novelist) ran the school with an emphasis on rote learning and the classics. Darwin detested it, and often ran the mile home to play. Towards the end of his school days, Charles and his brother rigged up a chemistry lab in a tool shed, which earned him the nickname "Gas". His performance at school gave no hint of his later genius, both his father and headmaster castigating him as a waster. His father proclaimed: "You care for nothing but shooting, dogs, and rat-catching, and you will be a disgrace to yourself and all your family."

granddaughter of Thomas Huxley (Angela Huxley) married a great-grandson of Charles Darwin (George Darwin, a computer pioneer). Fourth generation Darwin descendants include classical scholar, poet and journalist **Ruth Padel**; novelist **Emma Darwin**, whose *Mathematics of Love* was published to critical acclaim in 2006; **Carola Darwin**, chemist-turned-singer; **Sarah Darwin**, botanist; **Matthew Chapman**, writer and filmmaker; and **Randal Keynes**, writer and father of **Skandar Keynes**, the child-actor who plays Edmund in the *Chronicles of Narnia* films.

Other noteworthy Darwin relatives include English composer, **Ralph Vaughan Williams** (1872–1958), grandson of Charles's sister Caroline and Emma's brother Josiah, and two cousins of Charles: **Francis Galton** (1822–1911), founder of eugenics and **William Darwin Fox** (1805–88) amateur naturalist and life-long friend of Charles Darwin.

The Wedgwood/Darwin Dynasty Imelda Clift (Melrose, 2008)

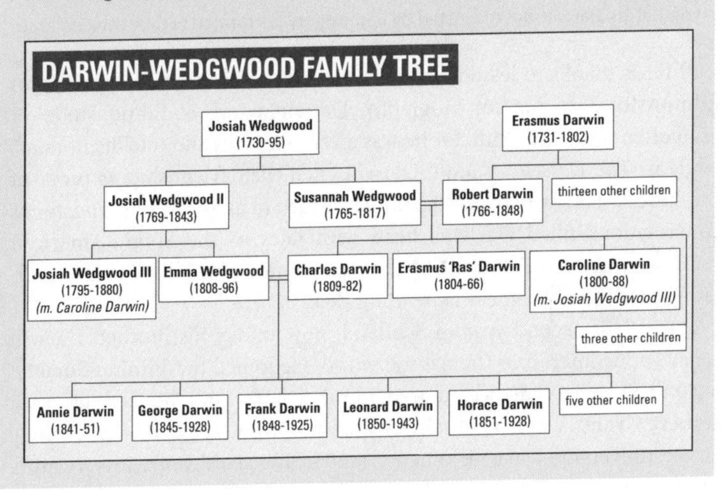

Savage surgery in Scotland

On leaving school at sixteen, Charles spent the summer as a medical assistant in his father's practice. In the autumn, he joined the **University of Edinburgh** in Scotland to study **medicine** alongside Erasmus. From the outset, neither brother was keen on the subject. In his first year, Charles was bored by his lectures, except for those on chemistry. He was disgusted by the sight of dissected corpses and botched surgical operations performed without anaesthetic. One highlight was learning how to

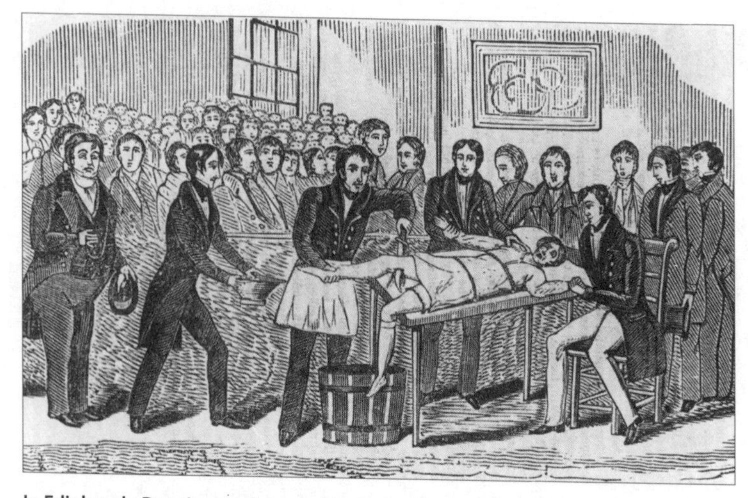

In Edinburgh, Darwin was disgusted by bad surgery performed without anaesthetic.

stuff birds thanks to lessons from a freed black slave from Guyana, **John Edmonstone**. In his autobiography, Darwin wrote of Edmonstone: "I used often to sit with him, for he was a very pleasant and intelligent man", while in *The Descent of Man*, Darwin cited their friendship as proof of the close similarity between the minds of men of all races. Edmondstone's conversations filled Darwin's head with tales of the South American rainforest. Around this time, Charles acquainted himself with his grandfather's evolutionary ideas by reading *Zoonomia*.

During his second year in Scotland, now minus his brother, Darwin renewed his interest in the natural world. He joined the **Plinian Society**, a student society dedicated to natural history, and hooked up with **Robert Grant**, a radical atheist evolutionist who taught marine biology at the university. Darwin joined Grant in his spare time, investigating the marine invertebrates on the rocky shoreline of the Firth of Forth (a river estuary near Edinburgh). Grant introduced Darwin to the concept of anatomical homology: the idea that deep similarities between body parts, for example in the bones of the forelimb, underlie superficial differences in form and function, for example between an arm, a wing or a flipper. Grant also took Darwin to a local natural history society run by Edinburgh geologist **Professor Robert Jameson**. Towards the end of his second academic year at Edinburgh, Darwin made his first scientific presentations, to the Plinian Society, on leech eggs in oyster shells and cilia on the larvae of sea mats.

Darwin and Lincoln: freedom evolves!

American freethinker and orator, Robert Ingersoll noted in the 1890s: "On the 12th of February, 1809, two babes were born – one in the woods of Kentucky, amid the hardships and poverty of pioneers; one in England, surrounded by wealth and culture. One was educated in the University of Nature, the other at Cambridge. One associated his name… with the emancipation of millions, with the salvation of the Republic. He is known to us as Abraham Lincoln. The other broke the chains of superstition and filled the world with intellectual light, and he is known as Charles Darwin." This concurrence of birth delighted evolutionary essayist Stephen Jay Gould, but spawned a long edit war on Wikipedia over whether it was notable enough for inclusion in the encyclopedia's entry on Darwin.

So, does anything else link the two men apart from their birthday? Both opposed slavery – Darwin with passion, Lincoln with action. Darwin supported Lincoln's war: "In the long run, a million horrid deaths would be amply repaid in the cause of humanity." Lincoln concurred with evolution (he read *Vestiges of the Natural History of Creation*) and opposed creationism. But it is unlikely that Lincoln ever read *The Origin* before his murder in 1865. Both men's rhetoric justified death: Lincoln with "these dead shall not have died in vain… this nation… shall have a new birth of freedom", Darwin with "Thus, from the war of nature, from famine and death, the most exalted object which we are capable of conceiving… directly follows". The two men were just a few handshakes apart – via abolitionist Moncure Daniel Conway and Lincoln's biographer James Russell Lowell who was a pallbearer at Darwin's funeral.

A hundred years to the day after Darwin and Lincoln's joint birthday, William du Bois and others established the National Association for the Advancement of Colored People. A hundred years after the Emancipation Declaration, Martin Luther King "let freedom ring" in Lincoln's "symbolic shadow". Ironically, a modern understanding of human evolution, which links former slaves and former slave owners together as the scatterlings of Africa, provides the strongest evidence that "All men are created equal". American history thus links Darwin and Lincoln in a simple dictum: freedom evolves!

Rebel Giants: The Revolutionary Lives of Abraham Lincoln & Charles Darwin David R. Contosta (Prometheus, 2008)

Cambridge University

After two sessions at Edinburgh, Darwin's father realized that his son was never going to make it as a physician, and encouraged him to study for the clergy. Having forgotten anything he ever knew about the classics, Darwin spent a few months with a private tutor in Shrewsbury before moving to **Christ's College, Cambridge** after the Christmas vacation, early in 1828. He spent three years there, but in his own words, the "time was wasted, as far as the academical studies were concerned". Darwin bunked off lectures. He fell in with a set of "dissipated low-minded men". He went hunting. He ate and drank to excess. However, Darwin's time at Cambridge was not entirely wasted. He acquired a taste for music (see box, below). He passed his exams, sometimes with good grades. He gained an intimate knowledge of Paley's *Natural Theology*, sharpening his intellect on the theologian's long lines of argument. Enraptured by the scientific travelogue of Prussian naturalist **Alexander von Humboldt**, he was inspired to travel.

Darwin developed an obsession with collecting beetles, which fueled his interest in natural history, and promoted some important friendships, for example with his second cousin **William Darwin Fox**. Through Fox, Darwin got to know the professor of botany, **John Stevens Henslow**, with

On Darwin's iPod

Although Darwin did once draw up his list of musical favourites, this has been lost to posterity. However, something of his musical tastes as student can be gleaned from his autobiography: " I acquired a strong taste for music, and used very often to time my walks so as to hear on weekdays the anthem in King's College Chapel. This gave me intense pleasure, so that my backbone would sometimes shiver." And from the reminiscences of his son Francis: "I never heard him hum more than one tune, the Welsh song 'Ar hyd y nos,' which he went through correctly; he used also, I believe, to hum a little Otaheitan song... He liked especially parts of Beethoven's symphonies, and bits of Handel... in June 1881, when Hans Richter paid a visit at Down, he was roused to strong enthusiasm by his magnificent performance on the piano. His niece Lady Farrer's singing of Sullivan's 'Will he come' was a never-failing enjoyment to him." So, as creative anachronism, here is a guess at what Darwin might have had on his iPod:

1 **AR HYD Y NOS** from ***We'll Keep A Welcome***
Darwin's favourite Welsh song, sung in a traditional arrangement by Bryn Terfel.

2 **ZADOK THE PRIEST by G.F. Handel** from ***Handel: Coronation Anthems***

3 **HALLELUJAH CHORUS by G.F. Handel** from ***Choral Favourites from King's College***
Two stirring pieces that Darwin could have heard in Cambridge, sung here by King's College Choir.

4 **HIMENE TARAVA TERO** from ***Coco's Temeava Vol.2***
Reminiscent of church singing Darwin heard on Tahiti, performed by Royal folkloric troupe, Coco's Temeava.

whom he took so many long walks discussing science and religion that he earned the nickname "the man who walks with Henslow". Through walks and country excursions, Darwin met other senior eminent men, including polymath and philosopher of science, **William Whewell**, who was a key proponent of Baconian inductive reasoning.

Darwin made good use of his summer and winter breaks away from Edinburgh and Cambridge. He made excursions to London and Birmingham. He travelled to North Wales, where he climbed Mount Snowdon. Shortly after leaving Edinburgh, he made his only trips to Ireland (visiting Belfast and Dublin) and to continental Europe, staying in Paris for several weeks. He spent much time at the family home of his uncle **Josiah Wedgwood II**, in Maer, Staffordshire and at the Owen family

5 **INITIATION SONG** from ***Bushfire: Traditional Aboriginal Music***
To recall the corroboree Darwin attended in Australia.

6 **DU MALHEUR AUGUSTE VICTIME by Antonio Sacchini** from ***Oedipe a Colone***
In a letter of 1829, Darwin enthused about a recital given by leading opera singers, which included an aria (from a different opera) by the now forgotten Sacchini.

7 **ODE TO JOY by Ludwig van Beethoven** from ***Symphony No.9 in D minor, Opus 125***
Even the tone-deaf Darwin would have been able to pick up the tune to the last movement of Beethoven's Choral Symphony!

8 **NOCTURNE NO. 8 IN D FLAT MAJOR by Frédéric Chopin** from ***The Romantic Generation/Charles Rosen***
Emma Darwin studied under Chopin, and might have played this to her husband. Charles Rosen, the pianist on this recording, studied with a pupil of a pupil of Chopin.

9 **TANNHAUSER OVERTURE by Richard Wagner (transcribed for piano by Franz Liszt)** from ***Siegfried-Idyll/Mikhail Rudy***
The great conductor and Wagner advocate Hans Richter may well have played this during his trip to Downe. This recording features Russian pianist Mikhail Rudy.

10 **WILL HE COME by Arthur Sullivan** from ***Clara Butt: The Acoustic Years***
A favourite song of Darwin's written by one half of what later became Gilbert and Sullivan, and sung here by a legendary Edwardian diva.

The young Charles Darwin riding a giant beetle; a cartoon drawn by fellow student Albert Way. At Cambridge, Darwin became obsessed with collecting these insects.

estate, Woodhouse, in the village of Rednal (thirteen miles northwest of Shrewsbury). From the summer of 1827, Darwin cultivated a relationship with **Fanny Owen**; he was devastated when she became engaged to local aristocrat Robert Myddelton Biddulph five years later.

Darwin took his final exams at Cambridge early in 1831, but was obliged by residence requirements to stay on at the university until June. From the April of that year, inspired by his reading of Humboldt, Darwin began to plan an ocean voyage to Tenerife. At first, he tried, without luck, to enlist Henslow, but then convinced a fellow Cambridge man, Marmaduke Ramsay, to travel with him. In preparation for the trip, Darwin brushed up on his knowledge of rocks, studying with Cambridge geologist **Adam Sedgwick** in North Wales during August 1831. Darwin's hopes for the Tenerife voyage were crushed when Ramsay died suddenly in July. But on returning to Shrewsbury from Wales in August, Darwin received a fateful letter from Henslow which was to exert a decisive effect on the course of his life and the history of science.

A geologizing Indiana Jones

In June 1831 **Robert FitzRoy** (see box, p.25) was re-appointed Captain of HMS *Beagle* and set about preparations for a second survey of South America. Worried about his own mental health during such a long voyage, FitzRoy asked his friend and sponsor Francis Beaufort to find a naturalist to provide him with companionship during the stresses of the voyage and to capitalize on the opportunities for scientific exploration. Darwin was not first choice: clergyman-naturalist Leonard Jenyns and botanist Henslow considered going, but only Darwin was free of domestic or professional obligations. He jumped at the chance, only for his father to object. After a tense few days and the felicitous intervention of Darwin's uncle, Josiah Wedgwood II, the old man relented. Darwin then had to face FitzRoy and convince the tetchy aristocrat that he was up to the job. At first, FitzRoy tried to put Darwin off, claiming that the position was already filled. Luckily, despite their political differences (Whig versus Tory) and FitzRoy's objections to Darwin's nose (the shape of which, he thought, revealed a weakness of character), the two men soon warmed to each other and Darwin was accepted on to the *Beagle*.

The voyage begins

After two months of delay and two false starts, Darwin and the *Beagle* set sail from Plymouth Sound the day after Boxing Day, 1831. The *Beagle* voyage that followed included island hopping through the Atlantic, surveys and explorations of **South America**, navigating across the Pacific (with trips to the Galápagos and Tahiti), visits to the new colonies in New Zealand and Australia, and then stops on the way home that included more islands, Cape Town and Brazil. Although often characterized as a sea voyage, Darwin in fact spent over three years on land. Threading through Darwin's voyage were the twin themes of adventure and discovery. Less than a year into his journey, Darwin was already comparing his escapades to the fantastical adventures of Baron Münchausen. To the modern reader, he is perhaps best envisaged as a geologizing Indiana Jones.

There was risk and privation. Afloat, the quarters were cramped, the seasickness unrelenting. On land and sea, Darwin faced storms. Three of his shipmates died of malaria; fellow crewman Edward Helyer drowned collecting specimens. Darwin himself was ill on several occasions, once seriously. The voyage brought Darwin face to face with war and rebellion. He saw Napoleon's tomb on St Helena. He visited the **Falkland Islands**

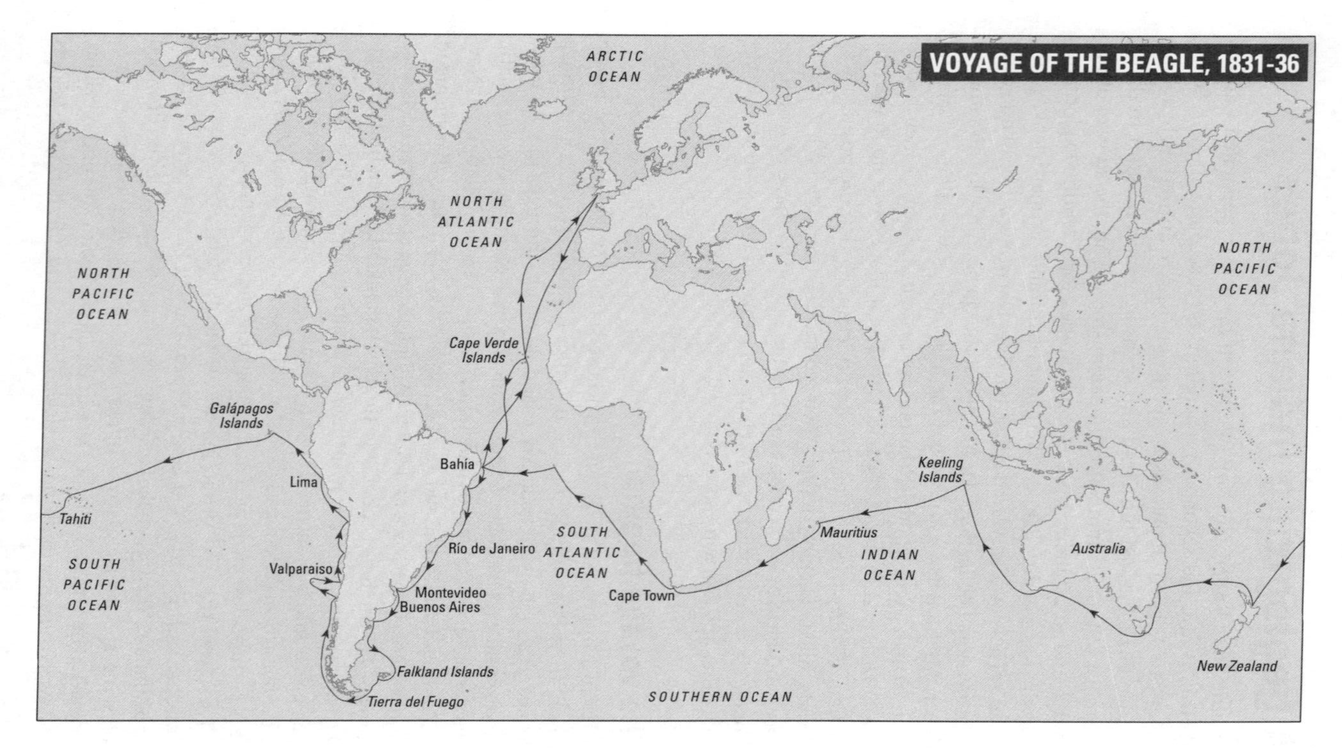
VOYAGE OF THE BEAGLE, 1831-36
ARCTIC OCEAN
NORTH ATLANTIC OCEAN
NORTH PACIFIC OCEAN
NORTH PACIFIC OCEAN
Cape Verde Islands
Galápagos Islands
Bahía
Lima
Tahiti
Río de Janeiro
SOUTH ATLANTIC OCEAN
Keeling Islands
Mauritius
INDIAN OCEAN
Australia
SOUTH PACIFIC OCEAN
Valparaiso
Montevideo
Buenos Aires
Cape Town
Falkland Islands
New Zealand
Tierra del Fuego
SOUTHERN OCEAN

Darwin's captain: Robert FitzRoy

The suicidal Captain FitzRoy.

Robert FitzRoy (1805–65) entered the Royal Navy at age thirteen and, after passing exams with full marks, moved quickly up the ranks. In 1828, he became temporary captain of the *Beagle*, returning the ship to England in October 1830. The following May, FitzRoy stood unsuccessfully as Tory candidate for Ipswich. A few weeks later, the *Beagle* and her captain were commissioned for a second South American Survey.

Fitzroy knew that he was prone to bouts of morbid depression and was haunted by two recent suicides. The first was that of his uncle **Viscount Castlereagh**, a brilliant but controversial politician who, as Foreign Secretary, had helped bring peace to Europe in the wake of the Napoleonic Wars. In 1822 Castlereagh fell victim to a real or imagined gay sex scandal, claiming to the king that he was being blackmailed. His mind unhinged, three days later he slit his own throat with a letter opener. The second incident occurred a few years later, during the *Beagle*'s first survey of the southern hemisphere under the command of **Captain Pringle Stokes**. In August 1828, during the gloomy southern winter, Stokes locked the door to his cabin, shot himself in the head, and then took an agonizing twelve days to die. Mindful of these dangerous precedents, FitzRoy took Darwin along as his gentleman companion and changed the course of history. Nonetheless, FitzRoy did succumb to despair part way through the second *Beagle* journey, resigning his captaincy for a short while before being persuaded to resume command.

But FitzRoy was more than just a bit player in Darwin's story. In the 1840s, he served as the Tory MP for Durham before serving disastrously as the second **Governor of New Zealand** – during his term, the colony almost became bankrupt and a new war broke out. However, FitzRoy is justly celebrated for his pioneering contributions to meteorology: he invented the **storm glass** (a device for predicting the weather), developed new and improved **barometers** and invented **weather forecasts** and gale warnings for fisherman. FitzRoy retired in 1863 with the rank of Vice-Admiral. But, at the age of sixty, the depression he had so feared aboard the *Beagle* finally caught up with him – one morning, FitzRoy got out of bed, went to his washroom and, echoing his uncle's demise, slit his own throat with a razor.

FitzRoy John and Mary Gribbin (Headline, 2003)

This Thing of Darkness Harry Thompson (Headline, 2005). A long and gripping novel about FitzRoy's life.

barely two months after their seizure by the British. In a feat of derring-do, Darwin escaped revolution in **Buenos Aires**. He helped suppress an uprising in **Montevideo**, but he was confined to ship during an insurrection in **Lima**. In the South American equivalent of the Wild West, Darwin witnessed the ruthless genocidal destruction of Native Americans. He met the Argentine dictator **Juan Manuel de Rosas** and rode and camped with the **gauchos** (the Hispanic equivalent of cowboys). He recoiled from the sight of drunken men belching blood and was disgusted by the bloody massacre of bullocks in Buenos Aires.

Darwin felt the full fire and fury of nature. He watched an Andean volcano erupt; he felt the earth move beneath his feet. At Concepción in **Chile**, he was fascinated by the devastation wrought by the shuddering

The voyages of the *Beagle*

Darwin's HMS *Beagle*, one of several ships to bear that name, was launched in May 1820 as a ten-gun brig. Later, in preparation for her use as a survey vessel, she was fitted with an extra mast (thus becoming a barque). On her first voyage from 1825 to 1830 she surveyed the southern regions of South America. Darwin's voyage, which began the following year, was the *Beagle's* second excursion. The ship made a third outing as a survey vessel in 1837 to 1843, mapping the coastline of Australia. During that trip, surveyor **John Lort Stokes** named an Australian harbour "Port Darwin" (now the City of Darwin) after his former shipmate. In 1845, the *Beagle* took on a new life as a **coastguard vessel**, guarding the Essex marshes from the River Roach, near Paglesham. She served as home to coastguards and their families for many years, but then disappeared from the historical record. However, recent investigations by Robert Prescott of the University of St Andrews, have confirmed that the *Beagle* survived intact until 1870, when parts of her timbers were sold to two local men, William Murray and Thomas Rainer. It is thought that the timbers were used to build a new farmhouse and boathouse. However, the hull probably just sank into the mire. In 2003, Prescott found traces matching the expected profile of the hull using an approach known as atomic dielectric resonance. Thus, it seems likely that all that remains of one of the most important vessels in history lies rotting below the Essex mud.

Beagle 2, named after Darwin's vessel by British scientist Colin Pillinger, was an ill-fated British spacecraft unfortunately lost on Mars on Christmas Day, 2003. Plans are now afoot for a successor, tentatively named **Beagle 2 Evolution**, to fly in 2009. That same year, Darwin's ship is scheduled to rise again, in the form of a £3.5 million replica of HMS *Beagle*, built in the Welsh port of Milford Haven. **The Beagle Project** (www.thebeagleproject.com) was initiated by David Lort-Phillips, a relative of John Lort Stokes, and Peter McGrath, a writer and yachtsman. The re-built *Beagle* will celebrate the Darwin bicentenary by earth and terrifying tsunami: "The earthquake must however, be to every one a most impressive event: the earth, considered from our earliest childhood as the type of solidity, has oscillated like a thin crust beneath our feet; and in seeing the most beautiful and laboured works of man in a moment overthrown, we feel the insignificance of his boasted power."

Afloat, beneath the southern stars, he watched the *Beagle*'s masthead and yardarm shine with St Elmo's fire, the ship afloat on a ghostly luminous sea. He saw waterspouts and precipitous blue-iced glaciers. In **Patagonia**, Darwin and his associates explored a desolate river valley, dragging their boats for a hundred and fifty miles against the raging stream. He climbed lofty mountains. He gloried in the exuberance of the tropical forests, the magnificent desolation of **Tierra del Fuego** and the boundless plains of

visiting sites of significance and acting as a venue for cutting-edge investigations of global biodiversity.

HMS Beagle The Story of Darwin's Ship Keith S. Thomson (Norton, 1995)

HMS Beagle Survey Ship Extraordinary Karl Heinz Marquardt (Conway, 2003). Includes plans for building a scale model of the *Beagle*.

HMS *Beagle* carrying Darwin through the Straits of Magellan.

Patagonia: "no one can stand in these solitudes unmoved, and not feel that there is more in man than the mere breath of his body". He surveyed lagoons lifted by coral and rode an elephant around **Mauritius**.

During his *Beagle* years, Darwin also encountered a rich variety of human life, both savage and civilized (to use his own "non-PC" terms). He delighted in the ladies of Buenos Aires ("the handsomest in the world") and of Lima ("like mermaids; could not keep eyes away from them"). He pitied the lot of the Chilean miner; he reveled in the colour and character of the Tahitians: "To see a white man bathing by the side of a Tahitian, was like comparing a plant bleached by the gardener's art, with one growing in the open fields". In **Tahiti**, Darwin even tasted the leaves of the local hallucinogen, kava, but avoided the psychoactive roots. In **New Zealand**, Darwin watched the Maori ceremony of nose pressing, while in **Australia**, he enjoyed Aborigines dancing and singing the corroberee.

Two facets of the human condition enflamed Darwin more than any other. Firstly, he abhorred **slavery** and found it impossible to reconcile the civility of a gentleman with the inhumanity of the slave-owner. Darwin quarreled with FitzRoy over this issue. In 1833, a few weeks before the Slavery Abolition Act was passed in Britain, Darwin proclaimed: "What a proud thing for England, if she is the first European nation which utterly abolishes it. I was told before leaving England, that after living in Slave countries all my opinions would be altered; the only alteration I am aware of is forming a much higher estimate of the Negro character". In 1836, on leaving Brazil for the last time Darwin exclaimed: "I thank God I shall never again visit a slave-country. To this day, if I hear a scream, it recalls with painful vividness my feelings, when… some poor slave was being tortured."

El'leparu (or York Minster) a Fuegian abducted by FitzRoy on an earlier voyage.

The second source of despair for Darwin was the bleakness of life for the inhabitants of **Tierra del Fuego**. The crude savagery that Darwin encountered at the tip of South America contrasted markedly with the civilized veneer acquired by the three Fuegians (nick-named York Minster, Fuegia Basket and Jemmy Button) who had traveled to England on the *Beagle*'s first voyage and returned to their homeland with Darwin. Pity mixed with abhorrence and disbelief: "a woman, who was suckling a recently-born child, came one day alongside the vessel, and remained there whilst the sleet fell and thawed on her naked bosom, and on the skin of her naked child…Viewing such men, one can hardly make oneself believe they are fellow-creatures, and inhabitants of the same world… What is there for imagination to picture, for reason to compare, for judgment to decide upon? To knock a limpet from the rock does not even require cunning, that lowest power of the mind." Darwin's gloomy evaluation was reinforced when the three "civilized" Fuegians reverted to "savagery" after resettlement in the land of their birth. Nonetheless, he saw a common thread of humanity, viewing the Fuegians not as separate species but potential ancestors: "One's mind hurries back over past centuries, and then asks, could our progenitors have been such as these?"

Darwin's *Beagle* voyage was more than just an adventure. He transported home an astonishing collection of thousands of botanical, zoological and geological **specimens**. But importantly, the *Beagle* voyage also fired the continuing mis-education of a creative free-thinker. Equipped with Lyell's books, Darwin geologized relentlessly. Everywhere he found proof of the principle that the present is the key to the past. In Chile, he saw first hand the changes in sea level produced by an earthquake; elsewhere he concluded that the land must have been lifted up after finding shells and marine fossils well inland, or even high up in the mountains. He observed erratic boulders transported through ancient glaciation, then came face to face with contemporary glaciers. He studied coral atolls, sensing how their gradual growth kept pace with the sinking rock beneath. He saw fossils of massive extinct mammals mixed with the remains of lineages still alive today.

During his time on the *Beagle*, Darwin was intrigued by the curiosities of **biogeography** – what different kinds of organisms were found where and how they got there – and reading his notes, one sees the themes emerging that would soon drive him towards evolutionary explanations. He puzzled over the curious spatial **compartmentalization** of the biological world: the two zones of life in the Americas (North versus South); the peculiar rodents (the capybara, the tuco-tuco, the agouti) found only in South America; a fox found only on the Falkland Islands; mice living on remote Chilean islands. In Australia he was mesmerized by the Alice-in-Wonderland

Darwin's heroes

Darwin hailed the Prussian naturalist and explorer **Alexander von Humboldt** (1769–1859) as "the greatest travelling scientist who ever lived". Three decades before Darwin's *Beagle* voyage, Humboldt explored Latin America, providing the first scientific descriptions of the region's natural history. During his adventures, Humboldt was met by meteor showers and electric eels; he scaled mountains and explored the Amazon and Orinoco river systems, laying the foundations of physical geography, biogeography and meteorology. Humboldt's multi-volume travelogue *Personal Narrative of Travels to the Equinoctial Regions of America* enthralled early-nineteenth-century readers, including Darwin. At his height, as naturalist, internationalist and diplomat, Humboldt was heralded across the world as the greatest European of his age. In his late seventies, he produced a five-volume masterpiece *Kosmos*, in which he attempted to unify all branches of scientific knowledge. Humboldt died in Berlin, where he received the full pomp of a state funeral. He is memorialized in the names of numerous species, towns, parks, geographical features and academic institutions in the New World.

Another of Darwin's heroes, **Charles Lyell** (1797–1875) was born in Scotland, but spent much of his childhood in England's New Forest. On graduating from Exeter College, Oxford, Lyell began professional life as a lawyer, before eyestrain led him to an alternative career in geology. In his most influential book, *Principles of Geology*, first published in three volumes in the early 1830s, Lyell forcefully defends the principle of **geological gradualism** or **uniformitarianism**, interpreting geological transformations as the accumulation of numerous small changes, of a kind still in operation, applied over vast spans of time.

FitzRoy gave Darwin the first of Lyell's volumes just before they set off together on the *Beagle*. On first landfall, in the Cape Verde islands, Darwin was thrilled to interpret local rock formations in Lyell's terms. He received the second volume in South America and became a firm disciple, incorporating Lyell's thinking into his own theories about the formation of coral atolls. On Darwin's return from the *Beagle* voyage, he and Lyell became close, lifelong friends. Lyell's support for the antiquity of geological time twinned with an emphasis on the influence of slow gradual changes had a profound influence on Darwin's ideas on evolution. Lyell is buried near to Darwin in Westminster Abbey.

weirdness of marsupials and of the duck-billed platypus: "lying on a sunny bank… reflecting on the strange character of the animals of this country as compared with the rest of the world… [a]n unbeliever in every thing beyond his own reason might exclaim: two distinct Creators must have been at work…" He saw how plants and animals imported by humans into regions far from their origins flourished in their new homes – cardoon and fennel on the pampas; many imported plants and animals on the island of St Helena – proving that climate alone could not explain the distribution of life. Encountering beetles, butterflies and grasshoppers on the open sea, he witnessed a mechanism for the natural dispersal of organisms en route from one land mass to another.

The Galápagos archipelago: a world within itself

In the popular imagination, one place on Darwin's voyage looms larger than anywhere else: the **Galápagos Islands**. This group of thirteen volcanic islands straddles the equator, 600 miles west of the South American mainland. The islands were annexed and settled by Ecuador in 1832. Darwin and the *Beagle* arrived on 15 September, 1835 and stayed for just over a month. He visited just four of the islands. Darwin quickly noticed the remarkable natural history of the archipelago: "it seems to be a little world within itself; the greater number of its inhabitants, both vegetable and animal, being found nowhere else", but he also noted that even the animals specific to the islands generally resembled their equivalents in the Americas. Black lava hillocks reminded him of the iron foundries of Staffordshire, while the "antediluvian" **giant tortoises** delighted him. Faced with such an abundance of tortoises and iguanas, he remarked that nowhere else in the world did reptiles replace mammals in so extraordinary a manner. He was told that one could tell which island a tortoise came from by its appearance, but was misled into believing that the tortoises had been introduced to the islands by buccaneers.

A giant tortoise, the largest native inhabitant of the Galápagos.

Contrary to popular belief, there was no eureka moment for Darwin in the Galápagos – while there, it never occurred to him "that the productions of islands only a few miles apart, and placed under the same physical conditions, would be dissimilar". His notes reveal that it was only months later, in the summer of 1836 during his journey home, that the Galápagos mockingbirds raised his first suspicions as to the fixity of species: "When I see these Islands in sight of each other… tenanted by these birds, but slightly differing in structure & filling the same place in nature, I must suspect they are only varieties… If there is the slightest foundation for these remarks the zoology of Archipelagoes will be well worth examining; for such facts would undermine the stability of Species."

Darwin returned to England at Falmouth, seventy miles from the point from which he had departed four years, nine months, five days earlier. The traveller arrived back in Shrewsbury late on 4 October, 1836 and surprised his family by materializing for breakfast the following morning. His father's first comment was on how the shape of Darwin's head was quite changed. After ten days in his old Shrewsbury home, Darwin left for a new life as a gentleman naturalist, with a reputation established by his specimen collections and accounts of adventures in far-off lands.

Conformity and creativity

In the months that followed his return from the *Beagle*, Darwin holed up in Cambridge, but made forays to London, Shrewsbury and Maer. At the end of October he made his final rendezvous with HMS *Beagle*, offloading crates at the Woolwich dockyard. He toured museums, finding homes for his collections. He spoke at learned societies. He made new acquaintances and refreshed old ones. In a fit of mutual admiration, Darwin and Lyell became firm friends. A few days later, while hosting a tea party, Lyell introduced Darwin to **Richard Owen**, rising star in the field of comparative anatomy.

London: the doubts start

At the start of March 1837, a few weeks after his twenty-eighth birthday, Darwin moved to London where, for a time, his life split in two. On the outside, he was a productive conformist, rising in scientific eminence and respectability. He wined and dined with the intellectual luminaries of the day: Charles Babbage (inventor of a prototypical computer), poet-historian Thomas Macaulay, astronomer John Herschel, and essayist Thomas Carlyle.

He even met his idol, Humboldt. In his first seven years back in England Darwin wrote fifteen papers, and within a decade had written up his account of the *Beagle* voyage, and published books on the geology of coral atolls, of volcanic islands and of South America. In 1838, he became vice-president of the Entomological Society and developed a lifelong interest in earthworms. Meanwhile, Darwin's finances were supported by investments from his father and the respectable young naturalist even secured a £1000 grant from the treasury to cover the costs of editing a massive five-volume *Zoology of the Voyage of HMS Beagle.*

Yet all the while, there was another side to Darwin: the mis-educated creative non-conformist, primed for radical reasoning. This Darwin read widely and dangerously: David Hume, whose *Dialogues concerning Natural Religion* banished the beneficent designer-deity, Adam Smith, whose "invisible hand" brought economic prosperity without planning (see p.223) and John Locke, who, in viewing the infant as "blank slate", saw the human mind shaped by sensation and reflection. In poetry, his tastes shifted from Milton to Wordsworth and Coleridge.

Darwin's intellectual landslide began just after his move to London, set in motion by the interpretation of his own specimens by others. Owen's work on Darwin's fossil mammals emphasized the geographical and temporal continuity of life by showing how extinct South American mammals were most closely related to the extant inhabitants of the same continent. Zoologist Thomas Bell established that the giant tortoises belonged to a species unique to the Galápagos. Ornithologist John Gould showed that finches from the Galápagos formed a distinct taxonomic group that included a dozen species previously unknown to science. Each species of mockingbird appeared to be unique to its own island, but it wasn't clear whether the same held true for the finches as Darwin's specimens were poorly labelled. Gould also showed that the two kinds of rhea (a flightless bird) that Darwin had collected from different parts of the South American mainland were distinct species.

The notebooks

Darwin quickly made the link between change over space and change over time. In his **Red Notebook** (a notebook held over from his *Beagle* days) he argues that the relationship between two contemporary species of rhea is the same as "extinct guanaco [a wild llama] to recent: in former case, position, in latter, time". A few lines later, Darwin speculates about one species turning into another and on a later page, he notes that the two species of rhea are "certainly different – not insensible change – yet one is urged to look to common parent?" In July 1837, Darwin began his **B notebook**, the first of four in which he recorded his private (and scandalous) thoughts on the **mutability of species**. In homage to his grandfather, he starts with the title *Zoonomia*. Page upon page of speculation follows on the tendencies of organisms to vary in response to a changing world and on the potential for **common descent**. Early on, he questions the link between evolution and progress: "each species changes; does it progress?" – a point echoed with "it is absurd to talk of one animal being higher than another". Eloquently, he debunks the importance of intellect: "When we talk of higher orders, we should always say intellectually higher, but who with the face of the earth covered with the most beautiful savannahs and forests dare to say that intellectuality is only aim in this world."

In consecutive sentences in the B notebook, Darwin makes two decisive breaks from Lamarck: "Changes not result of will of animal, but **law of adaptation** as much as acid and alkali. Organized beings represent a tree *irregularly branched*, some branches far more branched." He lets gradualism spill over to biology and outdoes geology in pursuit of deep time: "It leads you to believe the world older than *geologists* think." Darwin weaves in extinction: "The **tree of life** should perhaps be called the coral of life, base of branches dead; so that passages cannot be seen". The metaphor of the tree is his recurrent theme, culminating in the iconic "I think" sketch of an evolutionary tree – a poignant echo of Descartes' "I think therefore I am". And instead of a separate divine fiat for each creation, he has: "Let animals be created, then by the fixed laws of generation, such will be their successors."

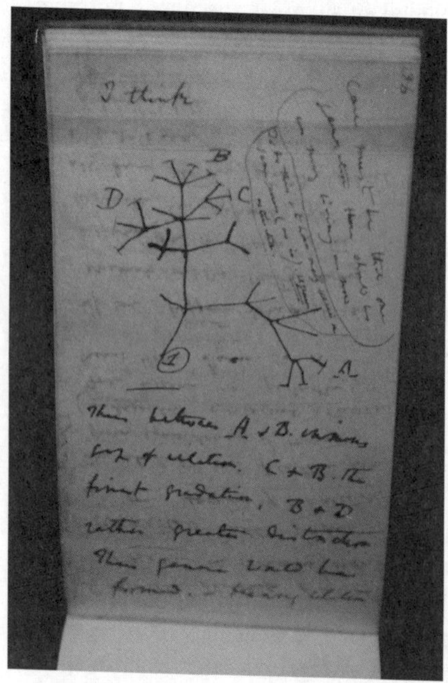

Darwin's first sketch of an evolutionary tree.

But Darwin's grasshopper mind was preoccupied with more than just transmutation – in his B notebook he intermingles body odour, slavery, animal liberation, mixed marriages, incest, male nipples, rudimentary bones, pigs that can open gates, mammalian mass extinctions, and floating trees. In an evolutionary echo of religious monotheism, Darwin states "there is but one animal", which passes "from worm to man highest". When classifying "man in a savage state", he first writes "species", then strikes it through and replaces it with "races". Like Alexander Pope, he places man in this isthmus of a middle state: "If all men were dead, then monkeys make men. Men make angels." He gropes at the need for a mechanism for evolution: "we are led to endeavour to discover *causes* of change, the manner of adaptation… what are the **Laws of Life**." Towards the end of the notebook, with a compassion worthy of the Buddha, he sympathetically proclaims the unity of humans and animals: "if we choose to let conjecture run wild, then animals our fellow brethren in pain, disease, death and suffering, and famine; our slaves in the most laborious works; our companions in our amusements. They may partake from our origin in one common ancestor; we may be all netted together."

Many of the themes raised in the B notebook reverberate through the three additional transmutation notebooks (C, D and E) from the late 1830s. In the **C notebook**, the phrase "my theory" is used repeatedly and Darwin recognizes the opposition that might greet his evolutionary ideas: "Mention persecution of early astronomers… I fear great evil from vast opposition in opinion on all subjects of classification." In the C and D notebooks, we see the first hints of the struggle for life and even natural selection: "Study the wars of organic being – the fact of guavas having overrun Tahiti, thistle Pampas… All this agrees well with my view of those forms slightly favoured getting the upper hand and forming species". There are the first hints of sexual selection: "Cock birds attract female by song (analogy of man)." And Darwin shamelessly speculates on the evolution of female sexuality: "In some monkeys clitoris wonderfully produced"; "In case of woman instinctive desire may be said more definite than with bitch… These facts may be turned to ridicule, or may be thought disgusting, but to philosophic naturalist pregnant with interest." Poetically, he sees the growth of women's breasts as more wonderful than the transition from caterpillar to butterfly. In the **D notebook**, domestic pigeons also receive their first mention.

Darwin ranges fearlessly over the continuum between humans and animals: "Now we might expect that animal half way between man & monkey would have differed in hair colour & form of head & features; but likewise in length of extremities." Time and again he unseats man from his haughty superiority: "Man in his arrogance thinks himself a great work worthy the interposition of a deity. More humble and I believe truer to consider him created from animals", "Let man visit Ourang-outang in domestication, hear expressive whine, see its intelligence… let him look at savage, roasting his parent, naked, artless, not improving, yet improvable and then let him dare to boast of his proud preeminence".

Darwin ponders the materialist origins of thought and even religious belief: "Love of the deity effect of organization, oh, you materialist!… Why is thought being a secretion of brain, more wonderful than gravity a property of matter?" Darwin considers the potential differences between the races of man, but poignantly places a question mark after "intellect".

Darwin's metaphysical meditations

Alongside the last two transmutation notebooks, Darwin kept two "**metaphysical notebooks**" (M and N). In these Darwin appears as a sympathetic psychologist ("everyone is insane at some time") and evolutionary philosopher of the mind ("He who understand baboon would do more toward metaphysics than Locke"). Within their pages, Darwin ruminates on the effects of stroke, the causes of madness, the naughtiness of children, the poetry of geology, the origins of language, the similarities in intellect and emotions between humans and animals, the arbitrariness of patriotism, and the conflicts between free-will, self improvement and predestination. Reflecting on the glory that was Greece and the subsequent dark ages, he makes clear that in "my theory there is no absolute tendency to progression". He notes that reading Dickens cured a headache brought on by an excess of philosophy.

Darwin anticipates subsequent thinkers: Freud and the unconscious ("The possibility of the brain having whole train of thoughts, feeling & perception separate from the ordinary state of mind"); Chomsky and his language-acquisition device ("it requires a far higher & far more complicated organization to *learn* Greek, than to have it handed down as an instinct"). And, as in his transmutation notebooks, he belittles the human intellect ("more fitted to recognize the wonderful structure of a beetle than a Universe") and berates mindless obedience to religion ("Weak people say I *know* it, because I was always told so in childhood, hence the belief in the many strange religions"). For those used to the avuncular image of Darwin as an old man, the young frisky Darwin of the metaphysical notebooks takes a surprisingly earthy interest in booze and sex: the student who, when drunk, thought everyone was calling him a bastard; the case of a Shrewsbury gentleman who tried to have sex with a turkey; dogs smelling each other's bottoms; the parallels between stallions licking the udders of mares and men's interests in women's breasts; the links between salivation and sex. Darwin wonders "why *all abnormal* sexual actions or even impulses… are held in abhorrence" and links the profane to the sublime: "A man shivers from fear, sublimity, sexual ardour. – a man cries from grief, joy & sublimity".

And in the C notebook, affirmations of common humanity are coupled with revolutionary cries for liberation: "Has not the white man, who has debased his nature by making slave of his fellow Black, often wished to consider him as other animal – it is the way of mankind & I believe those who soar above such prejudices yet have justly exalted nature of man… Educate all classes, avoid the contamination of caste, improve the women (double influence) & mankind must improve."

Reading Malthus

In September and October 1838, at the transition between the C and D notebooks, Darwin read **Malthus's** *Essay on the principle of population.* Malthus predicted that human populations would always eventually outgrow the means to support them, so that war and famine were inevitable. The force of Malthus's arguments gave a massive impetus to Darwin's ideas on the links between death, survival and selection, and provided him with, as he put it, "a theory to work by". Darwin generalized the Malthusian struggle for existence to the entire organic world. Grasping its violent but creative consequence, Darwin wrote in his C notebook: "One may say there is a force like a hundred thousand wedges trying to force every kind of adapted structure into the gaps in the economy of nature or rather forming gaps by thrusting out weaker ones." In the D notebook, he pits Malthus against Paley, "it is difficult to believe in the dreadful but quiet war of organic beings going on in the peaceful woods and smiling fields".

Thomas Malthus, whose gloomy view of population growth influenced Darwin.

Better than a dog anyhow

In 1838, at the age of 29, torn between work and the prospect of married life, Darwin jotted down notes weighing up the pros and cons of marriage. If he stayed a bachelor, he would be spared "the expense and anxiety of children", he would not be "forced to visit relatives" and would not suffer "loss of time" reading in the evenings. Marriage, on the other hand, promised an "object to be beloved & played with, – better than a dog anyhow" plus a "home, & someone to take care of house – Charms of music & female chit-chat... good for one's health". Darwin's logic finally settled on matrimony: "it is intolerable to think of spending one's whole life, like a neuter bee, working, working, & nothing after all. – No, no won't do... Marry-Mary-Marry QED."

A short while later he married his cousin, **Emma Wedgwood** (1808–96), who was worth far more than a dog! She was well off and well educated, speaking several European languages. She was an accomplished pianist (having taken lessons from Chopin) and she was also well travelled, having taken the grand tour of Europe. But the best was yet to come. In Emma, Darwin gained a lifelong companion, a confidante and gentle critic and a patient nurse during his periods of ill health. In a relaxed atmosphere of playful tolerance, she bore and brought up ten children (Darwin administered chloroform at some of the births). In old age, she even brought up one of her grandchildren. Emma took responsibility for the family's female staff. She entertained the numerous visitors, relatives and scientists to Down House. Starting life, like Charles, as a religious dissenter, she retained her faith while he lost his. Despite these differences in religious belief, she remained a faithful companion until Darwin's death. Emma is buried in Downe churchyard.

Emma Darwin Edna Healey (Headline, 2001)

But where Malthus set limits to improvability, Darwin saw endless variation, scribbling in his D notebook "it may be said that wild animals will vary according to my Malthusian views, within certain limits, but beyond them not – argue against this". In the same notebook, Darwin drew the analogy between natural and artificial selection: "It is a beautiful part of my theory, that domesticated races of organics are made by precisely same means as species – but latter far more perfectly and infinitely slower". He also made clear that the variation he had in mind was undirected: "the preservation of *accidental* hardy seedlings", "no *fortuitous* growth, yet... with infinitesimal advantage it would have better chance of being propagated" and reaffirmed his belief that evolution did not equate to progress: "if the simplest animals could be destroyed, the more highly organized would soon be disorganized to fill their places".

Marriage and a move to the country

While Darwin's inner creative demon was scribbling scandalous notes, his gentlemanly exterior moved forward in respectability. In the summer of 1838, Darwin rode out to the Wedgwood estate and started a brief courtship of his cousin, Emma (see box, p.38), proposing a few months

Darwin's illness

Shortly after returning from his voyage on HMS *Beagle*, Darwin was beset by **ill health** that lasted into old age. He consulted numerous doctors, but the medical science of his day was unable to provide a lasting cure or an authoritative diagnosis. In addition to malaise and tiredness, Darwin's symptoms spanned a range of systems: gastrointestinal (vomiting, bloating, colic, flatulence), neurological (headache, dizziness, vertigo, tinnitus, visual disturbances, muscle spasm and tremor; sensitivity to hot and cold), dermatological (blistering and pustular eruptions, oral lesions), psychiatric (insomnia, anxiety, depression, sudden feelings of impending doom, crying) and cardio-respiratory (palpitations, shortness of breath). Unable to examine the patient, even the best of today's doctors can never reach a definitive diagnosis. Speculation has filled the gap.

Partial explanations for Darwin's symptoms are easy to come by. Some of Darwin's episodes of ill health sound like **panic attacks**, but this does not rule out an underlying physical illness, nor identify a psychiatric cause. Agoraphobia has been suggested, although Darwin's trips away from home to scientific meetings or on holiday count against this. Meniére's disease could explain tinnitus and vertigo, but is unlikely in the absence of deafness; lactose intolerance could explain the gut symptoms; migraine could explain the headaches and visual disturbances; and Raynaud's phenomenon could explain the cold and numbness in the extremities.

The most parsimonious explanation has to link as many symptoms as possible to a single cause. One option is **arsenic poisoning**, but this is unlikely to have persisted for forty years. **Systemic lupus erythematosis** and **Crohn's disease** also fit the symptoms and the chronic up-and-down course of Darwin's disease. However, the most widely quoted explanation is the chronic parasitic infection, **Chagas' disease**, proposed by Israeli physician Saul Adler in 1959. In support of this idea, Darwin records being bitten by the relevant disease vector, the reduviid bug, while in the Andes (albeit at the margins of the disease's distribution), but Darwin's longevity and improved health in old age count against it, as does a pattern of alternating exacerbations and remissions. But the bottom line remains that, in the absence of physical evidence (which only a molecular examination of Darwin's remains could remedy), we will never know for sure.

later. In January 1839, Darwin moved into the future family home at 12 Upper Gower Street in the heart of London and on 24 January was elected a Fellow of the Royal Society – the country's premier scientific establishment. Five days later, he and Emma were married at Maer in Staffordshire, beginning a life together of domesticity that would last for 43 years. Emma spent most of the next dozen years pregnant, giving birth to nine children between 1839 and 1851 (a final child was born in 1856). Over the same period, Darwin was regularly plagued by chronic ill health (see box, p.39). This curtailed fieldwork and largely confined him to analyzing specimens already collected, or acquired from others.

Evolution outlined

By the summer of 1841, Darwin saw the need to escape from London to the countryside, but could not find a suitable house until the following year, when, with a heavily pregnant Emma and their two children, he moved to Down House (without an "e"), in the Kent village of Downe

Down House, the Darwin family home in Kent.

(now spelt with an "e"). The location suited them perfectly – less than fifteen miles from central London, but still isolated in "extreme rurality". After the move to Downe, Darwin's conventional career progressed nicely and within a year he had completed the task of editing *The Zoology of the Beagle Voyage*. In his letters to an ever-growing network of correspondents, Darwin comes across as a self-assured naturalist at home in scientific society. Around this time, he started a lifelong friendship with the botanist **Joseph Hooker** (1817–1911), who had, like Darwin, mixed foreign expeditions with the study of natural history.

But five years after his *Beagle* voyage, Darwin was still thinking about the mutability of species. Shortly before the move, in the summer of 1842, he had already scribbled out what is now known as his **Pencil Sketch**, a 35-page outline of his theory. It was scrawled on bad paper with a soft pencil, and is extremely difficult to read. But even a glance at the subject headings shows how much of what was to become *The Origin of Species* seventeen years later was already there: variation under domestication and in a state of nature; the natural means of selection; instincts; the evidence from geology; geographical distribution; classification; unity of type and abortive organs. The only serious omission is any substantial discussion of the principle of divergence.

Although Darwin is usually depicted as keeping his evolutionary hypotheses hidden during this time, his **letters** reveal that many of his established friends and contacts were in the know when it came to his unorthodox views on the mutability of species. In January 1844, Darwin wrote a now-famous letter to his new friend Hooker, admitting to his conversion to a transmutationist view: "I am almost convinced (quite contrary to the opinion I started with) that species are not (it is like confessing a murder) immutable." Quite what Darwin meant by the "confessing to murder" line is unclear. Some have fancifully suggested that it shows, quite literally, that Darwin felt a deep guilt over his theories. However, there are many other examples of Darwin using similar melodramatic language entirely in jest, so it was almost certainly intended merely as a joke. In any case, Hooker's reply was cautiously encouraging, accepting that there could have been "a gradual change of species". In the summer of 1844, Darwin expanded his pencil sketch into a much fuller essay of over 200 pages. With the breath of mortality on his neck after his own illness and the death of a third child as a baby, Darwin wrote a curious letter to his wife, asking that in the event of his death, she should have the essay edited and published.

A botched book, barnacles and bereavement

In October 1844, the idea of evolution, cosmic and biological, was thrust into public awareness by the publication of the anonymously written *Vestiges of the Natural History of Creation*. The book annoyed Darwin. In a superficial sense, it scooped him in promoting the idea of transmutation of species. But rather than a piece of tightly argued and well-supported scholarship, it was a work of shoddy popularization (later revealed to have flowed from the pen of Scottish author and publisher **Robert Chambers**). Tellingly, it said nothing about the mechanisms of evolution. The following January, Hooker and Darwin exchanged comments. Hooker thought it amusing, but remarked that the book seemed "more like a 9 days wonder than a lasting work". Darwin replied: "I have, also, read the Vestiges, but have been somewhat less amused at it... his geology strikes me as bad, & his zoology far worse." But it was the book's theological implications that drew the harshest comments. On reading it, Darwin's former geology teacher, **Sedgwick**, even went so far as to proclaim: "with a bright, polished, and many-coloured surface... the serpent coils a false philosophy, and asks them to stretch out their hands and pluck the forbidden fruit."

In 1846, Darwin allowed his research to veer away from geology towards a subject that would obsess him for eight years: **barnacles** (Cirripedia). Here his interest was so prolonged and so intense that Darwin's son George assumed that barnacling was the normal occupation of the head of any household – when visiting family friend John Lubbock, he asked, "where do you do your barnacles?"

Darwin's interest began with one particular barnacle, *Cryptophialus minutus,* which he nicknamed **Mr Arthrobalanus** – a tiny aberrant species that he had collected while in the Chonos Archipelago. It was odd for several reasons: rather than build its own home, it lived in a hole in the shell of a mollusc; stranger still, unlike most other barnacles, which are hermaphrodites, Darwin's specimen was a male, but with a degenerate body and an disproportionately large penis. In describing it, even the usually sober Darwin could not avoid an exclamation mark: "penis is wonderfully developed, so that in *Cryptophialus*, when fully extended, it must equal between eight and nine times the entire length of the animal!"

Fired up from his studies on Mr. Arthrobalanus, Darwin launched himself into a wider study on the taxonomy and natural history of barnacles. This interest culminated in the publication of four landmark volumes between 1851 and 1854 and the award of a medal from the Royal Society. He was delighted to discover other barnacle species in which his specimens consisted of a large female, with multiple tiny

Annie Darwin: lost joy of the household

Darwin's daughter **Annie** began life as the subject of an experimental comparison of baby humans and orangutans – Darwin even published a paper on his observations entitled *Biographical sketch of an infant*. Nonetheless, Annie grew into an outgoing, happy-go-lucky child linked by a strong affectionate bond to her father. Her health deteriorated as she approached ten years of age. In March 1851, Darwin took her to Malvern to try Dr Gully's water cure, which had brought Darwin himself so much benefit two years before. Her mother, Emma, stayed in Downe, too heavily pregnant to travel. The letters that flowed home from Charles to Emma in the days that followed provide a poignant description of the hopes and fears of the father of a dying child. Annie died on 23 April, 1851 from what was called "bilious fever with typhoid character" (some suggested this was tuberculosis but several alternative diagnoses remain plausible). Darwin was so devastated that he rushed back to Emma without staying for Annie's funeral. A servant went almost mad with grief. A few days later, the distraught Darwin writes in a twelve-page memoir: "From whatever point I look back at her, the main feature in her disposition, which at once rises before me, is her buoyant joyousness." He ends the memoir: "We have lost the joy of the household, and the solace of our old age: she must have known how we loved her; oh that she could now know how deeply, how tenderly we do still and shall ever love her dear joyous face." It has been suggested that Annie's death chimed the final death knell for Darwin's belief in Christianity, although there is no direct textual support for this from Darwin's own writings. However, modern psychiatrists see the loss of a child as one of the most devastating life events, so there is no doubt that Annie's death represented the lowest of the low points in Darwin's life.

Annie's grave in Malvern.

Annie's Box: Charles Darwin, His Daughter and Human Evolution Randal Keynes (4th Estate, 2001) Inspired by a writing box that belonged to Annie, Darwin descendant Randal Keynes takes an intimate look at Darwin the family man.

Darwin's bulldog

Thomas Huxley, confident young professor at 32.

Thomas Henry Huxley (1825–95), like his friend Hooker, left medical school for a journey of exploration – to northern Australia and New Guinea as surgeon's mate on HMS *Rattlesnake*. He soon became a self-taught expert on invertebrate comparative biology, authoring several papers that clarified some tricky taxonomy. In 1854 he took up a chair of natural history at the Royal College of Mines (now part of Imperial College), where for over thirty years he made valuable contributions to British science. His writings and lectures on human evolution together with his arguments with Owen on the anatomical differences between humans and apes earned him the nickname **Darwin's bulldog**. On the minus side, Huxley was by

"complemental males" living as mere bags of spermatozoa attached to the female. Although sometimes seen as a digression from his work on evolution, Darwin's barnacle work was a logical extension of earlier studies of marine invertebrates but, crucially, also provided him with a practical grounding in taxonomy, and comparative developmental biology. In addition, the shifts in form and function affecting evolutionarily related structures in different species provided new evidence for his theory of descent with modification. Instead of the speculative "darwinizing" that his grandfather was accused of, Darwin was finding the devil in the details.

Throughout his thirties, Darwin's health problems continued to sap his energy and make him miserable. His father died in November 1848, but Darwin was too ill to attend the funeral. But, for Darwin, life began again at forty, thanks to a trip to **Malvern**. He took the family for a four-month stay in this hilly spa town in March 1849, a few weeks after his fortieth birthday. Here, Darwin took Dr Gully's **water cure** – a regime of immersion in the local spring water, heating by steam lamp, scrubbing and spending most of the day wrapped in wet towels. Darwin was today's standards a racist, and he was reluctant to accept all of Darwin's views on natural selection (Darwin's bulldog, but not his poodle). On the plus side, Huxley's numerous achievements include his prescient classification of birds with dinosaurs (only recently recognized as correct), a treatise on the physical geography of the Thames valley, a classic book on crayfish and a biography of David Hume. He helped secularize schools, opened up adult education and transformed the academic activities of universities (viewing them as factories of new knowledge rather than storehouses of old). He coined the word "agnostic". Huxley was buried in St Marylebone (now East Finchley) Cemetery and left behind an intellectual dynasty that included grandsons Julian (evolutionary biologist and first director of UNESCO), Aldous (author of *Brave New World*) and Andrew (Nobel prize winner for his work on the physiology of nerves). Huxley also left behind a treasure trove of aphorisms: "After all, it is as respectable to be modified ape as to be modified dirt"; "Life is too short to occupy oneself with the slaying of the slain more than once"; "Science is organized common sense"; "The great tragedy of science is the slaying of a beautiful hypothesis by an ugly fact." From his biographer Edward Clodd, comes the greatest tribute of all: "It was worth being born to have known Huxley"!

Huxley: From Devil's Disciple to Evolution's Priest Adrian Desmond (Penguin, 1998)

sceptical of Gully's rationale for the treatment (drawing blood away from inflamed nerves of the stomach) and of his homeopathy. Nonetheless, Darwin's health improved dramatically in Malvern – probably as a result of enforced moderation in diet ("At no time must I take any sugar, butter, spices tea bacon or anything good"), snuff ("the cruel wretch has made me leave off snuff – that chief solace of life") and alcohol, twinned with plenty of exercise ("yesterday in 4 walks I managed seven miles"). Sadly, two years later, a more desperate visit to Malvern marked the low point in Darwin's life, concluding with the death of his ten-year-old daughter Annie in April 1851 (see box, p.43). Soon afterwards, Darwin met the young naturalist, **Thomas Huxley**, who subsequently became a firm friend of Darwin and Hooker (see box, above).

The chief work of my life

In September 1854, Darwin concluded his research on barnacles and, in his own words, "began sorting notes for Species Theory". Eager to prove that plants could cross oceans to settle new islands, Darwin became an avid experimentalist, embarking on a flurry of often-eccentric studies on the transmission and survival of seeds. In a creative game of intellectual tennis, Darwin batted ideas to and fro with Hooker. In one experiment, Darwin and his son Francis floated a dead pigeon in salt water for a month and then showed that seeds from its crop still germinated. In another, he cut the feet off a duck and used them to show that snails could hitch a ride from birds. He used his vast network of correspondents – Darwin's equivalent of the Internet – to seek out new facts and new specimens. He bred and studied domestic pigeons.

By 1856, Darwin was boldly discussing his ideas with his confidantes, by letter and face to face – that year Lyell, Huxley and Hooker all made visits to Down House. Although not convinced that Darwin was right, Lyell wrote to him saying: "I wish you would publish some small fragment of your data…" In his reply, Darwin dithered, but made a prophetic remark: "I rather hate the idea of writing for priority, yet I certainly should be vexed if any one were to publish my doctrines before me". Slowly, Darwin began work on a publication. Preparations for a short sketch were soon

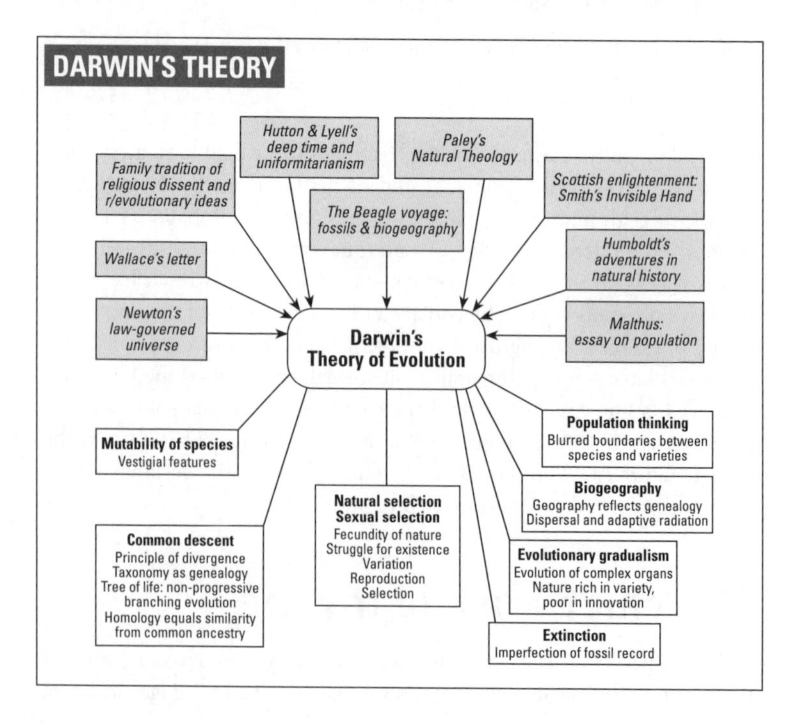

Darwin myth no.1: fear delayed publication

Many recent accounts of Darwin's life suggest that he delayed publication of his theory of evolution by natural selection and that, fearful of its reception, he kept it secret for many years. But none of this is mentioned in Darwin's autobiography, in his obituaries or in the early accounts of his life. Only in the mid-twentieth century did these myths of "Darwin's delay" begin to gain currency. In a recent, tightly argued paper, Darwin scholar John van Wyhe has highlighted the distortions and fabrications that have crept into the Darwin story and shown how these are derived from biased readings of a handful of passages plucked from Darwin's enormous written output. Darwin's apparent delay – between his essay of 1844 and *The Origin* in 1859 – was not down to any fears about the consequences of publishing on evolution. Instead, it was simply the accumulation of facts, ideas and arguments outrunning his ability to investigate, analyze and write about them, particularly when slowed down by illness. Yet, even when burdened with *Beagle* material and barnacles, Darwin was beavering away on his theory – as he says in *The Origin*: "from that period [ie 1837] to the present day I have steadily pursued the same object." There was no delay, no fear, no secrecy – nothing but perspiration racing to keep up with inspiration!

Mind the gap: did Darwin avoid publishing his theory for many years?
John van Wyhe (Royal Society, 2007) journals.royalsociety.org/content/gk6840u115705166

overtaken by something far more ambitious – a "big book" that was going to take years to write and might occupy multiple volumes. In September 1857, Darwin sent some notes on his theory to his new American friend, the Harvard botanist **Asa Gray**.

An explosive package

In June 1858, Darwin's lumbering complacency was blown apart by the intellectual equivalent of a parcel bomb. A package arrived at Down House, sent from the Malay Archipelago by 35-year old naturalist Alfred Russel Wallace. The parcel contained a twenty-page essay, written the previous February, which replicated much of Darwin's thinking. In an accompanying letter, Wallace asked Darwin to review it and pass it on to Lyell. Devastated at the thought of being scooped, Darwin sent the essay on to Lyell with a letter in which he lamented: "I never saw a more striking coincidence. If Wallace had my M.S. sketch written out in 1842 he could not have made a better short abstract!… So, all my originality… will be smashed".

Lyell and Hooker suggested an honourable compromise – that contributions from Darwin and Wallace should be published simultaneously at a meeting of the **Linnaean Society** in London on 1 July. The two men communicated a four-page extract from Chapter 2 of Darwin's 1844 essay, the notes Darwin had sent to Gray the year before and Wallace's paper. Neither Darwin nor Wallace was present. Wallace was half a world away, while Darwin was stuck at home recovering from a family crisis – first diphtheria, then scarlet fever had brought "death and severe illness and misery" to the children and their nurses. His last child, Charles Waring Darwin, died on 28 June, aged eighteen months.

The Darwin-Wallace papers elicited little interest. In September 1858, Darwin began work on an abstract of his "big book", the work culminating in his masterpiece, *The Origin of Species*. Darwin later wrote, "It cost me thirteen months and ten days' hard labour… It is no doubt the chief work of my life". In fact, Darwin had finished writing it by March, but he and Emma spent the summer checking over the proofs, so he was not free of it until October. Emma teased him over his poor use of commas.

Wallace: Darwin's rival or ambassador?

Alfred Russel Wallace (1823–1913) was born near the Welsh town of Usk and grew up in Hertfordshire. He worked as an apprentice **surveyor** for six years. During a brief spell as schoolmaster in Leicester, Wallace met entomologist **Henry Bates** and developed an interest in natural history. He worked for several more years as a surveyor and engineer. Then, inspired by Humboldt and Darwin, Wallace set off with Bates on an expedition to **Brazil**. In 1852, after four years collecting specimens and surveying the Rio Negro, Wallace set off back to England. At sea, a fire forced him to abandon his specimen collection and, adrift, he spent ten days in a lifeboat, awaiting rescue. Back safe in England, an insurance payment supported him while he wrote papers and forged links with naturalists, including Darwin. In 1854, Wallace embarked on an expedition to the **Malay Archipelago** (present-day Malaysia and Indonesia). During this six-year excursion, Wallace collected over 100,000 specimens, discovered the discontinuity between the kinds of plants and animals found in the western part of the archipelago and those found in the east (now called **Wallace's line**), and, crucially, hit upon the idea of evolution by natural selection independently of Darwin. Wallace's experiences were written up as a lively travelogue, *The Malay Archipelago*, selections from which were published by Penguin as *Borneo, Celebes, Aru* (2007). During his middle years Wallace was beset with financial problems, which were largely alleviated in 1881 by a government pension secured with help from Darwin. In late life, Wallace extended his

The Origin of Species

As Darwin described it, *The Origin* was "one long argument" for his theory of evolution by natural selection, weaving a large catalogue of facts into relentless, forceful lines of argument, all the time in language that remains accessible to the lay reader. Even a hundred and fifty years later, reading through the first chapter is to face the unrelenting thrusts of an intellectual fencing bout; at the end the reader is forced to concede, "Yes, I submit! I accept that organisms vary under domestication!" Throughout *The Origin*, Darwin constantly raises objections to his own hypotheses, only to demolish them. And what he doesn't say is equally important. Darwin leaves out any discussion of what his theory implies about God's relationship to the universe or to humanity. And he sneaks in just one cursory allusion to human evolution: "light will be thrown on the origin of man and his history". He says nothing about the evolution of the cosmos or of the origin of life.

The first edition of *The Origin of Species* was published on 24 November 1859 in London by John Murray. Exhausted from writing it, Darwin was away taking the water cure in Ilkley Spa in Yorkshire. The book sold

work on biogeography, became an early environmentalist and toured the US promoting evolution and natural selection. In old age, he settled in Broadstone, a suburb of Poole in Dorset, and is buried there in a grave capped with a fossil tree trunk and block of limestone.

Although often cast as Darwin's rival, Wallace remained a loyal and lifelong supporter and always accepted Darwin's claim to priority and even entitled his major book on evolution *Darwinism*. Wallace was an altogether more colourful character than Darwin, but also rather more flakey. He adopted **spiritualism** and, unlike Darwin, expounded a progressive, teleological view of evolution, with the universe working towards the birth of the human spirit. Wallace rejected natural selection as an explanation of the human mind, instead favouring interventions from the "unseen world of spirit". He became a **socialist** and an opponent of smallpox vaccination. He got tangled up in disputes as to whether the Earth was flat (he showed it wasn't) or whether there were canals on Mars (he argued there weren't). It is clear that, had Darwin died in South America, "Wallaceism" would have turned out quite differently from Darwinism.

The Alfred Russel Wallace web page www.wku.edu/~smithch/index1.htm

Alfred Russel Wallace: A Life Peter Raby (Princeton, 2001)

The Origin of Species: a one-page summary

Plants and animals under domestication show astonishing variation – compare the lively breeds of the domestic pigeon with the dull grey rock dove. This variation can be attributed to **artificial selection** (the conscious or unconscious actions of the breeder). Lesser but still appreciable variation occurs in nature, extensive enough to make it hard to distinguish species from mere varieties. Everywhere in nature we see an inevitable struggle for existence because more offspring are produced in each generation than can possibly survive to adulthood. This struggle is fiercest between individuals from the same species trying to exploit the same resources. Variants which are better suited to their environment will have a better chance of surviving and so will their offspring, whenever the variation is inherited. By analogy with artificial selection, this principle can be called **natural selection**.

Similarly, **sexual selection** is driven by struggles between members of the same sex in the same species. If some place in the web of nature is not well occupied, natural selection will preserve individuals that vary in the right direction and will fill the vacancy. Natural selection is always at work, but gradually, cumulatively and so slowly that we cannot see it in action, only its after-effects. The affinities of all living organisms can be represented as a great tree, which fills the crust of the Earth with its dead and broken branches (the extinct species), and covers the surface with its ever branching and beautiful ramifications (the existing species). Our ignorance of the causes and laws of variation is profound, but natural selection soldiers on nonetheless. We don't often see transitional forms because newer, better forms out-compete older ones, but gradations can be seen in nature (for instance, from ground squirrel to flying squirrel or from simple to complex eyes). It is unclear why hybrids between species are often sterile.

Geological time is so vast and the fossil record so imperfect that we can scarcely expect to find transitional fossils. But the theory of **descent with modification** provides the best explanation for extinction and for the succession of the same types of organism within the same areas. This theory also provides the best explanation for (1) curiosities in the geographical distribution of living organisms (such as differences between the Old and New World, or the strange inhabitants of oceanic islands); (2) the natural taxonomic hierarchy of groups within groups; (3) morphological and embryological similarities in form unrelated to function (why there are similar bones in wings, legs, flippers, etc); and (4) rudimentary organs. Although it stretches the imagination, descent with modification powered by natural selection is an inevitable consequence of the facts in front of us. What's more, it provides a nobler vision of the origin of species than special creation, and its creative power makes biology much grander and more interesting than the fixed laws and repetitive planetary orbits of physics.

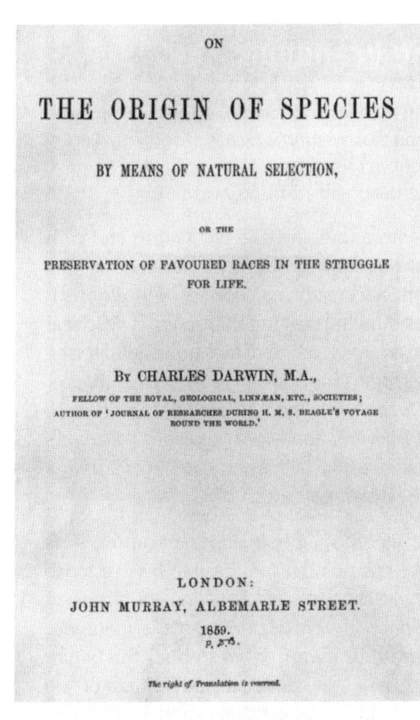

ON

THE ORIGIN OF SPECIES

BY MEANS OF NATURAL SELECTION,

OR THE

PRESERVATION OF FAVOURED RACES IN THE STRUGGLE FOR LIFE.

BY CHARLES DARWIN, M.A.,

FELLOW OF THE ROYAL, GEOLOGICAL, LINNÆAN, ETC., SOCIETIES;
AUTHOR OF 'JOURNAL OF RESEARCHES DURING H. M. S. BEAGLE'S VOYAGE ROUND THE WORLD.'

LONDON:
JOHN MURRAY, ALBEMARLE STREET.
1859.

The right of Translation is reserved.

Title page of the first edition of *The Origin*.

well and elicited keen interest from Darwin's associates and from naturalists, theologians and the general public. His brother Erasmus said it was "the most interesting book I have ever read". Lyell wrote that it was "a splendid case of close reasoning, and long substantial argument throughout so many pages…" Huxley exclaimed, "how extremely stupid not to have thought of that!", and wrote a glowing review for *The Times*. Even Owen, later cast as Darwin's rival, told Darwin that it offered the best explanation "ever published of the manner of formation of species". However, there were plenty of less positive comments. Astronomer Herschel described natural selection as "the law of the higgledy-piggledy". Adam Sedgwick, Darwin's mentor in geology, but wedded to a divine view of creation, received it with "more pain than pleasure". Henslow damned it with faint praise: "a stumble in the right direction". But people wanted to read it and *The Origin* soon went into a second edition.

X men, orchids and the human evolution

In June 1860, Darwin's theory was discussed at a meeting of the British Association for the Advancement of Science. Darwin was not present – he was taking the water cure, this time in Surrey – but Huxley and Hooker were there in his stead. A dull talk from American professor John William Draper was followed by an eloquent denunciation of Darwin's ideas by Samuel Wilberforce (Bishop of Oxford and son of the great abolitionist). What happened next is a subject for debate (see box, p.53) – but in the annals of legend, Darwin was left master of the field. The *Beagle*'s former

The Origin's culminating grandeur

"There is grandeur in this view of life, with its several powers, having been originally breathed into a few forms or into one; and that, whilst this planet has gone cycling on according to the fixed law of gravity, from so simple a beginning endless forms most beautiful and most wonderful have been, and are being, evolved."

Darwin once wrote: "With a book, as with a fine day, one wants it to end with a glorious sunset." Darwin's own glorious sunset, the final sentence of *The Origin*, shows him at his most eloquent and most memorable, with a not-so-hidden implication that biology is a lot more interesting than physics. **Michael Halliday**, an expert in discourse analysis, has pointed out how skilfully the rhythm and structure of Darwin's prose brings his book, his concluding chapter and his final sentence to a glorious climax, with the full force of hundreds of pages of argument bearing down on Darwin's final word "evolved" – Darwin's only use of the "e word" anywhere in the text. For Halliday, the power of Darwin's prose is matched only by the force of Beethoven's music.

Scratch beneath the surface and the sentence reveals another surprise – as English literature professor **Gillian Beer** has pointed out, despite his insistence on the action of impersonal laws, Darwin cannot avoid implying the action of personalized agents. The phrase "originally breathed" draws on the metaphor of the creator's divine breath and illustrates Darwin's desire to duck the origin-of-life question. His use of the passive voice – "are being evolved" instead of "are evolving" – implies that someone or something is doing the evolving (presumably Natural Selection), while the introduction of the present tense emphasizes that evolution is not limited to the distant past but is still ongoing.

The Origin's final sentence is a model of brevity and lucidity when compared to its much longer, less elegant precursors in Darwin's 1844 essay and 1842 pencil sketch. The sentence also survived virtually unscathed during the subsequent evolution of the book through six editions; a process which led to some other passages being rewritten four or five times and saw *The Origin* grow by a third. From the second edition onwards, Darwin merely appended the words "by the Creator" to "originally breathed" – ironic considering how he later expressed regret at having "truckled to public opinion & used Pentateuchal term of creation, by which I really meant 'appeared' by some wholly unknown process".

captain, FitzRoy, was also present at the debate, Bible in hand, proclaiming that *The Origin* had given him the "acutest pain" (fears from the *Beagle* days were realized when FitzRoy died by his own hand five years later).

In the years that followed the publication of *The Origin*, Darwin stayed out of the fray, through illness and probably also temperament. But through his contacts and correspondents, he kept a close eye on the commendations, critiques and caricatures that pertained to his work. In fierce arguments with Owen and others, Huxley earned the nickname

Darwin myth no. 2: the slaying of Soapy Sam

This myth asserts that science, personified by Huxley – Darwin's bulldog – slew religion, epitomized by "soapy Sam", **Samuel Wilberforce** (1805–73), Bishop of Oxford, at a meeting of the British Association in Oxford in 1860. According to popular accounts, in response to Sam's sarcasm ("Is it on your grandfather's or your grandmother's side that you claim descent from a monkey"), Huxley delivered a knock-out blow by saying that he would rather be descended from an ape than a bishop, winning the day for Darwin and leaving a resentful Wilberforce slithering off to nurse his wounds. But almost everything is wrong with this story. Contemporary accounts provide few details of what was said and done and most of the story is derived from biased accounts written twenty years after the event. In letters written shortly afterwards, Hooker, Huxley and Wilberforce all claimed to have gained the upper hand. Writing to Darwin, Hooker makes no mention of Huxley's repartee, even stressing that Huxley "could not throw his voice over so large an assembly, nor command the audience" – so Huxley's oft-quoted response was most likely mumbled under his breath, rather than barked across the room.

The learned Bishop Wilberforce.

Darwinians remained in the minority even after the debate. Wilberforce, rather than being resentful, remained gracious in all later dealings with Huxley. And Wilberforce was no scientific novice – he had been a Fellow of the Royal Society for fifteen years and based his arguments on science rather than religion. He wrote later, "we have objected to the views with which we are dealing solely on scientific grounds... We have no sympathy with those who object to any facts... or to any inference... because they believe them to contradict what... is taught by Revelation." Darwin himself admitted that Wilberforce's review of *The Origin*, which echoed in writing what was said in speech, was "uncommonly clever; it picks out with skill all the most conjectural parts, and brings forward well all the difficulties". Now that evolution is underpinned by so much new evidence, is there any need to celebrate this triumph of repartee over reasoned argument, of sarcasm over science? Today, the science speaks for itself!

Wilberforce and Huxley: a legendary encounter J.R. Lucas (1979) *History Journal* 22:313 or http://users.ox.ac.uk/~jrlucas/legend.html

Darwin's bulldog. In 1863, Darwin's supporters rallied against the view proposed by the Anthropological Society of London that "Negroes" were a separate, inferior species that deserved to be enslaved. In 1864, Darwin's disciples formed **the X Club**, a dining society dedicated to the defence of evolution and liberal rationalism; the X men's influence lives on in the journal *Nature*, founded by two of them, Huxley and Hooker, in 1869.

Although he made repeated revisions to *The Origin* throughout his life, Darwin's interests soon became more focused. Inspired by a holiday in Torquay, he became fascinated with **orchids**. As with barnacles, comparative studies on these plants revealed interesting adaptive variations in evolutionarily related structures. By 1862 he had enough material to publish a book. The following year, the discovery of the *Archaeopteryx*, the earliest known bird, provided fresh evidence for his theory of evolution. Darwin maintained his correspondence with Asa Gray during the American Civil War, proclaiming, "the destruction of slavery would be well worth a dozen years' war", and, through the 1860s, built up his friendship with Wallace.

It was in the late 1860s that Darwin started to grow his distinctive luxuriant beard and around this time he received visits from his German

Darwin myth no.3: Marx's dedication

This myth – that **Karl Marx** and Darwin were friends and that Marx offered to dedicate *Das Kapital* to Darwin – is the result of a simple error. A letter from Darwin to Edward Aveling, Marx's son-in-law, was mistakenly identified as a letter to Marx. In 1880, Aveling asked if he could dedicate his book *The Student's Darwin* to his hero. Darwin politely refused the offer. But when Aveling took over Marx's papers in the 1890s, Darwin's letter of response got mixed up with some papers from Marx, resulting in the later misassignment of authorship. The error was only picked up in the 1970s by Margaret Fay and Lewis Feuer.

Marx was, however, an admirer of Darwin and sent him a signed copy of *Das Kapital* (which, most of its pages uncut, Darwin never finished). Darwin's non-committal two-sentence acknowledgment of receipt was the only communication from the man of evolution to the man of revolution: "Dear Sir: I thank you for the honour which you have done me by sending me your great work on Capital; & I heartily wish that I was more worthy to receive it, by understanding more of the deep and important subject of political Economy. Though our studies have been so different, I believe that we both earnestly desire the extension of Knowledge, & that this is in the long run sure to add to the happiness of Mankind." The only other connection between the two men was Ray Lankester, a zoologist who corresponded with Darwin and who, after befriending the elderly Marx, was one of the few to attend his funeral.

A contemplative Darwin at sixty, sporting his characteristic bushy beard.

admirer Haeckel and from Gray. Re-drafted material from his "big book" formed the basis of *Variation of Plants and Animals under Domestication*, which went on sale in 1868. Here, Darwin demolished the idea that variation was divinely directed by drawing an analogy with an architect's use of rocks that had fallen off a cliff: "Can it be reasonably maintained that the Creator intentionally ordered… that certain fragments should assume certain shapes so that the builder might erect his edifice?"

The Descent of Man, and Selection in Relation to Sex followed in 1871, with Darwin applying the theory of evolution to human anatomical and mental features. In this weighty two-volume work, Darwin emphasizes structural and developmental homologies between humans and

What was Darwin really like?

The young Charles Darwin could not draw, but was a good shot. Although strong, he lacked poise – today's educationalists might have labelled him borderline dyspraxic. Nowadays, he might also have been signed up for speech therapy – he stammered, particularly over words beginning with "w". He was fond of quoting his grandfather, who, when asked if he found it inconvenient to stammer, replied: "No, sir, because I have time to think before I speak, and don't ask impertinent questions."

In adult life, Charles Darwin stood just less than six feet tall. As a youth, his blue-grey eyes were offset with a full head of dark brown hair, which receded in later life. His prominent brow ridge was decorated with bushy eyebrows, his cheeks with luxuriant sideburns. Darwin's blobby nose almost led FitzRoy to exclude him from the *Beagle* – in fact, Darwin himself later questioned whether such a nose could ever be the result of intelligent design! Unlike his father and grandfather, Charles Darwin never became obsese. In later life, he developed a ruddy complexion at odds with his ill health and from his late fifties, he sported an unruly whitish-grey beard.

Darwin was a **family man** who loved children; his daughter said: "To all of us he was the most delightful play-fellow, and the most perfect sympathizer." He played billiards with sons William and George. Darwin loved dogs, and on his return from the *Beagle* voyage, echoing Odysseus's return to Ithaca, Darwin's dog remembered his master. In middle age, his favourite was a fox terrier called Polly, whom he teased by making her catch biscuits off her nose. Darwin enjoyed snuff as a stimulant and smoked cigarettes to unwind. Although he drank too much at university, later in life Darwin consumed little, but enjoyed what he drank. He was an accommodating, unpretentious host, animated and bright-eyed in conversation and given to fits of hearty laughter. At Down House, he was a creature of habit whose daily routine included alternating periods of work and rest – working hard in the morning, writing letters and napping in the afternoons. Most days he walked the sandwalk, a circular gravel path next to Down House, where he did much of his thinking. His evenings were devoted to music, backgammon and reading: he admitted that he much preferred romantic novels with pretty women and happy endings!

other animals and notes the persistence of rudimentary organs within our bodies. He also emphasizes the continuities between the minds of "civilized" and "savage" humans and between humans, apes and dogs. He notes how civilization mitigates the effects of natural selection, but is ambivalent as to whether this is a good or bad thing. For his time, Darwin's views on race were progressive, seeing a common origin for all humans and stressing the superficial nature of many racial differences, which he saw as the effects of sexual selection rather than as adaptations.

However, he was clearly sexist in holding to the superiority of men over women. Curiously, from a modern perspective, more than half the book is devoted to describing sexual selection from molluscs to man. Like *The Origin*, *The Descent of Man* was a great success, although in a new intellectual climate, its conclusions ("Man still bears in his bodily frame the indelible stamp of his lowly origin") proved far less controversial. Darwin continued working through the next two years, producing the sixth (and last) edition of *The Origin* (the only edition to include the term "evolution") and completing the *Expression of Emotions in Man and Animals*. In 1874, Darwin produced a second edition of *The Descent of Man*, and then, at the age of 65, retired from writing on evolution.

An agnostic in the abbey

In his old age, Darwin remained stunningly productive, thanks, perhaps, to an unexpected liberation from a decades-long burden of ill health. His attention returned to botany, completing six books on the subject in four years. He wrote his own autobiography and a biography of his grandfather Erasmus. In 1877, the once and future prime minister, William Gladstone, visited Down House, prompting the self-effacing Darwin to comment: "What an honour that such a great man should come to visit me!" That same year Darwin received an honorary doctorate from the University of Cambridge.

For his final book Darwin returned to a theme first visited more than thirty years before: the action of **earthworms** in the production of soil. He continued with his eccentric experiments, playing music to worms and visiting Stonehenge to see how mighty megaliths were buried by the humble inhabitants of the soil. His final book was published in 1881. That same year, now an avuncular epitome of respectability, Darwin was invited to dine with the **Prince of Wales** (the future King Edward VII).

The humble earthworm, subject of Darwin's final book.

From late December 1881 Darwin suffered a series of

Darwin myth no.4: a deathbed conversion

The myth of Darwin's deathbed conversion arises from claims made by a Lady Hope, who said that she had visited a bedridden Darwin during his final illness. To an American audience, Lady Hope alleged that she found Darwin reading Hebrews, that she heard him express concern as to how his theory of evolution had been communicated, a wish to hear the local Sunday school sing and a desire to call a meeting of servants to discuss religion. Darwin's children denied these claims. His daughter Henrietta wrote: "I was present at his deathbed. Lady Hope was not present during his last illness, or any illness... He never recanted any of his scientific views, either then or earlier... The whole story has no foundation whatever". Even if Lady Hope's story were true – say she snuck in when Darwin's children were not there – her claims say nothing about any deathbed confession or conversion; these must count as embellishments to an already highly questionable story. According to accounts from his family, Darwin faced death bravely, stating that he was not afraid to die and at one point even saying, "if only I could die". In his autobiography, Darwin described himself as an agnostic and, according to his son Frank, Darwin was still an agnostic when he died.

angina attacks. He died at Down House at 4pm on 19 April, 1882. His last paper, on the subject of the dispersal of freshwater molluscs, was published in *Nature* just thirteen days before his death. By a curious coincidence that Darwin would have relished, the paper was co-authored by a certain **William Drawbridge Crick**, a shoe manufacturer from Northampton, who turned out to be the grandfather of Francis Crick, whose own *Nature* paper of April 1953 – on the structure of DNA and co-written with James Watson – represented the greatest advance in biology since *The Origin of Species*.

Darwin had expected a private burial next to his brother, Erasmus, in the country churchyard in Downe. However, in the week following his death, a groundswell of opinion among his friends and colleagues, fanned by exhortations and eulogies in the newspapers, diverted arrangements to a state funeral in **Westminster Abbey**. A horse-drawn hearse took all day to carry Darwin from Down House to Westminster. At noon on the 26 April, the funeral began. Nine pallbearers, including Huxley, Hooker and Wallace, carried Darwin to his final resting place, a few yards from the grave of Isaac Newton. A week later an obituary in the *Times* stated, "his work and character... have filled the minds of thinking people of all countries, classes, creeds and occupations..." The naughty boy from Shrewsbury had, in the words of the *Pall Mall Gazette*, become the "greatest Englishman since Newton".

Darwin's evolution revolution

In 1859 the world changed, changed utterly. The publication of *The Origin of Species* marks one of the greatest discontinuities in human thought, comparable only to the Copernican revolution. Modern biology was born, and most of what has followed in the life sciences can be considered footnotes to Darwin. But the implications of Darwin's work for the human condition were even more profound. As Sigmund Freud wrote in his *A General Introduction to Psychoanalysis* (1920): "Humanity has in the course of time had to endure from the hands of science two great outrages upon its naive self-love. The first was when it realized that our earth was not the center of the universe… The second was when biological research robbed man of his peculiar privilege of having been specially created, and relegated him to a descent from the animal world."

Darwin's big ideas

Darwin was no one-idea man, no one-hit wonder. Curled up inside what we now call Darwin's Theory of Evolution are several distinct ideas, each of them important and influential. In the introduction to *The Origin of Species*, Darwin himself highlights the importance of three ideas (the **mutability of species**, **common descent** and **natural selection**): "I am fully convinced that species are not immutable; but that those belonging to what are called the same genera are lineal descendants of some other and generally extinct species, in the same manner as the

If evolution is the answer, what is the question?

Biology is different from physics or chemistry. If a time machine transported a biologist, a chemist and a physicist back one hundred million years, the chemist and the physicist would find the laws of physics and chemistry the same. But the biologist would find a world populated by collections of plants and animals quite different from what we see today. Life changes while the world of physics and chemistry stays the same. One of the functions of evolutionary biology is to document and explain these striking changes that characterize the biological world.

Aside from its protean mutability, several other features of the biological world require explanation. Why is there such variety in the outward appearance, inner workings and behaviour of living organisms, whether extant or extinct? But beneath the superficial diversity of life, why such deep similarities? Why should a whale living in the sea, on close inspection, look more like a cow that lives on the land than like the fish that lives with it in the sea? Why should the bones in the wing of a bird resemble those in the arm of a man? Why should organisms look so alike when their lifestyles are so different? Why should organs and body parts look so alike when their functions are so different?

Living organisms are dramatically more complex that inanimate objects. Compare a stone to a stork, a pebble to a penguin, a grain of sand to a bacterium, or a pocket watch to a wild flower, and the vibrant complexity of life reigns apparent. How is it that much of life's gear, tackle and trim looks like an **adaptation**, an apparently designed contrivance, crafted with a clear purpose in mind – the white hair of polar bear to camouflage it against the snow, the eye of the eagle able to spy a rabbit a mile away, a fake insect built by an orchid to seduce a crane fly into polli-

acknowledged varieties of any one species are the descendants of that species. Furthermore, I am convinced that Natural Selection has been the main but not exclusive means of modification." In an essay from 1982, the great evolutionary biologist **Ernst Mayr** teased out two more threads in Darwin's thinking: **gradualness** and an emphasis on **variation** and **populations** rather than on archetypes.

Fraternity: the great tree of life

Darwin acknowledged the influence of his *Beagle* voyage in priming his first great idea, the **mutability of species**: "When on board H.M.S. *Beagle*, as naturalist, I was much struck with certain facts in the distribution of the inhabitants of South America, and in the geological relations of the present to the past inhabitants of that continent. These facts seemed to me

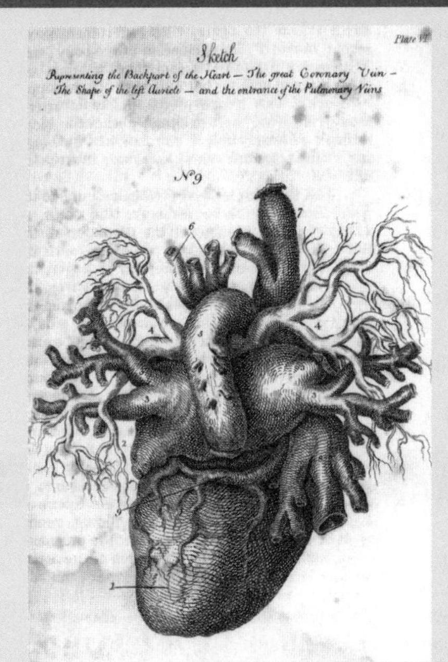

The human heart, a contrivance of nature, a living pump.

nating it? Why this appearance of design in nature? Why do contrivances of nature seem so perfectly to resemble the designed artifacts of man: the heart, a pump, the eye, a camera, our tendons and ligaments, ropes and pulleys. And why should "why" questions work at all in biology?

Until Darwin, the best explanation for all these features of the biological world was the "argument from design". Darwin's revolutionary theory of evolution by natural selection provided an explanation of the appearance of design in nature without the need for intelligent design. To quote Daniel Dennett: "In a single stroke, the idea of evolution by natural selection unifies the realm of life, meaning, and purpose with the realm of space and time, cause and effect, mechanism and physical law."

to throw some light on the origin of species." As we have seen, Darwin was not the first to suggest the possibility of evolutionary change. However, unlike previous authors, in the two decades leading up to *The Origin*, he marshalled an unprecedented array of arguments and facts to support the concept. Thanks to Darwin, the fact that evolution has happened has become so widely embedded in our society that even proponents of Intelligent Design, like Michael Behe (see box, p.284), have no trouble accepting it.

A crucial novel feature of Darwin's great idea was the notion of **non-progressive branching evolution**. Some of his predecessors saw evolution as organisms advancing up a ladder of progress in a single direction – the evolutionary equivalent of a straight unbranching bamboo cane rushing heavenwards. Darwin, on the other hand, borrowing a metaphor that harks back to Genesis, saw "a great **Tree of Life**", with its branches reaching out undirected in multiple directions: "As buds give rise by growth to fresh buds, and these, if vigorous, branch out and overtop on all sides many a feebler branch, so by generation I believe it has been with the great Tree of Life, which fills with its dead and broken branches the crust of the earth, and covers the surface with its ever branching and beautiful ramifications."

Indeed, it is no accident that the only figure to appear in the first edition of *The Origin* was also an evolutionary tree. And Wallace, co-discoverer of evolution, also hit upon the same metaphor, writing in his 1858 essay: "a complicated branching of the lines of affinity, as intricate as the twigs of a gnarled oak or the vascular system of the human body. Again, we have only fragments of this vast system, the stem and main branches being represented by extinct species of which we have no knowledge, while a vast mass of limbs and boughs and minute twigs and scattered leaves is what we have to place in order."

The species problem

One of biology's trickiest concepts is that of a "species" (a word which is the same in the singular and plural). In classical Latin the term just meant "kind" or "sort". However, by the seventeenth century it was being applied to biological groups. In his *Systema Naturae* (1735) the Swedish taxonomist **Carl Linneaus** (1707–78) placed *species* as the lowest taxonomic unit in a hierarchical classification that went **kingdom**; **phylum** (plural phyla); **class**; **order**; **family**; **genus** (plural genera); **species**. Linnaeus also introduced what are called "binomial" species names, in which a capitalized genus name is followed by a lower-case species names, both in italics. For example, humans belong to the species *Homo sapiens*.

In Linnaeus's time, species were thought to have been individually created by God and to have some sort of a real status, which could be defined by essential qualities, just as one can define a circle or triangle in geometry. However, even Linnaeus recognized that there were sometimes important sub-divisions that could be made below the species level, which were variably termed "races" or "varieties". Darwin argued vigorously in *The Origin* that the distinction between species and variety was "entirely vague and arbitrary" – in both cases, one was looking at populations of organisms that shared similar characteristics because of shared ancestry.

For plants and animals, the species concept received a new lease of life in the 1940s, when Ernst Mayr championed the **biological species concept**, namely that species consist of individuals who can interbreed and produce fertile offspring. However, Mayr's definition applies only to sexually reproducing organisms (ie most plants and animals), but fails in the world of bacteria, where exchange of genes across wide taxonomic distances is well recognized and different kinds of genome evolution apply in different bacterial lineages.

Implicit in Darwin and Wallace's Tree of Life is the concept of common ancestry or **common descent** – the notion that all members of any taxonomic group (for example, all primates, all mammals, all vertebrates) share a common ancestor. Darwin, however, could only speculate as to whether all life had a common origin: "Analogy would lead me one step further, namely, to the belief that all animals and plants have descended from some one prototype. But analogy may be a deceitful guide. Nevertheless all living things have much in common, in their chemical composition, their germinal vesicles, their cellular structure, and their laws of growth and reproduction… Therefore I should infer from analogy that probably all the organic beings which have ever lived on this earth have descended from some one primordial form, into which life was first breathed."

Even Mayr discussed several other species concepts and lively arguments continue among biologists and philosophers as to which definition is the best. Although sometimes these arguments take on an arcane angels-on-a-pinhead quality, in conservation biology, much often hangs on whether a threatened population represents a separate species or "merely" a sub-species of an otherwise non-endangered species.

Below is a table showing the scientific classification for an animal and a plant, both of which Darwin knew well. He identified the flightless **Lesser Rhea** (*Rhea pennata*) as a separate species from the American Rhea following his travels in South America, and the **Fly Orchid** (*Ophrys insectifera*) figured in his research into plant fertilization by insects, carried out in the 1870s.

Kingdom	Animalia	Plantae
Phylum	Chordata	Magnoliophyta
Class	Aves	Liliopsida
Order	Struthioniformes	Asparagales
Family	Rheidae	Orchidacae
Genus	*Rhea*	*Ophrys*
Species	*R. pennata*	*O. insectifera*

Common descent: from Darwin to DNA

Darwin, in proposing common descent, had no proof that all life had a common origin. However, in the last fifty years, abundant evidence from molecular biology has amply confirmed Darwin's suspicions making it dramatically clear that all living organisms on this planet – from humans to hydrangeas, from molluscs to microbes – are descended from a single organism (the last universal common ancestor: LUCA). All living organisms share the same building blocks of life and the same "genetic alphabet" – the same four bases in their DNA, the same (or very nearly the same) genetic code to turn information stored in DNA into proteins. Furthermore, we have now found hundreds of genes that have relatives in all forms of life, from bacteria to humans. Analyses of the sequences of these genes has provided clear proof of the common origin of all life and has even allowed us to draw up a tree of life that links humans with all known plants, animals and even microorganisms.

The honeybee, a distant cousin of humanity.

As with many important ideas, the idea of common ancestry risks losing its impact through familiarity. In an attempt to jolt you out of complacency, let us begin a brief meditation on the reality of common ancestry. Let us imagine you are reading this book sitting in an armchair in the sitting room of a suburban house. Even in such a domestic environment, you are surrounded by the DNA of other living organisms. Using the kind of approaches employed by crime scene investigators, a molecular biologist could obtain not just DNA but even DNA sequences from thousands of different organisms or their products within a hundred yards of where you sit.

But think about the strange implications of the theory of evolution. However alien these organisms seem to you, Darwin's notion of common ancestry means that you are related to them in the same way that you are related to your human relatives – your brothers, sisters, cousins, uncles and so on. The sequence of a very slowly evolving protein – say, a protein called EF-Tu, – is the same whether you are human or a chimpanzee and 99 percent the same in a dog or cat. Your pet zebrafish's EF-Tu is 94 percent the same as yours, the honeybee buzzing around your garden 85 percent, while the yeast in your kitchen cupboard is 81 percent the same as yours. But EF-Tu is just one of hundreds of sequences we could have chosen to illustrate one of the central messages of evolution: that all life on this planet is one!

Equality: evolution, the great leveller

Darwin's notion of evolution is a great leveller. His tree-like pattern of repeated branching speciation brings an **egalitarianism** to biology quite at odds with any notion of a ladder of inevitable progress with humans at the apex of the cosmos. Instead, Darwin builds on the Copernican revolution, removing any privileged status for humans among other organisms. Buried within the model are several important and still unappreciated concepts. Firstly, a tree of life model allows for extinction. As Darwin makes clear, not all branches of life survive until the present day: "From the first growth of the tree, many a limb and branch has decayed and dropped off; and these lost branches of various sizes may represent those whole orders, families, and genera which have now no living representatives, and which are known to us only from having been found in a fossil state." Indeed, the number of extinct species vastly outweighs the number of extant species, so today's biosphere is but a small subset of all the life forms that have ever existed.

Secondly, just as the twigs are all more or less equidistant from the trunk of the tree, any group of extant organisms can be considered equally remote from their most recent common ancestor. Kicking away the ladder of human arrogance, implicit in Darwin's tree is the notion that all organisms are "equally evolved". Although Darwin himself sometimes used value-laden language in his descriptions of evolution (for example, "higher" and "lower" animals), in his notebooks and elsewhere, it is clear that he also glimpsed the egalitarian consequences of branching evolution.

Clearly, if producing the greatest number of individuals is the best measure of evolution, then insects can also be considered to be more "evolved" than humans – a point emphasized by the evolutionary biologist Haldane, who, when asked if he could conclude anything about God from the study of nature, replied that the Creator must have "an inordinate fondness for beetles." William Paley in his *Natural Theology* (1802) saw divine attention to detail in the wings of an earwig. But post-Darwin, one could argue that these insects are more "evolved" than human gentlemen as lovers – the male earwig has two identical, fully functional and independently operable penises. Even the humble slug trumps us humans in the evolutionary love stakes, with a penis seven times the length of its body!

Underlining the counter-intuitive nature of branching evolution as the great leveller, the following recent discoveries have met with a great deal of surprise: the idea that chimpanzees are more evolved than humans (in

Survival of the aptest: evolutionary turns of phrase

The terms **adaptation** and **evolution** pre-date Darwin. The term "adapt" comes from the Latin *adaptare*, to make fit. William Paley used the terms "adapted" and "adaptation" repeatedly in *Natural Theology* (1802): "The eyes of fishes also, compared with those of terrestrial animals, exhibit certain distinctions of structure, *adapted* to their state and element." The word "evolution" stems from the Latin *evolutio* meaning "unfolding", particularly "the unrolling and reading of a scroll, the reading of a book". The term has acquired a general meaning covering any process of formation or growth or development; in biology it was first used as a term to describe embryological development. The word "evolution" was first used in connection with the development of species in 1762 by Swiss naturalist **Charles Bonnet**, who developed a theory of pre-formation (females carry within them all future generations in a miniature form) and catastrophism. Curiously, Darwin, in the first edition of *The Origin*, never used the term "evolution"; instead his preferred phrase for the idea was **descent with modification**. He did use the term "evolved", however, as the very last word of his text.

The term **natural selection** originates with Darwin, but was criticized as being too anthropomorphic, breathing agency into an inanimate process – selection implies a selector. Darwin in a letter to his geologist friend Lyell, a year or so after completing *The Origin*, states that if he were starting afresh he would

the sense that more of their genes appear to be under selective pressure); the fact that many plants, animals and even some unicellular organisms have larger genomes than humans – the salamander genome is ten times larger than the human genome, while the record-holder is an amoeba, *Amoeba dubia*, with a genome over 200 times larger than ours! And even in terms of the number of genes, rice has more genes than the humans that cultivate it.

Natural selection: freedom from the cosmic watchmaker

Darwin was not content with showing that species changed over time. As he notes in the introduction to *The Origin*, "such a conclusion, even if well founded, would be unsatisfactory, until it could be shown how the innumerable species inhabiting this world have been modified so as to acquire that perfection of structure and co-adaptation which most justly excites our admiration".

have used the term "natural preservation". However, the phrase that caught the public's imagination, then and now, is **survival of the fittest**, which originates not with Darwin, but with his contemporary **Herbert Spencer**. Alfred Russel Wallace regularly urged Darwin to dump the term natural selection and replace it with Spencer's phrase. Darwin went half way – in the fifth edition of *The Origin* he added "or Survival of the Fittest" to "Natural Selection" in the title of Chapter 4 and used the phrase several times in the text.

Despite its popularity with the public, the phrase "survival of the fittest" is now seldom if ever used by professional biologists and has been eliminated from any serious presentation of Darwin's ideas. There are several problems with it. A modern reading misunderstands Darwin's meaning: in Darwin's time, the word "fittest" primarily meant "best suited" or "most appropriate" rather than, as now, "in best physical shape". But more troublesome, the phrase has helped fuel the excesses of Social Darwinism (see p.264), erroneously suggesting that evolution provides moral justification for "might makes right" and for the mistreatment and even murder of those designated "unfit". In addition, if the fittest are defined as those best equipped to survive, the phrase becomes an uninformative tautology that obscures the essential features of natural selection. Instead, when it comes to the survival of the aptest, natural selection has emerged as clear winner.

Believing that the universe was governed by natural laws, he also wanted to find an explanation for evolutionary change and for adaptation – loosely defined as the appearance of design in nature. In his own words, he wanted to integrate in one coherent theory, **Unity of Type** (similarities that occurred between related organisms that were not simply tied to function) and **Conditions of Existence** (the ways in which species appeared to be adapted to their environments). Darwin and, some years later, Wallace both hit on a powerful explanation for adaptation and evolutionary change – **natural selection**. Darwin eloquently summarizes the principle of natural selection in *The Origin*: "Owing to this struggle for life, any variation, however slight and from whatever cause proceeding, if it be in any degree profitable to an individual of any species… will tend to the preservation of that individual, and will generally be inherited by its offspring. The offspring, also, will thus have a better chance of surviving, for, of the many individuals of any species which are periodically born, but a small number can survive. I have called this principle, by which each slight variation, if useful, is preserved, by the term of Natural Selection, in order to mark its relation to man's power of selection."

American philosopher Daniel Dennett in his book *Darwin's Dangerous Idea*, claimed natural selection was the best idea anyone ever had! Like many great ideas, it seems obvious after the fact. Yet, it represents one of the most potent conceptual advances in humankind's attempts to explain the universe and our place in it, banishing forever Paley's cosmic watchmaker. As Richard Dawkins puts it: "Never were so many facts explained by so few assumptions!" Three major concepts come together in this revolutionary idea: **reproduction**, **variation** and **selection**. As we have seen, Malthus's *Essay on the Principle of Population* was a major influence on both Darwin's and Wallace's thinking about reproduction. Darwin noted the wasteful **fecundity of nature**: the idea that most organisms tend to produce far more offspring than can possibly survive. For example, a single female codfish can produce up to five million eggs in a breeding season, while just one male human produces over 300 million spermatozoa in a single ejaculation – potentially enough to re-populate most of Western Europe!

"If I were to give an award for the single best idea anyone has ever had, I'd give it to Darwin... In a single stroke, the idea of evolution by natural selection unifies the realm of life, meaning, and purpose with the realm of space and time, cause and effect, mechanism and physical law."

Daniel Dennett

Logic took Darwin one step further: he recognized the consequences of geometric increase (the reproductive equivalent of compound interest) even in slowly reproducing organisms. A pair of elephants would eventually fill up the world with their descendents, even if progeny only marginally outnumbered parents in each generation. And yet the most superficial glance at the real world confirms that most organisms do not increase geometrically over time. Therefore, Darwin reasoned, growth must be checked by the hostile conditions of life. All organisms must be engaged in a **struggle for existence**, sometimes literally (in the form of a physical fight), more commonly metaphorically (for example, a plant struggling against the drought). Of the cod's five million eggs, no more than two survive to adulthood. Darwin hit on another key insight: that organisms didn't struggle only with their obvious adversaries, predators; they also struggled with members of their own species, in competing for the best opportunities in life.

The second key prerequisite for natural selection is variation. In most sciences, variation is seen as merely uninteresting or even as a nuisance. In contrast, Darwin placed biological variation centre stage, as the very seed corn for evolution, underpinning a population-based view of life that liberates us from the essentialism of Plato. Darwin spent much time documenting the tremendous variation that exists in domesticated and wild populations. It is a measure of Darwin's genius that he sought the answer to a question of cosmic significance in the study of an everyday animal, the **domestic pigeon**. In cataloguing the astonishing variety of breeds of domestic pigeon – all derived, Darwin argued, from a single wild progenitor – he became convinced of the remarkable variability inherent in all living organisms.

The third ingredient in the Darwinian mechanism of evolution is the idea of **selection** itself. In framing the term natural selection, Darwin drew an analogy with artificial selection, where the cumulative actions of breeders, generation upon generation, slowly shaped domesticated plants and animals to human needs and wants. Darwin recognized that some of the variations present in a population might give their owners a slight advantage in the struggle for existence: for example, if they allowed more efficient use of resources or greater success at avoiding predation. The variants that survive best will leave more offspring and these offspring are likely to benefit from the useful variations passed down from their parents. Over time, favorable variations will thus tend to accumulate within a population, and cause the character of a species to change, particularly in a changing environment.

A female frog produces far more offspring than can reach adulthood.

Key concepts that Darwin didn't have

With hindsight, it is clear that Darwin was handicapped by the lack of a coherent theory of inheritance and blinded by the widespread acceptance in his time of **blending inheritance**: the idea that organisms behaved like pots of coloured paint, so that once two extremes were mixed, only the intermediate remained (yellow and blue paints when mixed become green, and the original colours can never be recovered subsequently). If true, blending inheritance would severely hamper the ability of natural selection to effect change, as any new variation would quickly be diluted out of the population. Instead, as **Gregor Mendel** showed, biological inheritance is particulate, in that genetic information comes in discrete packets, now called **genes** (see p.105). What is mixed during inheritance are not pots of paint but bags of coloured marbles, or – in twenty-first century parlance – inheritance is digital rather than analogue.

Weismann recognized the distinction between body cells and germ cells.

Towards the end of the nineteenth century, in the years following Darwin's death, the German biologist **August Weismann** formulated another key concept, the **soma-germline distinction**, which rendered inheritance of

The Zen of natural selection

Although Darwin's description of natural selection has stood the test of time, our understanding of the processes involved has been sharpened up since Darwin first articulated the principle. It is now abundantly clear that the generation of variation in a population and the action of natural selection in optimizing fitness are independent, sequential processes. Heritable variation is generated spontaneously by the random processes of mutation and recombination – it is not directed towards any future outcome and arises in the same way in the absence or presence of selection.

acquired characteristics a logical impossibility. Weismann proposed that multicellular organisms possessed two types of cell, the germ cells, which conveyed heritable information from one generation to the next (the germ line), and somatic cells, which were used to build the body (or soma). Although environmental effects during the life of the individual might influence somatic cells, the germ cells were immune to external influences, due to the **Weismann barrier**, the principle that information moves only from germ line to the soma, and never from somatic cells to germ cells.

Early in the twentieth century, Danish botanist **Wilhelm Johannsen** described another important dichotomy: the **genotype-phenotype distinction**. He sought to clarify the difference between an organism's hidden, heritable genetic constitution (its genotype) from its observable morphological, developmental or behavioural features (its phenotype). We now recognize that an organism's phenotype is the result of an interaction between its genotype and the environment and that biological variation can be sub-divided into a heritable genotypic component and a non-heritable component, derived from environmental influences. Natural selection can act only on the heritable variation.

Also in the early twentieth century came the solution to another conundrum that greatly troubled Darwin. In the first edition of *The Origin*, Darwin estimated the Earth's age at several hundred million years, long enough for evolution by natural selection to generate the abundant diversity of species. However, physicist William Thomson (later **Lord Kelvin**), disagreed with Darwin's estimate. Thomson calculated the age of the Sun – based on the idea that gravity generated solar radiation – as, at most, only a few tens of millions of years; he also estimated that the temperature of Earth would have been too high even a million years ago to allow for life. Thomson's calculations troubled Darwin. But missing to Thomson and Darwin was the real source of solar energy and heat on earth, **radioactivity**, which enabled a much longer-lived Sun and Earth. Darwin turned out to be right after all.

Putting it bluntly: **variation precedes selection**. But although variation is random, natural selection as a whole is not a random process: as Dawkins has put it, natural selection is the non-random survival of random variants. But natural selection is also not entirely predictable or deterministic. An advantageous mutation may be so rare as to occur in one population but not another. Similarly, fitness depends on heritable features, but also on variation caused by the environment. Survival depends on luck as well as genes: those with a favourable trait may be faced with better chance of surviving, but not complete certainty. In a phrase coined by Jacques Monod: evolution is chance caught on the wing.

Natural selection is shortsighted, without any drive towards some pinnacle of progress or Promised Land. **Cancer** provides the ultimate proof of the shortsightedness of evolution. A tumour cell population evolves progressively, with natural selection favouring mutations that benefit the survival and reproduction of the tumour cells, culminating in a transition to malignancy and the eventual death of the body that supports them.

We see nothing of… slow changes in progress, until the hand of time has marked the long lapse of ages, and then so imperfect is our view into long past geological ages, that we only see that the forms of life are now different from what they formerly were.

Charles Darwin, from *The Origin of Species*, Chapter 4

Darwin imported the idea of uniformitarianism into evolution from geology. Evolutionary change is a **gradual process**: thus, to account for the huge differences between all existing organisms and the divergence between existing and extinct life forms requires a huge expanse of time. The fact that biological change was gradual rather than occurring in sudden jumps is often summed up by a Latin phrase from Linnaeus: *natura non facit saltum* (nature does not make jumps). Evolution has to proceed gradually, drawing on the cumulative power of natural selection, without any chance of ever returning to the drawing board; and yet each stage has to remain viable in the unbroken chain of life. As Darwin put it:

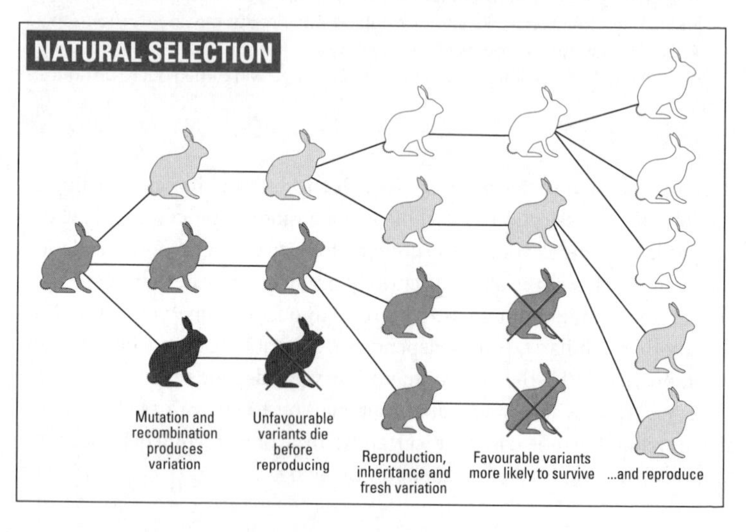

A selection of natural selections

Evolutionary biologists now recognize various types of natural selection: for example, stabilizing, directional, disruptive and balancing selection. **Stabilizing selection**, also called purifying selection, acts to remove genetic diversity from the population, keeping things centred on an adaptive average and selecting against extreme values for a given character trait. For this kind of selection, life is a trade off between the costs and the benefits of a trait. Stabilizing selection is natural selection as a conservative force, thwarting deviations in form or function, silently at work most of the time in most populations. Stabilizing selection explains why the morphologies of some plants and animals, such as sharks and ferns, or the sequences of some important proteins have varied little over millions of years. It also explains why, despite the immense potential for variation inherent within the dog family, as evidenced by the diversity of breeds of domestic dog, wolves and foxes remain much the same generation after generation.

Directional selection occurs when one natural selection favours a phenotype at one extreme of the current range of variation. For example, in a population of bacteria exposed to antibiotics, directional selection might drive the population to become progressively more resistant. Similarly, over-fishing year on year tends to select for progressively smaller fish. To take just on example, the average size of pink salmon in the Canadian North-West has plummeted by a third in just 25 years in response to the introduction of gill netting. Directional selection is also likely to drive an organism's evolution when introduced to a new environment or new niche.

Disruptive selection occurs when individuals at the extremes of a character distribution are preferred over those closer to the average. Thus, on a landscape where dark volcanic outcrops interdigitate with light sandy soil, extremely dark or extremely pale mice will be selected because they are well camouflaged against one or other background. Intermediate forms, which are visible against either background will be picked off by owls and other predators. Disruptive selection is believed to be the driving force behind the birth of new species in the same location, a process called **sympatric speciation** (see p.115).

Balancing selection is a form of natural selection that maintains genetic variation within a population via a number of different mechanisms. In a situation called **heterozygote advantage** (or heterosis), an individual who has two different versions of a gene (a heterozygote) will be favoured over individuals with two identical versions of the gene. The best known example here is sickle cell trait, where people who have a mixture of diseased and normal red blood cells are better protected against malaria than normal people, while those with entirely diseased cells suffer from full-blown sickle-cell disease. Another type of balancing selection is **frequency-dependent selection**, where the more common a character is, the more it is selected against. An extreme example here is the profusion of human tissue types (MHC in the jargon), which have been driven by the action of infectious micro-organisms.

"Whatever the cause may be of each slight difference in the offspring from their parents… it is the steady accumulation, through natural selection, of such differences, when beneficial to the individual, that gives rise to all the more important modifications of structure…" Evolution relies on compound interest, rather than a windfall on the lottery.

Sexual selection: survival of the jiggiest

Although conventional natural selection explains adaptations, there are many contrivances of nature that are highly complex but also highly costly, or even deleterious to survival, that cannot be accounted for in this way: the peacock's tail, the complex and highly decorated follies built by bower birds, the antlers of stags, the horns of male stag beetles or the reggae music of Bob Marley. Darwin wrote to Asa Gray, "The sight of a feather in a peacock's tail, whenever I gaze at it, makes me sick!"

To account for these kinds of phenomena, Darwin proposed **sexual selection**. Although often contrasted with natural selection, sexual selection is in fact a specialized form of natural selection, (the term **ecological selection** is sometimes used to describe natural selection minus sexual selection). Sexual selection is sub-divided into what are now called **intrasexual selection**, typically involving male-male competition, in which males compete aggressively among themselves for access to females, and **intersexual selection**, in which males compete for the attention of the females. As Darwin puts it in *The Descent of Man and Selection in Relation to Sex*: "The sexual struggle is of two kinds: in the one it is between the individuals of the same sex, generally the males, in order to drive away or kill their rivals, the females remaining passive; while in the other, the struggle is likewise between the individuals of the same sex, in order to excite or charm those of the opposite sex, generally the females, which no longer remain passive, but select the more agreeable partners".

Male-male competition drives the evolution of biological weaponry and behaviours peculiar to males (horns, antlers, tusks, spurs, etc). Direct fighting may occur, but less aggressive interactions may also result. Male-male competition cannot explain the peacock's tale, which has no conceivable use in fighting. Instead, it is explained by intersexual selection, which places female choice centre stage. Although many naturalists, including Wallace, balked at the idea that something as fickle and intangible as female aesthetic sensibilities could explain biological phenomena, Darwin fearlessly defended this idea.

Sexual selection in action

Canadian biologists **Barrette and Vandal** have studied **antler combats** in wild male woodland caribou (*Rangifer tarandus caribou*). There are two kinds of combats: fighting, which is rare, violent, and potentially fatal, and sparring, which is common, gentler and seldom dangerous. Barette and Vandal found that in 713 matches between males of different antler size, males with smaller antlers withdrew 90 percent of the time. They suggested that sparring serves as a means to assess a partner's weight and strength relative to one's own. A more subtle approach to male-male competition is shown by males of the **damselfly**, *Calopteryx maculata*. Brown University's **Jonathan Waage** has shown these males have two uses for their penises: not just to transfer sperm, but also, using a series of brushes and hooks, to scrape out sperm already in the female's sperm storage organs.

In *The Descent of Man and Selection in Relation to Sex,* Darwin briefly discusses dancing in birds as a courtship signal driven by sexual selection. A recent study initiated by two Rutgers anthropologists, **Lee Cronk** and **Robert Trivers**, has provided evidence that dance in humans might play a role in sexual selection by providing clues to the dancer's general fitness. Through an unconventional coupling of Darwin and the dancehall style of reggae music, they showed a correlation between ability to dance well and body symmetry in a population of young Jamaicans.

Male caribou lock antlers during a sparring match.

Runaway effects and the handicap principle

In the early twentieth century, English evolutionist **Ronald Fisher** proposed the **runaway principle** as an explanation for sexual selection of traits, like the peacock's tail, that provide no obvious survival advantage. According to Fisher, female preferences initially evolve because the preferred trait is favoured by natural selection and likely to improve the fitness of their offspring. However, once female preferences exist, males with the trait become doubly fit, with advantages both in terms of natural selection and sexual selection. Fisher reasoned that such a link between the female choice and male phenotype could drive a runaway effect in which the preference and the effect become progressively more exaggerated. Such effects may be not limited to features in males – it has been argued that women's breasts are so much larger than is required for lactation because of a link between a male preference for large breasts and **female breast size**, although proponents of contrary views argue that breasts are heavily sexualized only in Westernized cultures and more conventional adaptationist explanations apply.

Another recently developed theory of sexual selection is the **handicap principle**, championed by Israeli evolutionary biologist **Amotz Zahavi**, as a way of explaining traits that are not just not adaptive, but which bring an obvious disadvantage. According to Zahavi, if the male of the species is able to survive to reproductive age with such an obvious handicap, the female considers this verification of his global fitness. Such handicaps might establish he is free of disease, or show that he possesses more than enough speed or strength to overcome the negative effects of the trait.

The evidence for evolution

When Darwin published *The Origin of Species*, it was criticized as a work of speculation, rather too thinly underpinned by facts. However, in the 150 years since its publication, new lines of evidence have opened up and huge masses of confirmatory facts have been slotted into the Darwinian paradigm. Biochemistry and molecular biology have confirmed the unity of life, gaps have been filled in the fossil record and evolution has been observed before our eyes.

Fundamental unity of life

Geneticist Jacques Monod once said that all that is true of the bacterium *E. coli* is also true of the elephant. By that he meant that all organisms run according to the same molecular design rules and all are fashioned from the same building blocks: all cells use DNA for information storage, all cells use proteins to get things done. Darwin knew none of this, because although DNA was discovered in surgical pus in 1869, it was not recognized as the basis of inheritance until the 1940s and 50s. Proteins were discovered earlier, in 1838, but their central role in living organisms was not understood until 1923, when the enzyme urease was shown to be a protein.

It may be argued that building life needs "something like" DNA and "something like" protein, just as written communication needs "something like" the Roman alphabet. But there are many ways in which the "information molecule" and the "doing molecule" could have been built differently. There are hundreds of naturally occurring amino acids that could be used to build proteins (but all living organisms exploit the same twenty or so). There are over a hundred naturally occurring bases that could be incorporated into DNA (but the same four are always used in nature). And recently scientists like Peter Nielsen, Ichiro Hirao, Floyd Romesberg and Peter Schultz have all created exotic yet still functional variants of DNA.

So why should all living organisms use the same alphabet? Imagine discovering a tribe previously unknown to the outside world and then discovering that tribe happened to use the Roman alphabet. Rather than accept that they independently invented the same alphabet that we use, we would be forced to accept that the missionaries must have got there first. To try and explain life's molecular universals without recourse to a common evolutionary origin would be just as absurd. And the existence of such universal standards is not just of academic interest: the very success of gene cloning depends on the inter-operability of the molecular apparatus of microbes and men. Millions of diabetics depend on insulin made by a human gene expressed in a bacterium.

"We can plainly see why nature is prodigal in variety, though niggard in innovation. But why this should be a law of nature if each species has been independently created, no man can explain."

Charles Darwin

Natural selection observed

While evolution over geological time periods (macro-evolution) can only be inferred, evolution in action on a human timescale (micro-evolution) has been observed many times. Time and again, a few years after the introduction of new antibiotics, we see their utility blunted by the emergence of **antibiotic resistance**. Similarly, mosquitoes and head lice have evolved resistance to pesticides, rats to rat poison, rabbits to myxomatosis. In some human infections, such as tuberculosis or AIDS, we even see natural selection at work within the individual patient, helping micro-organisms evade drug treatment or the patient's immune response. Rapid bacterial evolution can even be a force for good in providing micro-organisms that can remove environmental pollutants.

Sparrows, moths and guppies

The humble **house sparrow** provides a compelling example of evolution in action in a well-known animal. Originally from Europe, this bird was introduced to Brooklyn in 1852 and spread explosively across North America, reaching Vancouver by 1900 and Mexico City by 1933. Geographically distinct populations rapidly acquired differences in size, colour and shape.

In Hawaii, a population of sparrows derived from England via New Zealand in the late nineteenth century diverged the most, with the birds turning brown and losing their dark patches. That such changes can occur in a single species within a single human lifespan is testament to the variability inherent in natural species. Human predation has driven evolution even since Darwin's time: fish populations have responded to over-fishing by evolving smaller body sizes and shorter times to maturity. Similarly, the average size of elephant tusks has declined steadily over 150 years in response to hunting and poaching.

The evolution of the **peppered moth** over the last two hundred years provides the classic textbook example of evolution in action. Originally, most peppered moths in England were light-coloured, camouflaged against the bark and lichens on which they rested. During the Industrial Revolution the lichens died off and the trees became blackened by soot. Light-coloured moths were replaced by dark-coloured "melanic" moths – a phenomenon termed **industrial melanism**. Improvements in environmental standards then reversed the changes, so that light-coloured peppered moths have become common again. In the 1950s, English biologist **Bernard Kettlewell** showed that natural selection accounted for these changes. He bred populations of light and dark

A light-coloured peppered moth (*Biston betularia*) on a dark background. Field studies on the peppered moth have illustrated evolution in action.

moths in the laboratory. He marked the underside of their wings to allow identification of laboratory-bred moths. He then released a mixture of light and dark moths in two separate wooded areas: a polluted wood on the Bournville estate in Birmingham and a rural unpolluted wood in Dorset. Kettlewell found that the moths that matched the colour of the tree trunks were the ones that survived – in Birmingham the black moths, in Dorset the light moths.

American ecologist **John Endler** carried out experiments on **Trinidadian guppies** that clearly show evolution at work in real time. Colourful male guppies are attractive to female guppies and to predatory fish, as mates or as food sources. In natural environments, where there are few predators, male guppies tend to be colourful to attract females, but where predators are plentiful, the male guppies are duller. Endler introduced predatory fish to an environment with brightly coloured male guppies. The result was evolution in action: the population of males rapidly shifted from colourful to dull, as natural selection exploited the existing variation within the population as the raw material for change.

Bacteria, flies and mice

Laboratory experiments have also provided powerful illustrations of evolution at work. For more than twenty years, American evolutionary bacteriologist **Richard Lenski** has been running an experiment on the model laboratory bacterium, *E. coli*, in which he has documented numerous adaptive changes in bacterial lineages that were derived from the same ancestral stock, but have then been propagated in parallel. In biotechnology laboratories, a process known as **directed evolution** is now used routinely to generate interesting and useful variants of naturally occurring enzymes.

In 1980, **Michael Rose** initiated a series of experiments in fruit fly evolution, starting with ten populations, split into two sets of five. One set was allowed to breed when young, the other only when old. After a few generations, Rose had a colony of **Methuselah flies**, so called because they lived 10 percent longer than normal flies. After more than two decades of selection, Rose has now evolved fruit flies that live more than twice as long as normal. In 1993, **Ted Garland** started breeding lab mice that are highly active on running wheels. His "High Runner" mice are now three times more active than unselected mice and, like children with attention-deficit hyperactivity disorder, they respond to Ritalin by returning to normal!

The fossil record

The fossil record provides several lines of compelling evidence for evolution. Firstly, the appearance of so **many extinct forms**, from trilobites to *T. rex*, is readily explained from an evolutionary viewpoint. Secondly, the **geological succession** of organic beings (the times at which groups first appear in the geological record) by and large shows a progression from simpler to more complex forms, and matches our expectations from the study of morphology. Thus, invertebrates appear before vertebrates, fish before frogs, reptiles before mammals. The English evolutionary biologist J.B.S. Haldane once stated that all that would be needed to falsify his belief in evolution would be the discovery of a fossil rabbit from the Precambrian! (See www.talkingsquid.net/archives/133 for a modern humorous view of this.)

Thirdly, fossils on occasion provide evidence of **speciation in action**. For example, the fossil record of diatoms (single-celled photosynthetic marine organisms) from the genus *Rhizosolenia* is so complete between 3.4 and 1.6 million years ago that we can see one species split into two. However, as Darwin himself noted, fossilization is such a rare event that much more commonly we see gaps between distinct successive forms. Given that these gaps can often span millions of years, it is no surprise that fossils from before and after the gap show dramatic changes.

Missing links

This brings us on to the key fourth point: despite its shortcomings, the fossil record provides abundant evidence of **missing links** that vividly illustrates evolutionary transitions between taxonomic groups. What's more, our knowledge has progressed steadily since Darwin's day (see boxes on pp.82 and 84). New fossil discoveries have illustrated numerous evolutionary transitions: from fish to four-legged animals, from dinosaurs to birds, from reptiles to mammals, from land-dwelling mammals to whales and sea cows, from limbed reptiles to snakes, from apes to man. We even have insects caught in amber, illustrating the transitions from wasps to ants.

Missing links occur as intermediate and transitional forms. **Transitional forms** appear in the fossil record as organisms that resemble an ancestral form, but possess derived features characteristic of a newly emerging taxonomic group. **Intermediate forms** possess a mixture of traits that puts them at the interface between two taxonomic groups, illustrating how a transition might have occurred, even if, as contemporaries, the intermediate cannot be ancestral to either group. For example, a platypus shares egg-laying with reptiles and milk production and hair with all other mammals, illustrating how the last common ancestor of mammals might have looked. But as our contemporary, it cannot be our ancestor. In addition, it possesses specific adaptations, for example to aquatic life, that were never present in the common ancestor.

Titktaalik, a fishapod from the Canadian Arctic

For a long time evolutionary biologists had assumed that tetrapods (vertebrates, like us, that have four limbs) must have evolved from lobed-finned fish. However, until recently, the gap in the fossil record between the two groups remained frustratingly wide. On the fishy side of the gap, from 385 million years ago, sat *Eusthenopteron*, a lobed-finned fish with tetrapod-like internal nostrils and *Panderichthys*, perhaps the first walking fish, with a pectoral fin skeleton and shoulder girdle intermediate between the equivalents in lobed fish and tetrapods. Closer to the present, on our side of the gap, were two primitive tetrapods from 365 million years ago, known from specimens from Greenland: *Ichthyostega* and *Acanthostega*. These animals are unmistakably tetrapods, with limbs that bear digits (seven or eight, rather than five), but they retain fish tails with fin rays. In the gap between *Panderichthys* and tetrapods lay only enigmatic fragmentary finds – that is, until a recent discovery in Nunavut, in Canada's frozen north. Neil Shubin, Ted Daeschler and their co-workers in the Nunavut field project (tiktaalik.uchicago.edu) went to Ellesmere Island in the Canadian Arctic in the late 1990s with the clear aim of finding a "fishapod", a new intermediate between fish and tetrapods. They focused their search on sediments from just the right time (375 million years ago) and just the right environment (rivers), and, in a beautiful example of predictive palaeontology, they struck gold with three fossilized skeletons of ***Tiktaalik roseae*** (named by Inuit elders after a local word for burbot). *Tiktaalik* proved to be a missing link with a remarkable mixture of fish and tetrapod characteristics: fish gills and scales, ears and limbs of intermediate form, but with a tetrapod neck, rib cage and lungs. Particularly striking is *Tiktaalik*'s flat crocodile-like skull, with its eyes on top. *Tiktaalik* has rapidly become a classic example of a transitional fossil.

A classic example of a transitional fossil.

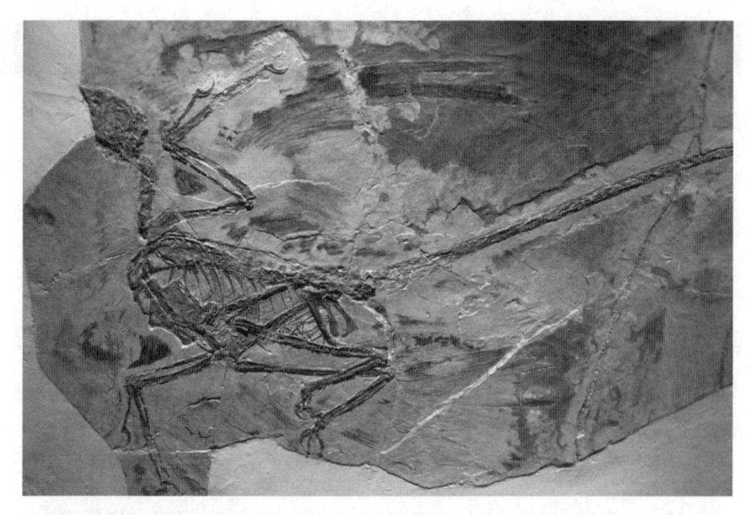

***Microraptor gui*, a feathered four-winged dinosaur recovered from fossil beds in Liaoning, China.**

The 150-million-year-old ***Archaeopteryx*** is the most famous missing link. Similar in size and shape to a magpie, this primitive feathered bird displays a mixture of features associated with birds and dinosaurs. But in recent years, the cutting edge in the study of avian evolution has shifted elsewhere, to the **Yixian Formation**, a geological site in China that has yielded striking transitional fossils of **feathered dinosaurs**. The most primitive is *Sinosauropteryx*, from 150–120 Ma, whose body was covered with feather-like structures resembling hollow tubes, or hairs (whether these really are related to feathers is disputed by some palaeobiologists). More convincing bird-like feathers are seen on *Protarchaeopteryx* and *Caudipteryx* (135–121Ma), but their feathers were probably used for display or insulation rather than flying. *Microraptor,* a small bird-like dinosaur from the same period carries the first known feathers clearly adapted for flying. Curiously, *Microraptor* had feathers on both fore and hind-limbs, earning it the name the **four-winged dinosaur** – like the Wright brothers, *Microraptor* took to the air with two sets of wings. Among feathered dinosaurs, the *Jurassic Park* favourite, **Velociraptor**, provides a telling example of the predictive power of evolutionary thinking; the presence of **bird-like feathers** on Velociraptor was first predicted because of its close relationship to other feathered dinosaurs, but in 2007 was confirmed from direct fossil evidence of quill knobs (points at which feathers are anchored to bone).

Whippos and walking whales

Until the 1990s, cetaceans (whales, dolphins and porpoises) were thought to be out on a limb taxonomically with an extinct group of hoofed carnivores, the mesonychids. However, molecular studies have shown that cetaceans belong with the **artiodactyls** (even-toed hoofed mammals such as pigs, cows, camels and giraffes). Surprisingly, their closest land-living relative is the hippopotamus, so the last common ancestor of whales and hippos is sometimes called a **whippo**!

Darwin noted the lack of fossils illustrating the evolution of whales from land mammals. However, in recent decades, spectacular fossil finds from Pakistan and India have illuminated the transitions that turned land dwellers into exclusively aquatic mammals. *Indohyus*, a small deer-like mammal from the Eocene, sits at the first branch point on the path leading to modern cetaceans. Recent studies on *Indohyus*, led by Hans Thewissen, have shown that a wading aquatic lifestyle predated the origin of cetaceans, but unlike cetaceans, *Indoyus* remained a vegetarian. *Pakicetus inachus*, the earliest known cetacean (from the early Eocene, 56–34 million years ago) looked rather like a dog with hoofs and illustrates the shift to a carnivorous lifestyle. *Pakicetus* was originally identified as a close relative of whales through the distinctive position and contents of its middle ear, shared only with other cetaceans. Its teeth also resemble those of later fossil whales. Complete fossil skeletons have confirmed that these mammals were shore-living land-dwellers.

Biogeography

The study of biogeography – the question of what lives where and why – provides a rich source of evidence for evolution. Without evolution, it would be reasonable to assume that a species should live wherever the climate and environment provided a suitable habitat. But then, why, until humans introduced them, were there no wallabies in England or rabbits in Australia, even though each can live wild in either setting? Why is geographical proximity a better predictor of biological similarity than similarity of climate? Evolution provides an explanation: **geography reflects genealogy**. According to the evolutionary view, if species diverge in time and space from common ancestors, the most closely related species should be found close together geographically, irrespective of their habitat or specific adaptations. Even Oceanic islands obey this rule: the Galápagos and Cape Verde Islands have similar climates, but have a distinctive collection of plants and animals, more closely allied to that of the adjacent continental land mass than to that of the other archipelago.

Ambulocetus ("walking whale" in Latin) is perhaps the most remarkable find from Pakistan: a ten-foot-long, mammalian crocodile. It was amphibious, with hind legs well adapted for swimming and probably hunted like modern crocodiles. Other adaptations to a partially aquatic existence include changes in the structure of the nose and ear. Related walking whales from the remingtonocetid family resembled modern sea otters and probably hunted for fish in the shallows. *Rhodocetus* was another amphibious whale-like mammal that retained features of a terrestrial existence, including hind legs. Its anklebones share distinctive features with those of other artiodactyls, confirming the links predicted from molecular evidence.

The first known exclusively marine whale-like mammals – *Basilosaurus* and *Dorudon* – are known from several sites in North America and Egypt and lived together 38 million years ago. *Basilosaurus* had an elongated body shape: at 18 metres long, it was initially thought to be a reptilian sea monster, and it represents the closest a whale ever came to a snake. It features in *Moby Dick* as the archetypal fossil whale. Despite its sleek aquatic profile, *Basilosaurus* retained two tiny hind legs that probably functioned as love handles during mating. *Dorudon* was more like a modern whale in size (around 5 metres long). Although both *Basilosaurus* and *Dorudon* look very much like modern whales, they were smaller-brained and both lacked the "melon organ" that equips modern toothed whales and dolphins with the ability to use ultrasound. In addition, both groups show a transitional arrangement of nostrils mid-way between the snout and the top of the skull.

Contradictory islands

Ever since Darwin visited the Galápagos Islands, scientists have been fascinated by the **contradictory biology** of oceanic islands. On the one hand, islands remote from any continental land mass show a remarkable paucity of certain plants and animals – for example, most remote oceanic island lack mammals. As Herman Melville wrote of the Galápagos: "Little but reptile life... The chief sound of life here is a hiss. No voice, no low, no howl is heard". Even a large island territory like New Zealand is home to only two native species of land mammal, both bats. And yet, islands feature as the hottest of hot spots for **biodiversity** for many groups of animals or plants. New Zealand has more flightless birds than any other country. In the Galápagos, islands within sight of each other are home to distinct species of birds or reptiles found nowhere else. Hawaii is another island group that abounds in biodiversity (see box, p.87).

Evolution provides a ready explanation for the rareness of some groups on islands but the abundance of others. The rareness is explained by the fact that the islands have appeared relatively recently in geological terms, while remaining remote from any continental source of new species. Even though the most remote islands are occasionally colonized through rare events – storms blowing airborne insects and birds or rafts of vegetation washed out to sea carrying plants and smaller animals – these will tend to exclude larger terrestrial animals or inland plants. But any new island populations descended from the lucky colonists will, over time, diverge from the parent population, often rapidly spawning many new species through a process known as **adaptive radiation**. Through this process island incomers rapidly diversify to fill vacant niches, occupied by other groups on the continental homeland. The island continent of **Australia** provides the most extensive example of adaptive radiation among mammals. Because placental mammals evolved elsewhere and never reached Australia, marsupials diversified to occupy a wide variety of niches occupied by placental mammals elsewhere in the world – from marsupial wolves to marsupial moles.

The marine iguana (*Amblyrhynchus cristatus*) is unique to the Galápagos and the only lizard to live and forage in the sea.

Hawaii: evidence of evolution from the fiftieth state

The Hawaiian islands provide a remarkable example of the peculiarities of island biology. These volcanic islands are over two thousand miles from the nearest continent. They have no native species of reptiles, amphibians, or conifers and only two native species of mammal (a bat and a seal). Before the arrival of humans, there were no termites, ants, or mosquitoes.

However, these islands are home to an enormous number of species of flies: around **800 species** of *Drosophila* and *Scaptomyza*, about a quarter of the worldwide total! The only plausible explanation is an evolutionary one – adaptive radiation. Many other species of plants and animals from Hawaii also originate from adaptive radiations. These include the fifty or so species of **Hawaiian honeycreepers** that have evolved from a single finch-like colonist. This ancestor had a small bill, but today's honeycreepers possess a wide variety of bill shapes, each specialized for a specific diet. Then there are the 240 or so species of **Hawaiian crickets**, evolved from as few as three progenitor species. The grandly named **silversword alliance** is a set of thirty species of plants native to Hawaii that taxonomists group together on account of shared, but obscure, features of leaves and flowers that are accompanied by riotous variability. The glorious flowering silverswords live high on volcanic slopes, but their relatives include trees, shrubs, mats and vines that occupy elevations ranging from sea level to the high mountain or habitats as diverse as desert and rainforest. Molecular phylogenetic studies have shown that all members of this alliance are descended from a single ancestral species that arrived in Hawaii millions of years ago. Their closest living relative outside Hawaii is the tarweed, a daisy-like plant from the west coast of North America. Tellingly, tarweed comes equipped with an evolutionary smoking gun: a sticky fruit just perfect for hitching a lift with a migratory bird.

Evolution in Hawaii Steve Olson (National Academy of Sciences, 2004)

But why should continents as well separated as Australia and South America share similar organisms (such as marsupials) while regions as close as the northern and southern islands of Indonesia (on either side of Wallace's line) have quite different faunas? An explanation came with the acceptance of **continental drift** and **plate tectonics** in the late twentieth century: Australia and South America were once joined in a single southern continent, **Gondwana**, while the two parts of Indonesia have separate tectonic histories, only recently coming into contact. The emerging field of **vicariance biogeography** has shown that organisms as diverse as midges, fish, frogs, turtles and birds show patterns of distribution and divergence explained by the movement of continents.

Variation under nature

What is a biological species? The most widely used definition for plants and animals – the **biological species concept** – views a species as a population of individuals that can reproduce with one another but which cannot reproduce with members of other populations (from which they are said to be "reproductively isolated"). But if new species emerge gradually through branching evolution, rather than through sudden creation, we can expect to find intermediates along the path of speciation, from partially interbreeding populations via populations that can interbreed only with reduced or loss of fertility, to fully isolated populations. Darwin himself noted that it is impossible to distinguish **varieties** and **species** merely by observing their morphological features, particularly when one looks at the immense variation under domestication. If a Chihuahua and a Great Dane occurred in the wild, their appearance would suggest that they should be classified as separate species, but as with all other breeds of dog, they can interbreed and produce fertile offspring.

Pairs of closely related species – lions/tigers or donkeys/horses – illustrate an intermediate stage in reproductive isolation, in that they can cross-breed in captivity and produce viable offspring (ligers or tigons; mules or hinnies), but the offspring are generally, but not always, infertile. Similar inter-species **hybridization** occurs in nature; for example, the West European carrion crow (*Corvus corone*) and the Asian hooded crow (*Corvus cornix*) have distinct ranges but meet in narrow hybrid zones in Germany, Denmark, Britain, northern Italy and Siberia, where they inter-breed.

The hybrid offspring of a horse and a zebra. Darwin discussed such equine crosses in *The Origin*.

Ring species provide the most striking demonstration of how small cumulative changes can lead to the formation of new species. Ring species form as an ancestral species expands along two geographical fronts, looping around an area of unsuitable habitat. At one point in the ring of populations, two distinct forms coexist without interbreeding, behaving as different species. However, around the rest of the ring, the traits of one of these species change gradually, through intermediate populations, into the traits of the second species. Around two dozen ring species have been described, but two in particular have been the focus of intensive scientific scrutiny: *Ensatina* salamanders and greenish warblers. *Ensatina* **salamanders** live in the mountains along the west coast of North America. Two distinct varieties, differing dramatically in colour, coexist in southern California and interbreed there only rarely. However, a chain of interbreeding populations, encircling the Central Valley of California, connects the two varieties. Similarly, two populations of **Greenish warblers** overlap in a narrow zone of central Siberia where they differ in colouration, in genetic characteristics and in the songs males use to attract mates. However, the traits that distinguish the two kinds of warbler change gradually through a ring of populations encircling the inhospitable Tibetan Plateau to the south.

> **"... I was much struck how entirely vague and arbitrary is the distinction between species and varieties."**
>
> **Charles Darwin**

Speciation observed

For plants and animals, speciation can be defined as the production of a novel population of individuals that can no longer breed with members of the progenitor community or with any other population. The birth of a new species has been observed repeatedly, many times under the hand of man and several times in nature. The commonest route to speciation involves doubling of the number of chromosomes, a process known as **polyploidy**. As chromosomes have to pair up during reproduction, the resultant mismatch between parent and offspring precludes inter-breeding and signals the foundation of a new species. For example, in the twentieth century, Dutch botanist **Hugo de Vries** discovered a variant of the evening primrose (*Oenothera lamarckiana*) with twice the usual number of chromosomes. He was unable to breed this variant with the parent species and so named it *Oenothera gigas*. In addition to providing a mechanism for speciation, chromosome doubling provides a route for rescuing otherwise sterile inter-species hybrids; a new species of primrose, *Primula kewensis*, was generated this way in Kew Gardens in 1912. **Goatsbeard** (*Tragopogon*) plants provide an example of natural hybrid speciation. Three species were introduced into North America in the twentieth century: *T. dubius*, *T. pratensis* and *T. porrifolius* (this latter the source of the vegetable salsify). Within a few decades all three had became established in the USA. In the 1950s, botanists discovered two new species in the parts of Idaho and Washington where the original species overlapped. One new species, *T. miscellus*, is a polyploid hybrid of *T. dubius* and *T. pratensis*, while the other, *T. mirusis*, is derived from *T. dubius* and *T. porrifolius*. Polyploidy is also found in many crop plants, including cabbage cereals, such as the wheat-rye hybrid **triticale**. It has also been seen, albeit rarely, in mammals: in two species of South American rodent, the red and the golden viscacha rats.

Reproductive isolation without polyploidy has been observed in plants and fruit flies, particularly under artificial selection. The American **apple maggot fly** (*Rhagoletis pomonella*) provides an interesting example of incipient speciation in the wild. This species is native to eastern North America, where it feeds on the native hawthorns. However, a variety of this species that feeds on apples spontaneously emerged in the early nineteenth century, following the introduction of the apple from England.

Galápagos finches. Darwin commented on the perfect gradation in the size of their beaks in *The Voyage of the Beagle*.

Darwin's finches: evolution in action

The Galápagos Islands are home to thirteen distinct but closely related species of **finches**. These birds are all of similar size and colouring (brown or black), but differ in the size and shape of their beaks: the ground finches have deep, wide beaks, the cactus finches have long, pointed beaks, while the beaks of the warbler finches are more slender. They represent a classic example of adaptive radiation, occupying niches as diverse as those of the tool-using woodpecker finch or the vampire finch that supplements its diet with the blood of sea birds.

Despite playing only a minor role in Darwin's thinking, Darwin's finches have in recent decades provided one of the most compelling and exciting examples of evolution in action. Since 1973, the husband and wife team of Princeton ecologists **Rosemary** and **Peter Grant** has been tracking thousands of individual finches generation after generation, documenting natural selection in action. They have shown that during a year of drought on the Galápagos in 1977, while finch populations plummeted, natural selection wrought tangible evolutionary changes on the finches. Among medium ground finches (*Geospiza fortis*) there was an increase in body size and beak size. This was a consequence of only the biggest and toughest seeds surviving the drought, nuts that are usually too difficult for medium ground finches to crack. The Grants estimated that if such droughts were to occur once every ten years, a new species of finch could arise in just over two centuries. However, they have also shown that in years when rainfall is plentiful, the tables are turned: natural selection then favours smaller birds. In recent years, Harvard developmental biologist **Cliff Tabin** has unravelled some of the biochemical basis of beak variation, showing that natural selection has been acting on the expression of a signalling protein, calmodulin.

Darwin's Finches: Evolution of a Legend Frank Sulloway (*Journal of the History of Biology*, 15, 1982) Available at www.sulloway.org/Finches.pdf

The Beak of the Finch: A Story of Evolution in Our Time Jonathan Wiener (Knopf, 1994)

Today there is an apple-feeding race that avoids feeding on hawthorns and a hawthorn-feeding race that does not normally feed on apples. The two varieties seldom inter-breed and have now begun to diverge genetically, despite similar appearances.

Groups within groups

One of the most striking features of living organisms is that they fall into a **hierarchical organization** of groups within groups (lions and tigers within carnivores, carnivores and primates within mammals, mammals and birds within vertebrates and so on). Although a higher taxonomic group can contain multiple lower groups, no lower group is ever a member of more than one higher group. This hierarchical arrangement was recognized as the basis of biological classification by the eighteenth-century Swedish naturalist, **Linnaeus**, and has coped admirably with the addition of thousands of newly discovered species. Long before Darwin, naturalists felt that there was something non-arbitrary, something natural, about this approach to biological classification but, until the discovery of evolution, they had no plausible explanation for it.

Carl Linnaeus (1707–78) laid the foundations of modern taxonomy.

The natural, hierarchical organization of living organisms stands in stark contrast to what we see when we try to classify many other things. Take the book that you are holding. You might want to shelve it with similar books so you can find it again. But how do you decide what kind of similarity to use? You could shelve it with volumes of similar size. But you might prefer to separate hardbacks from paperbacks. You could use an alphabetical system. But then do you catalogue according to author or title? You could try classifying by subject and put it with the other science books. Why not put it with your other *Rough Guides*? But then how much really links evolution, reggae, the Gambia and climate change? Hierarchical systems for classifying books by subject do exist, such as the Dewey Decimal System. However, such systems are dogged by arbitrary, subjective choices to the point of bigotry: in the Dewey system, a hundred numbers are set aside for religion, eighty of them for Christianity, one each for Jews and Muslims, while Buddhists (294.3), Sikhs (294.6) and Jains (294.4) have to hide behind the decimal point of a single number. And whether one classifies a book on "Jews for Jesus" under Christianity or Judaism remains an arbitrary choice!

Biologists don't have this problem, biological classification remains largely objective and depends on non-arbitrary choices. The layman may have difficulties, after superficial consideration, with the **classification of whales** – are they mammals, fish or both? But the expert has no such difficulty. Well before any evolutionary explanation had been proposed, Linnaeus recognised that whales should be classified as mammals, on the basis of the structure of their hearts and lungs, their warm-bloodedness, and because they nurse their young and possess hollow ears and moveable eyelids. Darwin's theory of evolution – descent with modification twinned with repeated branching giving rise to new lineages – provides a ready explanation of life's natural hierarchy, because "all true classification is genealogical".

"From the first dawn of life, all organic beings are found to resemble each other in descending degrees, so that they can be classed in groups under groups. This classification is evidently not arbitrary like the grouping of the stars in constellations."

Charles Darwin

The great tree of life: molecules and morphology

Since Darwin's time, biologists have drawn countless phylogenetic trees. For over a hundred years after *The Origin*, these trees almost always depended on morphological characters. However, in the last forty years or so, biologists have gained access to a second, independent line of evidence, trees drawn from the sequences of DNA and protein molecules. A linguistic analogy clarifies why sequence evidence is independent of morphological evidence. Just as there is an almost infinite number of ways to create a love poem conveying (more or less) the same message, there is an astronomically large number of ways to build a gene (or a protein, or even a limb) with identical function from different sequences of the same molecular letters. Thus, if organisms were all independently created, there would no reason to assume that phylogenetic trees based on these two independent lines of evidence – morphology and molecular sequences – would agree. However, if, as we predict from the theory of evolution, there is one true tree that connects all species in an objective genealogy, the two lines of evidence should converge on the same tree.

As predicted, carefully constructed phylogenetic trees based on morphology and molecular sequences match each other very closely, at least for animals (bacteria don't play by the same rules). It is worth re-emphasizing that the odds of this happening in the absence of evolution are infinitesimal. For as few as ten taxa, the odds of arriving at two similar trees just by chance are millions to one. More recently genome sequencing has provided additional kinds of evidence that can be used to draw trees (for example, repertoires of genes or proteins, or positions of genes). As expected these trees concur with those produced by earlier approaches, providing yet more molecular evidence for evolution.

The baggage of history

As organisms evolve, their features are subject to the maxim of "use it or lose it": useless features are discarded. However, their disappearance is seldom instantaneous; instead, such features decay gradually, leaving vestiges of ancestral forms. Blind cave-dwelling animals with degenerate eyes provide a simple illustration of vestigiality. The Mexican **blind cavefish** (a popular inhabitant of the domestic aquarium) belongs to the same species as the river-dwelling variety. Both varieties develop a

Triplophysa xiangxiensis, a blind cave fish from Hunan province in China. The eye develops early in life but degenerates before adulthood.

precursor lens and rudimentary optic cup during the embryo's first 24 hours. But in the blind form, the lens degenerates, normal development of the retina is ablated and the cornea and iris never appear. The eyeball sinks down, roofed over by a flap of skin. Something similar happens in the olm, a **blind cave salamander** that lives in subterranean lakes in the regions around Trieste. While eyes develop in the larvae, they regress later in development, to be covered by skin in the adult. Why build an eye and then bury it? Degenerate eyes have also evolved in many burrowing animals, not just in the true moles of the northern hemisphere, but independently in the golden moles and naked mole rats of Africa and the marsupial moles and blind mole rats of Australia. The eye of the blind mole rat *Spalax ehrenbergi* represents an interesting intermediate in vestigiality: the animal cannot see, but its subcutaneous retina captures enough light to drive circadian rhythms.

"Rudimentary organs may be compared with the letters in a word, still retained in the spelling, but become useless in the pronunciation, but which serve as a clue in seeking for its derivation."

Charles Darwin

Vestigial limbs and genes

The **wings** of flightless birds, such as ostriches, emus, cassowaries, rheas and penguins, are also vestigial structures. Although ostrich wings still perform simple functions, in balance or courtship, they can be considered vestigial because they are useless for their original function – as wings built for flight. Additional examples of vestigial organs of flight include the wings of flightless beetles, often hidden under fused wing covers, and the non-functional wings sometimes seen in female Gypsy moths. **Leg bones** in legless vertebrates also illustrate vestigiality. Whales, dolphins and porpoises possess small vestigial leg bones concealed within the bulk of the body, which represent remnants of the legs of their land-living ancestors. Some whales also have undeveloped, unused, pelvis bones. Many species of pythons and boas carry hidden vestigial pelvises floating free in the abdominal cavity. In addition, some lizards carry rudimentary, vestigial legs underneath their skin.

Vestigial characters are not restricted to anatomical features, they can include behavioural traits and broken genes. **Whiptail lizards** from the all-female species *Cnemidophorus uniparens* show complex mating behaviours, even though they now reproduce exclusively by parthenogenesis, i.e. without fertilization by males. Non-functional **pseudogenes** ("broken genes") abound in genome sequences. Thus, although most mammals can make vitamin C, we humans cannot because we have lost a key enzyme, L-gulono-gamma-lactone oxidase. However, we have not lost the gene for this enzyme, it is there in our genome, but merely non-functional. Similarly, our genome possesses many non-functional remnants of genes involved in the sense of smell. In fact, comparisons between the human and chimpanzee genomes have identified dozens of genes that have stopped working since our divergence from the human-chimp ancestor.

Atavism

An **atavism** (or "throw-back") is the reappearance of a lost character, last seen in a remote ancestor but not observed more recently. Atavisms are similar to vestigial structures, but occur only in rare individuals, rather than in the whole species. As Darwin put it: "these characters, like those written… with invisible ink, lie ready to be evolved whenever the organisation is disturbed". Julius Caesar owned a horse with **extra toes**, a throw back to the arrangement of digits in an ancestral horse. In October 2006, Japanese fishermen captured a bottlenose dolphin with an extra pair of **hind limbs**, atavistic evidence of evolution from a terrestrial precursor. In fact, hind limb parts have been observed many times in sperm whales, and occasionally in other whales: the vestigial limbs include femurs (attached to the whale's vestigial pelvis), tibias, fibulas and, rarely, even feet with digits. **Teeth** in birds provide another example of atavism. Although their dinosaur ancestors possessed teeth, birds lost them at least 70 million years ago. However, laboratory experiments have shown that bird tissues can be induced to produce teeth (an **induced atavism**) and in 2006, **John Fallon** even found crocodile-like teeth in the embryos of a mutant chicken.

Development

The form and function of an adult organism is the product of numerous cumulative developmental steps. Evolutionary changes in the appearance of the adults must result from changes in the basic developmental processes. Among distantly related organisms, one would expect to see conservation in the early stages of development, but divergence later on. With organisms that are more closely related, we should expect to see them share their developmental programme for progressively longer.

Human vestigial features and atavisms

Your **appendix** is a sometimes troublesome vestige of the tip of the caecum, a structure used to digest plants in other mammals. In parallel with its degeneration in humans, the caecum is similarly diminished in other mammals that have switched away from a diet rich in tough plant matter (for example, dogs and cats). Your **coccyx**, or tailbone, consists of four fused vertebrae at the base of the spine that represent remnants of a lost tail. Even though the coccyx still serves minor functions, as an attachment point for parts of the gluteus maximus muscle, it is clearly vestigial as a tail, having lost its original functions in facilitating balance, mobility or grasping. In rare cases of atavism, humans can be born with **tails**; over a hundred cases have been reported in the medical literature. Although these human tails contain a complex arrangement of connective tissue muscles, blood vessels and nerves, they usually lack skeletal structures. However, several have been found with cartilage and up to five, well-developed, articulating vertebrae. The **semilunar fold** is a small fold of conjunctival tissue on the inside corner of the human eye. It is the vestigial remnant of the nictitating membrane, a translucent or transparent "third eyelid" that sweeps horizontally across the eyeball in many other vertebrates. The three **extrinsic muscles** of the **human ear** (the *Auricularis superior*, *Auricularis anterior* and *Auricularis posterior*) are greatly underdeveloped compared to those of monkeys (where they are used to re-orientate the ears without moving the head). Some people can use the muscles to move the ears; in most people, they are entirely useless. **Darwin's tubercle** is a small point of skin toward the top of the outer ear, found in a minority of humans. It is thought to represent an atavism, recalling the pointed top of an ancestral ear.

"Man still bears in his bodily frame the indelible stamp of his lowly origin."

Charles Darwin

In mammals with fur, raising the hair is used to trap air, which helps keep the animal warm. One could just about argue that **goose bumps** in the cold serve the same purpose in humans. But many mammals also raise body hair under stress, to make them appear larger in the hope of scaring off predators or adversaries. In humans, the formation of goose bumps under stress must now surely count as a vestigial reflex.

When we compare vertebrate embryos, this is more or less what we see. As Baltic-German biologist **Karl Ernst von Baer** (1792–1876) noted in the early nineteenth century, at an early point in their development the embryos of different species are much harder to distinguish from each another than are the equivalent adult forms. In addition, comparative embryology reveals a branching pattern in development that is consistent with, and even evidence of, Darwin's branching evolution.

Disappearing limbs

Comparative embryology provides another compelling reason for believing that evolution has happened. To see why, lets make a comparison between the developmental evolution of human artifacts and living organisms. Cars have changed over the years. When I was a boy, I learnt that a carburettor was an essential component of all cars. A few years ago I was surprised to discover that carburettors had been universally abandoned in cars built since the late 1980s and early 1990s. In human design processes, one can go right back to the drawing board and make radical changes in design. Thus, one would expect to see no trace that the carburettor had ever existed in a modern production line.

Contrast this with what we see in biological evolution, which is governed by gradualistic transformation, where sudden radical changes in design and development are seldom permitted. Instead, in biological organisms,

Men's tits, women's clits and the female orgasm

Men have nipples and vestigial breast tissue. Darwin once argued that these were evidence of an evolutionary past in which both sexes suckled their young – a phenomenon still evident in the **Dayak fruit bat**, the only mammal in which males routinely assist females in breast-feeding their young. An alternative explanation relies on the interplay between development and evolution. The default mammalian body plan is female; a switch to the male body plan requires the action of male hormones from the seventh week of development. However, by this time the tissue that gives rise to nipples and breasts has already been formed, making it hard for evolution to devise a mechanism that specifically removes these tissues from males. Whatever the explanation, the fact that thousands of men develop breast cancer every year means that male breasts rate as an example of "unintelligent design".

In his 1991 essay, *Male Nipples and Clitoral Ripples*, **Stephen Jay Gould** raised similar arguments over the existence of the female clitoris and the female orgasm – the developmental and behavioural homologues respectively of the male penis and ejaculatory orgasm. A key problem for any adaptationist explanation for these features of female sexuality is that most human sexual intercourse occurs without female orgasm. After berating Freud and heaping praise on sexologists Kinsey and Hite, Gould concluded that these features of the human female have no adaptive function and, like nipples, are just a byproduct of a common human developmental programme. As the clitoris is made from the same foetal tissue as the penis, it can't help but precipitate orgasms too. Such views have been echoed recently by philosopher-biologist **Lisa Lloyd**, who, in her book *The Case of the Female Orgasm*, has exhaustively examined every other possible explanation for this phenomenon, including the "phallacies" of male vanity.

when an ancestral character is abandoned in the adult form, it is often retained in the early stages of growth: the character is not struck off the drawing board, but instead its later development is merely suppressed or modified so that it never reaches its full-blown ancestral form.

Imagine how strange it would be to see carburettors inserted during the manufacture of a modern car, only to be removed further down the production line! Yet, in **legless vertebrates**, such as snakes and slow worms, **limb buds** appear during embryonic development, only to be reabsorbed before hatching – clear evidence of evolutionary descent from a four-legged ancestor. Similarly, although outside the womb, whales and dolphins lack hind legs, **hind limbs** appear in the cetacean foetus, complete with bones, nerves, and blood vessels, only to disappear before birth. The human embryo develops a **tail**, which is subsequently re-absorbed.

Looking further back in our evolution, for a time early in the development of mammalian embryos, we see **pharyngeal arches**, barely distinguishable from the **gill pouches** of fish. In fish embryos, the pharyngeal pouches eventually develop into gills, whereas in mammals they give rise to structures that evolved from gills, such as the Eustachian tube, the middle ear, the tonsils, the parathyroid, and the thymus. In fish, several branches of the tenth cranial nerve (the vagus nerve) make a quick loop around the arterial arches that supply the gill slits. The **recurrent pharyngeal nerve** represents the mammalian equivalent of one of these branches, connecting the brain to the larynx (a derivative of the sixth pharyngeal arch). But during mammalian development, arterial components of the sixth pharyngeal arch migrate into the thorax to become the pulmonary artery. Tied in a developmental and evolutionary knot around the artery, the recurrent laryngeal nerve loops down into the thorax, avoiding what any rational designer would see as the shortest route to its target. In the giraffe, at fifteen feet in length, this nerve stands as the ultimate in undesigned biology!

Homology

Richard Owen coined the term "homology" to account for the similarity of form seen in anatomical structures from different organisms. In 1843, he defined **homology** as "the same organ in different animals under every variety of form and function". While one would expect structures that performed precisely the same function to be built in exactly the same way, inherent in Owen's concept of homology was the fact that deep similarities in structure and organization are sometimes seen in organs that performed different functions. For example, very similar arrangements of very similar bones and muscles occur in primate arms, bat wings, bird wings, pterosaur wings, dolphin flippers, horse legs, pig legs, the digging forelimbs of moles, and webbed amphibian legs. Owen's rather feeble explanation for this homology was that there must be a common structural plan for all vertebrates, which he called the archetype. But, as Darwin notes, this amounts to little more than a restatement of the facts. The only compelling explanation is an evolutionary one: homology is similarity that results from descent from a **common ancestor**.

Through the careful study of comparative anatomy, morphological homology can be seen in many other contexts: homology can be glimpsed in the **mouthparts** of insects, despite numerous modifications that enable insects to exploit a wide variety of food sources and in the numerous derivatives of **arthropod limbs** (the wings of a fly, spinnerets of a spider, book gills of the horseshoe crab). Homology can also be established, through scrutiny of the fossil record, between the three small bones of the middle ear of mammals and bones in the jaw of fish.

Detailed studies of structures independently co-opted to the same function ram home the point that similarities in function cannot explain structural homology; instead, due to the contingencies of evolutionary history, even when homologous parts are independently recruited to the same function on repeated occasions, they are re-deployed in different ways. For example, as **Pat Shipman** describes in her book *Taking Wing*, the wings of bats, birds and pterosaurs all function for flight, but each has been modified from the ancestral pentadactyl limb in a different way: in the "finger wing" of a pterosaur, the fifth digit supports a membranous wing; in the "arm wing" of a bird, the flight feathers are mounted all along the arm and the digits are short; in the "hand wing" of a bat, elongated digits 2–5 support the wing membrane. These different wing designs illustrate one of the most curious features of homologous structures: that they often look as if they have been jerry-rigged into new functions in an ad hoc fashion through bricolage.

Sequence homology

In the last two decades, molecular biology has also provided stunning evidence of **sequence homology** among DNA and protein molecules from a wide range of organisms. Eugene Koonin and others have identified over a hundred proteins that are common to all living organisms, providing evidence of descent from a universal common ancestor. However, as with morphological homology, it is the strange cases of molecular bricolage that

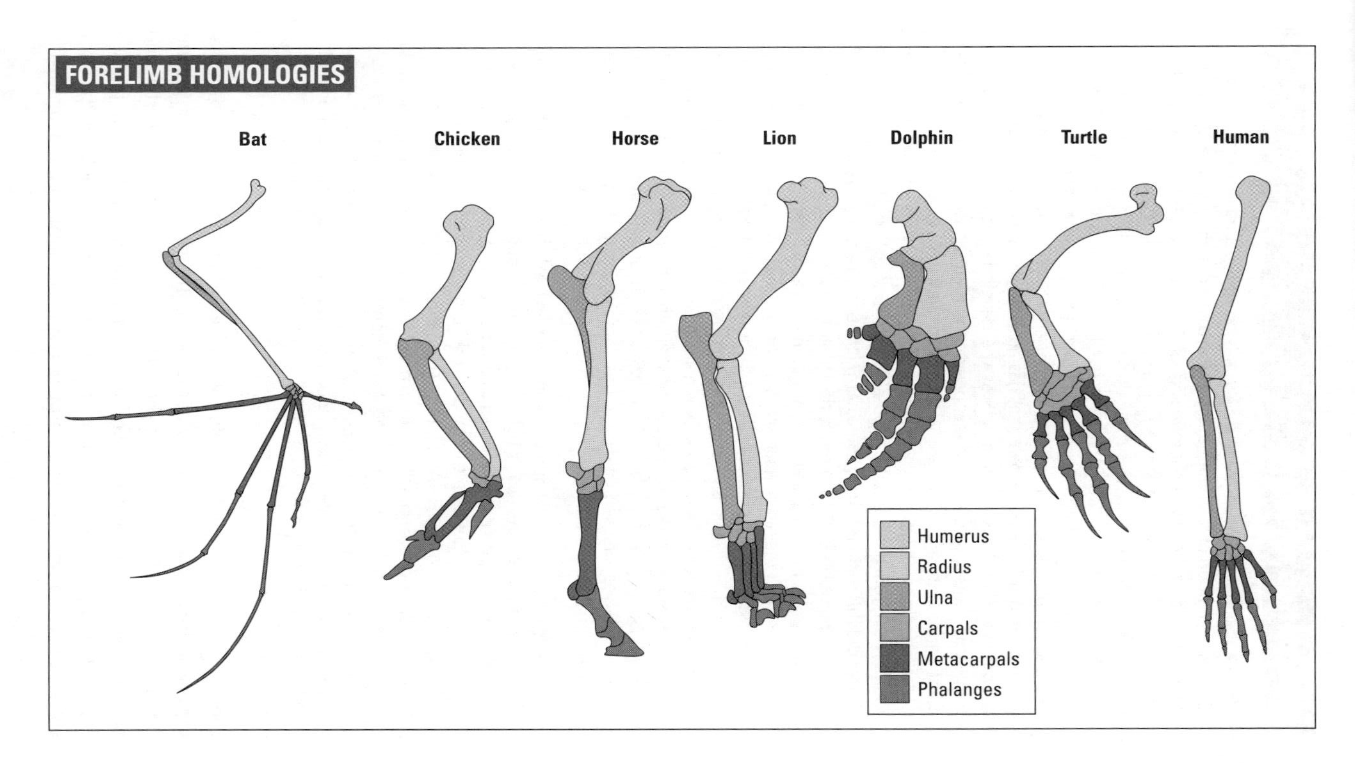
FORELIMB HOMOLOGIES
Bat
Chicken
Horse
Lion
Dolphin
Turtle
Human
Humerus
Radius
Ulna
Carpals
Metacarpals
Phalanges

The panda's thumb

The giant panda, *Ailuropoda melanoleuca*, is an unusual member of the order Carnivora (a group of mammals that subsist wholly or largely on meat), because its diet consists primarily of bamboo shoots. As a plant eater equipped with the digestive system of a meat eater, the giant panda struggles to glean enough calories from its diet. Just to survive, it has to munch its way through 10–15 kilograms of bamboo shoots, day in, day out. In 1964, Chicago anatomist **Dwight Davis**, after 25 years of study, published a meticulous description of the anatomy of the giant panda. Davis argued that, contrary to a supposed link to racoons, the giant panda is, in fact, a bear that has adapted to eating plants. One of the adaptations that Davis highlighted was the development of a sixth digit, an opposable thumb, to grasp and manoeuvre the bamboo while eating. But the giant panda's "thumb" is not a true thumb! In the panda's hand, there are the usual five digits – four, like your fingers, built from four bones; one, like your thumb, built from three. But the opposable sixth digit contains but a single bone. Where did this panda's thumb come from? If one compares the hand of the giant panda to that of bears, it becomes clear that the bone in the panda's "thumb" is homologous to a tiny bone, found in the wrist of bears and some other mammals, called the radial sesamoid. Sesamoid bones add strength to tendons as they cross joints. The panda's "thumb" is a greatly enlarged sesamoid bone. Thus, rather than evolve an opposable digit from the true thumb, as humans – and even racoons – have done, the giant panda has, instead, jerry-rigged a wrist bone as a workable contraption to strengthen its grasp. Curiously, even in the panda's foot, the equivalent sesamoid bone is enlarged compared to that of bears. Is this a separate adaptation to tree climbing, or an incidental side effect of a developmental programme common to fore and hind limbs?

One other carnivore has adapted to eating bamboo: the red panda. But the red panda is not a bear; instead, it belongs in a large group with weasels, skunks and racoons. Despite its independent adoption of a herbivorous lifestyle, the red panda also possesses a thumb derived from a sesamoid bone: a remarkable instance of convergent evolution. The panda's thumb, as an evolutionary co-option of a bone, originally there to prevent wear and tear in a tendon, to a novel function in handling bamboo, has become an iconic example of evolution's blind attempts to make the best of the materials to hand. It has leant its name to a book by Stephen Jay Gould, to a blog for discussing evolution and even to the controversial creationist "textbook", *Of Pandas and People*.

provide the best evidence of evolution. For example, one might expect that the **crystallin proteins** used to build the lens of the eye in different organisms might be unique to lens tissues, show evidence of common ancestry and have evolved specifically for use in the lens. Instead, crystallins provide multiple examples of molecular opportunism, where metabolic enzymes or the cell's fire-fighting and repair protein responses have been independently

The skeleton of the giant panda's paw reveals how the "thumb" is, in fact, a modified wrist bone.

co-opted into the role of crystallins in many different lineages. Indeed, in some cases, the very same proteins are deployed as a crystallin in the eye and as an enzyme elsewhere; in the duck eye, the enzyme lactate dehydrogenase becomes one kind of crystallin, while another enzyme, alpha-enolase, becomes another. Similar evolutionary opportunism is the only explanation as to why the crystallin of a squid eye should closely resemble an enzyme, aldehyde dehydogenase, used in humans to recover from a hangover!

5 Evolutionary biology

It would be nice to say that after Darwin's death, his reputation soared, his ideas were universally accepted and evolutionary biologists lived happily ever after. Sadly, none of this is true. Although Darwin's idea of descent with modification was accepted, natural selection was not. Even during his lifetime, some scientists – including Darwin's American friend, Asa Gray – persisted in seeing a role for divine intervention in guiding evolution.

The eclipse of Darwin

The English biologist **St George Jackson Mivart** (1827–1900) was an early supporter of Darwin, but later argued against natural selection on the grounds that it could not account for the emergence of complex structures (the "what use is half an eye" argument). Another contemporary, **Herbert Spencer**, propounded a version of cosmic and biological evolution that entailed universal progress via natural evolution. Darwin's German admirer, **Ernst Haeckel**, elaborated a misguided view of evolution that is often encapsulated in the phrase "ontogeny recapitulates phylogeny": the view that the development of the individual from single cell to adult recapitulates the evolutionary history of the taxonomic group to which the individual belongs.

After Darwin's death, things got worse: evolutionary biology entered a phase dubbed by Julian Huxley "the eclipse of Darwin". German zoologist Theodore Eimer and several American evolutionists (Osborn, Hyatt and Whitman), held to the idea of **orthogenesis**, that life exhibited an inherent tendency to evolve in a progressive, directional fashion along channels of internal constraint, even to the point of overshooting the mark (for example, deer evolving antlers so big that their owners became extinct).

Gregor Mendel: minding the peas and bees

Gregor Mendel (1822–84) was a German-speaking priest from a part of the Austrian Empire now in the Czech Republic. He entered an abbey in Brno in 1843, took time out to study at the University of Vienna in 1851–53, and then settled into the monastic life for his remaining thirty years. Between 1856 and 1863, Mendel carried out a heroic series of experiments on garden peas, cultivating nearly thirty thousand plants in the monastery's garden. He presented his work in a paper read at two meetings of the local natural history society in 1865 and published in 1866. Mendel's paper elicited little immediate interest from the scientific or the wider community. The focus of his experiments shifted from peas to bees, but then his research petered out in favour of administrative responsibilities. His funeral in 1884 was noteworthy in that Czech composer Janáček played the organ.

In his experiments with garden peas, Mendel wisely chose varieties showing clear-cut differences that bred true when self-fertilized, for example, differences in seed colour (green versus yellow). When crossing such varieties he found that, in the first generation, all plants carried the same character (for example, all seeds were yellow), which Mendel called the "dominant" character. But when these first generation peas were allowed to self-fertilize, something curious happened: both ancestral states appeared, but in a ratio of three of the dominant state (yellow) to one of the other, "recessive" state (green). Mendel's interpretation of his results was that:

1. Inheritance was not blending but particulate.

2. Each of the particles (today called **genes**) that carried inherited information existed in two versions (now termed **alleles**), one dominant (A), the other recessive (a).

3. Each plant had two copies of a gene, which segregated as single copies in the sex cells (pollen or ovum). The true-breeding parent plants carried two identical copies of the gene (AA or aa). The first-generation plants carried one of each type (Aa), with the outward appearance of the plant (today called its **phenotype**) conforming to the action of the dominant version.

4. Each member of the second generation had an equal chance of receiving either version of the gene from either parent, so four combinations of genes (or "genotypes") appeared: AA, Aa, aA, aa. As the first three combinations all had the phenotype of the dominant gene, this accounted for the observed 3:1 ratio.

These conclusions are often summarized as **Mendel's first law of inheritance** or **the law of segregation** (two genes in plant segregated randomly as single copes into sex cells). Mendel also went on to show, in experiments on variants in two or more distinct characters (for example, colour of peas *and* shape of peas) that different genes were sorted from generation to generation *independently* of each other, a principle often summarized as **Mendel's second law of genetics** or **the law of independent assortment**.

A Monk and Two Peas: The Story of Gregor Mendel and the Discovery of Genetics Robin Marantz Henig (Weidenfeld & Nicholson, 2000)

Many naturalists also preferred **Lamarckism**, placing the lively desires of living organisms in the driving seat of evolution, rather than cold, calculating natural selection. Haeckel favoured this view, as did English novelist Samuel Butler (grandson of Darwin's old headmaster) and American fossil hunter Edward Drinker Cope. In response, one of Darwin's disciples, **George Romanes** (1848–94), coined the term **neo-Darwinism** to describe evolution by natural selection shorn of any Lamarckian influence (and of Darwin's own misguided theories of inheritance).

One major problem with Darwin's view of evolution was his lack of a coherent theory of inheritance. Although the Austrian monk **Gregor Mendel** (1822–84) had quietly laid the foundations of modern genetics during Darwin's own lifetime, Mendel's results were not widely known or understood in the decades that followed. However, as the twentieth century arrived, three botanists rediscovered Mendel's work: **Carl Correns** (1864–1933), **Hugo de Vries** (1848–1935) and **Erich Tschermak** (1871–1962). In each case the "rediscovery" was the consequence of independent investigations reported in near-simultaneous publications. Correns' paper, published in January 1900, was the first to cite Mendel and Darwin in the same bibliography. In it, Correns describes Mendel's results and Mendel's laws of segregation and independent assortment. Later in the spring of the same year, de Vries used Mendel's terms "dominance" and "recessive", but neglected to mention Mendel, until reprimanded by Correns. Tschermak's paper from June 1900 also refers to Mendel but conveys only a hazy grasp of his ideas. A few years later, American geneticist Walter Sutton (1877–1916) and German biologist Theodor Boveri (1862–1915) proposed what is now known as the **Boveri-Sutton Chromosome Theory**: the idea that the genes underlying Mendelian inheritance were located on chromosomes (organized structures that had been seen in the nuclei of plants and animals).

Genetics pioneer, Gregor Mendel.

Darwin and Mendel: the great what if?

Darwin and Mendel were contemporaries. One of the great "what ifs" in the history of science is "what if Darwin and Mendel had met to discuss each other's work, or, at least, had exchanged notes?" The closest they came to meeting was in the summer of 1862, when Mendel visited England to attend the International Exhibition, a world fair held in South Kensington. Charles Darwin was less than twenty miles away, but their paths never crossed as the Darwins were stuck at home, nursing their son Leonard through scarlet fever. Mendel read a German translation of Darwin's *The Origin* before publishing his seminal paper in 1865, but he did not see any connection between his work and Darwin's. It has been claimed that Mendel's paper sat on a shelf at Down House, unread, but this is just a myth. Although Darwin possessed two books that briefly referred to Mendel's work, there is no evidence that he read the relevant sections; in one of the books, the pages are clearly uncut. Darwin leant one of these two books to his friend George Romanes, who used it to write an encyclopedia entry, priming another myth: that Darwin wrote about Mendel in the *Encyclopaedia Britannica*.

How close was Darwin to discovering Mendel's laws of inheritance? As early as 1838, Darwin scribbled in his notes a question that, in retrospect, seems pregnant with potential: "Do races of peas become intermixed & gardener have hybrid seedlings?" In a letter written to Wallace in February 1866, Darwin recognizes that inheritance can be non-blending: "My dear Wallace... I do not think you understand what I mean by the non-blending of certain varieties... I crossed the Painted Lady and Purple sweetpeas, which are very differently coloured varieties, and got, even out of the same pod, both varieties perfect but not intermediate." Furthermore, as Chinese plant scientist Yongsheng Liu has pointed out, Darwin describes experiments that are uncannily similar to Mendel's, in his 1868 work *Variation Under Domestication*: "Now I crossed the peloric snapdragon... with pollen of the common form; and the later, reciprocally, with peloric pollen. I thus raised two great beds of seedlings, and not one was peloric. The crossed plants, which perfectly resembled the common snapdragon, were allowed to sow themselves, and out of a hundred and twenty-seven seedlings, eighty-eight proved to be common snapdragons, two were in an intermediate condition between the peloric and normal state, and thirty-seven were perfectly peloric, having reverted to the structure of their one grandparent..." The ratio, at 2.4:1, is close enough statistically to conform to an expectation of 3:1, so this might look like a glimpse by Darwin of **Mendel's first law** (although we now know that at least two genes are involved). But given that Mendel himself did not recognize the universality of his own work, it is unfair to expect Darwin or anyone else to do so, particularly in the face of less easily interpreted results from crosses in other species of plants and animals.

However, even as genetics gained a firmer footing, some eminent biologists abandoned Darwin's gradualism. Thus, English naturalist **William Bateson** (1861–1926), who coined the term "**genetics**", and de Vries, who coined the term "**mutation**" (a sudden genetic change) instead preferred **saltationism**, the view that new species arise primarily through sudden large changes (saltations), rather than through the steady accumulation of small changes. Bateson gathered impressive evidence for discontinuous variation in biology, classifying unusual morphological variations in animals according to changes in the number of structures (for example, extra fingers, ribs or nipples in humans, extra oviducts in crayfish) or situations where one body part has been replaced by another (for example, bees with legs instead of antennae, an example of what he called **homoeosis**). Support for saltationism persisted well into the twentieth century, encapsulated in 1940 in the phrase **hopeful monster** by German-American geneticist **Richard Goldschmidt** (1878–1958) to describe what he saw as a sudden dramatic evolutionary change that could give rise to a new group of organisms.

Nonetheless, even in the early 1900s some scientists held true to Darwinian gradualism. In a discipline dubbed **biometry**, zoologist **Walter Weldon** (1860–1906) and socialist statistician **Karl Pearson** (1857–1936) adapted and developed statistical approaches to the study of biological variation, emphasizing the usually continuous nature of such variation and arguing that "the problem of animal evolution is essentially a statistical problem". Pearson and Bateson clashed in arguments as to how Mendelism could be reconciled with continuous variation. The answer, spelt out as early as 1902 by Scottish statistician **Udny Yule** (1871–1951), was simple: if a character is influenced by many different genes (not just by one, as with Mendel's peas), then the effects of the genes are blended together in a large population to give continuous variation for that character.

Genetic breakthrough

Genetics took a great leap forward during the second decade of the twentieth century through the work of **Thomas Hunt Morgan** (1866–1945), professor of experimental zoology at Columbia University in the USA. Morgan adopted the **fruit fly**, *Drosophila melanogaster*, as a model organism for studying genetics. Through his early experiments on flies, Morgan was able to confirm and extend Mendel's theory of inheritance, confirming that traits were linked to chromosomes and that those linked to sex chromosomes showed sex-linked patterns of inheritance. Later studies in Morgan's "Fly Room" demonstrated that genes could be linked on the same chromosome (i.e. could break Mendel's law of independent assortment), that chromosomes could swap genes by recombining (through a process known as

J.B.S. Haldane: from war hero to dissecting table

John Burdon Sanderson Haldane (1892–1964) was born in Oxford into a well to do Scottish family. Educated at Eton and Oxford, he fought in the trenches during World War I and took to war with enthusiasm, earning the nickname "Bombo". Twice wounded, first in France and then in Mesopotamia, he convalesced in India. In 1919 he became a fellow of New College, Oxford, moving to Cambridge four years later. In 1924, Haldane befriended a married woman, a 31-year-old reporter for the *Daily Express*, Charlotte Burghes. After helping her obtain a divorce by spending the night with her at the Adelphi Hotel, Haldane was sacked from his university position by the university's sextet of academic lawyers, the "sex viri" (or the "sex weary" as Haldane called them). Haldane successfully appealed against the decision and was reinstated in 1926. He married Charlotte the same year. Nineteen years later the two divorced and Haldane married one of his research associates, Helen Spurway, who remained with him until he died.

> **"A satisfactory theory of natural selection must be quantitative… we must show not only that it can cause a species to change, but that it can cause it to change at a rate which will account for present and past transmutations."**
>
> **J.B.S. Haldane**

As well as making seminal contributions to the mathematics of evolution, Haldane was an enthusiastic popularizer of science, often with ideas ahead of his time. In 1924, he wrote a provocative paper that proposed the idea of test tube babies, and it was he who coined the word "clone" in a 1963 essay. In response to the rise of fascism in the 1930s, Haldane became active in leftwing politics, visiting Spain during the Civil War to advise the revolutionary government on civil defence. By the late thirties, he was an open supporter of the Communist Party and from 1940 chaired the editorial board of *The Daily Worker*. Two years later he joined the Party. However, like many left-leaning scientists of his generation he became disillusioned with the Soviet Union in the 1950s and quietly left the Party. In response to the invasion of Suez in 1957, Haldane and his wife left the UK for Calcutta, becoming Indian citizens in 1961. From 1962 until his death in late 1964, Haldane headed up a research institute in Bhubaneswar in Orissa. Shortly before his death, he wrote a satirical poem, "Cancer's a Funny Thing", and then on dying, having no further use for it, left his body to the local medical school.

crossover) and that chromosomes could be mapped by determining how often the genes on them showed linked inheritance.

Two great advances primed the merger of Mendel's work with Darwin's in the first half of the twentieth century. The first was the foundation of population genetics by a trio of scientists: two British, **Ronald Fisher**

(1890–1962) and **J.B.S. "Jack" Haldane** (1892–1964), and one American, **Sewall Wright** (1889–1988). A pioneer in statistics as well as genetics, Fisher was educated in Cambridge and in his early years was an associate of Darwin's sons Horace and Leonard. In a groundbreaking paper in 1918, and twelve years later in his landmark book *The Genetical Theory of Natural Selection*, Fisher showed how continuous phenotypic variation could result from the additive effects of many discrete genes, proving that Mendelian genetics was entirely consistent with evolution by natural selection. He also elaborated new mathematical formulations of the idea of **evolutionary fitness** (which, in the new synthesis, was defined in terms of the ability of an individual to maintain its share of genes in the next generation).

Haldane published ten papers from 1924 to 1934 under the title *A Mathematical Theory of Natural and Artificial Selection*. In them, he showed how real-world examples of natural selection could be explained in mathematical terms (for example, melanism in Mancunian moths) and how natural selection could exert its effects far more quickly than anyone supposed. The two key concepts that Sewall Wright brought to evolutionary biology were **genetic drift** and **fitness landscapes**. According to his theory of genetic drift, in small enough populations, the frequencies of genes could change quite dramatically, even through random effects, without any input from natural selection. With his concept of **fitness landscapes**, Wright

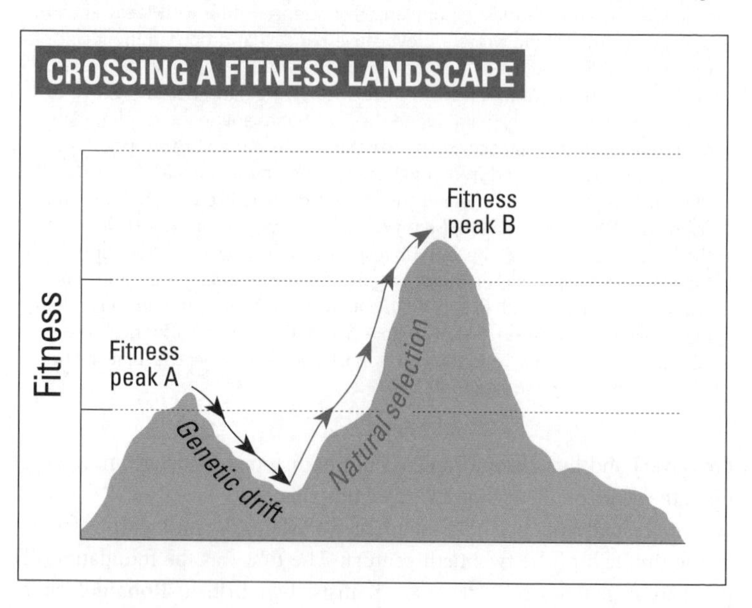

graphically described the relationship between genotype/phenotype and fitness as a surface where the horizontal axes represent frequencies of genes or characters and the vertical axis represent fitness. Thus, optimal populations would appear as fitness peaks rising up out of the plains. Natural selection would drive populations to climb the nearest peak, while genetic drift would cause them to wander across the surface, so that they could sometimes reach a new higher fitness peak.

The modern synthesis

The second great advance, now known as the **modern synthesis**, took place during the 1930s and 40s, chiefly through the work of another trio: **Theodosius Dobzhansky** (1900–75), **Ernst Mayr** (1904–2005) and **George Gaylord Simpson** (1902–84). Ukrainian-American Dobzhansky took Morgan's work on fruit flies out of the lab and into the field. In his 1937 book *Genetics and the Origin of Species*, Dobzhansky made the arcane maths of evolutionary population geneticists accessible to a wider audience. He also pithily redefined evolution in genetic terms as "a change in frequency of an allele within a gene pool", where the term **gene pool** was used to mean all the genes, including all their allelic variants, found within a given population.

Ernst Mayr was born in Germany, but spent most of his long life in the USA. In his 1942 book *Systematics and the Origin of Species*, Mayr introduced a new definition of "species" (as a population linked by ability to inter-breed rather than purely by similar appearance). Drawing on his extensive field studies of birds, Mayr also explained how natural selection and genetic drift could drive the emergence of new species whenever a population within an existing species became isolated, particularly if the population was very small.

Palaeontology's contribution

George Gaylord Simpson's contribution to evolutionary biology was to bring palaeontology to the modern synthesis with his book *Tempo and Mode in Evolution* (1944). Here, through detailed scrutiny of extinct mammals, he overturned simplistic notions of linear progression in the fossil record (for example, in the evolution of the horse) and confirmed that evolution was messy, branching and non-directional, thus signalling the death knell for neo-Lamarckism and orthogenesis. Simpson also argued that the large-scale changes seen in the fossil record (**macroevolution**) could mostly be

The long life of Ernst Mayr

Ernst Walter Mayr (1904–2005) was the philosopher-king of twentieth-century evolutionary biology, bringing speciation to the modern synthesis and championing the philosophical distinctiveness of biology. Plus as a centenarian, he personified the link between Darwin's age and our own: Wallace was still alive when Mayr was born! As tropical explorer and naturalist, Mayr was heir to the legacy of Darwin, Wallace, Hooker and Huxley; within his lifetime, he named over two dozen new species of birds and over three dozen orchid species.

Born in Bavaria, Mayr moved to Dresden in his teens. He became a keen ornithologist, impressing experts such as Erwin Streseman, who became his mentor. Like Darwin, Mayr tried his hand at medicine, studying at Greifswald on the Baltic. But within two years he switched to a doctorate in ornithology at the Berlin Museum before, at the age of 21, accepting a position there. A year later, with backing from the second Baron Rothschild, Mayr set off on a three-year expedition to New Guinea, which included an excursion to the Solomon Islands. Despite the physically gruelling terrain and the threat of disease, drowning and death at the hands of natives, Mayr triumphed in collecting ornithological specimens and studying rare birds of paradise. He even achieved linguistic mastery of Malay and Neo-Melanesian. In 1931, a year after the end of the expedition, Mayr became a curator at the American Museum of Natural History in New York. In 1935, he began his 55-year marriage to Gretel Simon.

Mayr's landmark book, *Systematics and the Origin of Species*, was published in 1942. In 1953, Mayr moved to Harvard University, where he went on to serve as director of the Museum of Comparative Zoology. Following his retirement in 1975, Mayr published more than two hundred articles and over a dozen books. In 1999, at the age of 95, he was joint recipient of the prestigious Crafoord Prize of the Royal Swedish Academy of Sciences. His twenty-fifth book, *What Makes Biology Unique?*, was published in his hundredth year. On his 100th birthday, interviewed by *Scientific American* magazine, he revealed an intellect as razor sharp as ever: his discussions ranging from Kant to Crick and Watson, from creationism to the Cambrian explosion. His beautifully understated last words in the interview serve as a fitting epitaph: "And I hope you find my provocative ideas sufficiently useful that you will at least be gentle in your criticism."

explained by an accumulation of the smaller changes (**micro-evolution**) observed by population geneticists, although he did accept that something he called "quantum evolution" might account for the origins of the high-level taxonomic groupings. For Simpson, even when it is gradualistic, not all evolution occurs at the same tempo.

The term "modern synthesis" was coined by Thomas Huxley's grandson, the biologist and internationalist polymath **Julian Huxley** (1887–1975) who wove several strands of evidence together in his book *Evolution: the Modern*

Cellular selection

Although natural selection is usually thought of as a force acting on whole organisms, similar phenomena operate at the cellular level. Over a century ago, German microbiologist Paul Ehrlich suggested that cellular selection might explain how it was possible to raise antibodies against substances never encountered in nature. Ehrlich speculated that antigens bound to antibodies on the surface of cells and this resulted in the secretion of excess receptors.

Danish immunologist **Niels Jerne** (1911–94) hit on a similar idea in the 1950s, in what he termed the natural selection theory of **antibody production**. In 1957, Australian virologist **Frank Macfarlane Burnet** (1899–1985) published a brief but lucid paper, in which he incorporated Ehrlich's and Jerne's ideas into the **clonal selection theory** of antibody production. According to his theory (which Macfarlane Burnet called Darwinian), antigens counter-intuitively play no role as templates in the construction of antibodies. Instead, a huge repertoire of antibody-producing B cells exists dormant in the body, each cell carrying one type of antibody on its surface. When an antigen binds to the antibody on the surface of a cell, the cell is selected from the silent majority and stimulated to proliferate, spawning many progeny cells that produce large amounts of identical soluble antibody. Such clonal selection is also now known to operate in other parts of the immune system (for example, in T cells).

American biologist **Gerald Edelman** (b.1929) went on to elucidate the mechanisms underlying clonal selection, showing that the signal to increase antibody production was a change in shape of the antibody as it combined with the antigen. Similarities between the immune and nervous systems led Edelman to formulate what he called **neural Darwinism** (the title of his 1989 book). According to Edelman, frequently used connections between nerve cells are selected from a pre-existing repertoire of cell-cell links in the brain, with developmental selection occurring before birth and experiential selection occurring later in life.

In recent years, American cell biologist Carlo Maley has argued that natural selection drives the **evolution of cancer cells**. In a cellular struggle for existence, cancer cells compete for space and resources, evade predation by the immune system and evolve to disperse and colonize new organs. In addition, cancer cells evolve in response to treatment, often becoming resistant to anti-cancer drugs. Fortunately, ideas from evolution also promise a cure for cancer: Cambridge molecular biologist **Greg Winter** has used a Darwinian selection strategy to evolve novel anti-cancer antibodies.

Understanding Evolution website http://evolution.berkeley.edu/evolibrary/news/071001_cancer

Synthesis (1942). Over the decade that followed, other strands were added, most notably plant science through the book *Variation and Evolution in Plants* (1950) by American botanist **G. Ledyard Stebbins** (1906–2000).

Now over sixty years old (and so not so modern), the modern synthesis, with its reconciliation of genetics and evolution and its reaffirmation of the Darwin's gradualism and natural selection, stands as one of the greatest scientific achievements of the twentieth century. However, as we shall see in the coming sections, evolutionary biology has not stood still during the last six decades.

A modern view of speciation

How do new biological species arise? Evolutionary biologists have defined two principal modes of speciation: *allopatric* (from the Greek for "different homeland"), where geographical separation is a prerequisite for speciation, and *sympatric* (from the Greek for "same homeland"), where new species form while occupying the same place; two minor variants are also recognized: *peripatric* and *parapatric* speciation. **Allopatric speciation** is the easiest to understand. An original single continuous population is split into two by the facts of geography: population movements, destruction of an intervening population or new geographical barriers (mountains, deserts, rivers, etc), or by a mixture of all three. Over time, the two populations diverge sufficiently in their mating behaviours and/or genetic composition that they become reproductively isolated, i.e. no longer capable of exchanging genes, even if contact were to occur. For example, when the Isthmus of Panama closed about three million years ago, populations of marine organisms inhabiting the Atlantic and Pacific coasts became isolated. Studies on snapping shrimps from both sides of this new barrier have shown that, although still superficially similar, they will no longer mate.

A variant of allopatric speciation termed **peripatric speciation** occurs when a small sub-population (maybe even a single individual) colonizes a new habitat at the periphery or even outside of the original geographical range. In these circumstances, there is drastic loss of genetic diversity in the colonizing population (sometimes called the **founder effect**), which, combined with genetic drift, may be enough to ensure reproductive isolation from the original population. Founder effects and peripatric speciation account for many of the oddities of island biology, the most famous example being the Galápagos finches.

In **parapatric speciation**, two diverging populations occupy and adapt to separate zones but come into contact at the boundary between the two zones. Here the driving force in speciation might be a reduced fitness for hybrids, which will be less well adapted to either environment than members of either

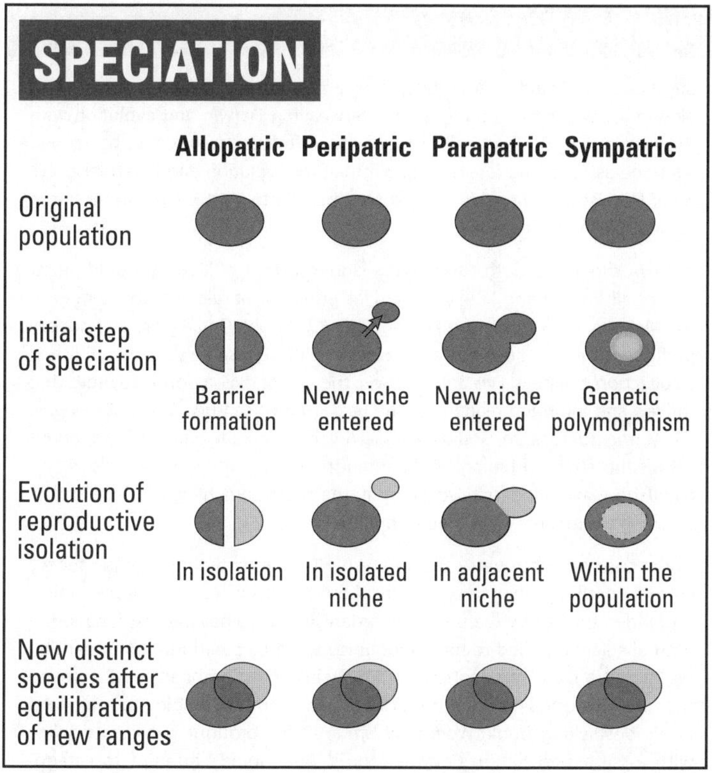

SPECIATION

	Allopatric	Peripatric	Parapatric	Sympatric
Original population				
Initial step of speciation	Barrier formation	New niche entered	New niche entered	Genetic polymorphism
Evolution of reproductive isolation	In isolation	In isolated niche	In adjacent niche	Within the population
New distinct species after equilibration of new ranges				

parent population (for example, light-coloured lizards will be camouflaged on light rocks, dark-coloured lizards on dark rocks, but grey-coloured lizards will be disadvantaged on both types of rock). The ring species described in an earlier chapter represent a special case of parapatric speciation.

In **sympatric speciation**, populations diverge into species while inhabiting the same locality. The importance of sympatric speciation in generating biodiversity remains controversial and Ernst Mayr was opposed to the very idea. However, two credible cases have already been discussed in a previous chapter: polyploidy represents an unequivocal route to sympatric speciation, while the incipient speciation of apple maggot flies resulting from different choice of food renders the concept plausible. Sympatric speciation accompanied by divergence in breeding time, **allochrony**, has been described in salmon and periodical cicadas. Additional plausible instances of sympatric speciation have been documented in plants, plant-eating insects, fish and birds, but not so far in mammals. Examples

Stephen Jay Gould: Darwin's essayist

Stephen Jay Gould (1941–2002) was an American palaeontologist who eloquently wrote essay after essay, interweaving Darwin and evolution with the details of history, baseball and light opera. For over twenty years he reigned supreme as America's foremost spokesman for evolution and defender of his hero, Darwin. His status as the people's palaeontologist even earned him a cameo role in *The Simpsons*!

A New Yorker by birth, upbringing and sporting affiliation, Gould spent almost all his working life at Harvard, as professor of geology and curator at the Museum of Comparative Zoology. His 1977 book *Ontology and Phylogeny* prefigured the evo-devo movement (see p.140). That same year, he published a collection of essays, *Ever Since Darwin*, the first of over a dozen volumes that thrilled and informed readers for the next quarter century. In his 1981 book, *The Mismeasure of Man*, Gould fearlessly defended human equality. Throughout his life, he attacked biological determinism, particularly in sociobiology and its offshoot evolutionary psychology. In the courtroom, he fought against the teaching of creationism in American schools.

Gould courted controversy at every turn. His punctuated equilibrium theory stirred up things in palaeontology. Gould's opposition to adaptationism – the idea that every feature of an organism has to have an explanation in natural selection – led to him being misquoted by creationists and derided by his "ultra-Darwinist" critics. Gould's belief that Cambrian fossils proved that life's history depended on luck and was non-repeatable – popularized in his bestselling book, *Wonderful Life* (1989) – brought him into dispute with fossil expert Simon Conway Morris. And Gould's attempt to categorize religion and science as two separate fields of human experience, as "non-overlapping magisteria" has been seen as excusing religion from intellectual scrutiny.

In 1982 Gould was diagnosed with mesothelioma, a usually fatal form of cancer. After two years of aggressive treatment, sweetened by non-recreational use of cannabis, he recovered to write a magazine article, "The Median isn't the Message". Here he highlighted the perils of hasty conclusions when faced with the statistics of cancer survival, giving hope to many subsequent cancer sufferers. He died twenty years later from an entirely different kind of cancer, surrounded by books in an attic room in an apartment in his beloved Manhattan. He lived just long enough to see his 1400-page magnum opus *The Structure of Evolutionary Theory* published. Creative in his controversy, Gould's legacy is perhaps best summarized by the title of a paper written by two fellow palaeontologists shortly after his death: "Wonderful Strife"!

The Richness of Life: A Stephen Jay Gould Reader Stephen Jay Gould (Norton, 2006)

American palaeontologist and essayist Stephen Jay Gould.

among birds include the **indigobirds** of Africa, who get other birds to raise their young and have experienced sympatric reproductive isolation when colonizing new hosts; **buntings** on the remote islands of Tristan da Cunha; and the **band-rumped sea-petrel**, a tropical island-dwelling sea bird, that has undergone speciation by allochrony.

To seal the separation of populations into new species, they must undergo **reproductive isolation**. The barriers to inter-breeding can be **pre-zygotic** (they prevent fertilization from ever happening) or **post-zygotic** (genetic impediments to the production of fertile offspring). Pre-zygotic impediments can be classified as temporal (for example, plants flowering at different times), behavioural (for example, changes in courtship rituals), structural (for example, anatomical changes that preclude mating) or biochemical (different mating hormones, incompatibility of egg and sperm). An imbalance in chromosome number between the two parents (for example, through polyploidy) precludes normal development of the offspring in most settings.

Invading planet earth: alien species

Across the globe, humans have deliberately or by accident allowed thousands of **introduced species** (also called alien, exotic or immigrant species) to run free in new locations and new environments. In many situations, these incomers cause harm to native species, to local biodiversity or to human health and wealth, making them **invasive species**. Island ecologies have been decimated by such introductions and invasive species are second only to habitat destruction as a threat to global biodiversity. Deliberate release provides repeated proof of the short-sightedness of human intervention. The motivation may be economic gain: for example, the **apple snail** was introduced to Taiwan and Hawaii as a food source; instead, it became a serious threat to crop production and to local ecosystems. Sometimes whimsical motives are at work, such as ecological nostalgia. In the nineteenth century, there was even an Acclimation Society of North America. One of its members, **Eugene Schieffelin,** purportedly wanted New Yorkers to see all the birds mentioned by Shakespeare. As a result, from 120 released birds, the **starling population** of North America grew to over 200 million; the number of **house sparrows** on the continent is not far behind. Both species are now serious pests. Two-dozen **rabbits** introduced to Australia in 1859 by Thomas Austin led to the most explosive growth ever recorded of any mammalian species, giving a population at its peak of over 600 million and causing utter devastation to indigenous plants and animals.

Misguided efforts to control pests can also prime deliberate release. The **cane toad** was introduced to Australia in 1935 in an attempt to control the cane beetle, but is now spreading relentlessly across the continent. Plants escaping from a horticultural setting provide another route to invasion: in Britain, **giant hogweed** and **Japanese knotweed** have become serious problems. Accidental passage of species along trade routes can also have devastating effects. The Russian freshwater **zebra mussel** has colonized many canals, rivers and lakes elsewhere, causing billions of dollars worth of damage in the US alone.

As Darwin noted, introduced species provide compelling proof that local climate and conditions are insufficient to explain the distribution of species. Ecological invasions also illustrate the remarkable volatility of the living world and provide evidence of evolution in action. In Australia, cane toads have evolved larger bodies and longer legs as they invade the continent and have driven indigenous snakes to evolve smaller gapes and longer bodies. On Gough Island in the South Atlantic, house mice have tripled in size as they have evolved over two centuries into vicious carnivores of albatross chicks.

Out of Eden: An Odyssey of Ecological Invasion Alan Burdick (Farrar, Straus and Giroux, 2005)

Invasion Ecology Julie Lockwood, Martha Hoopes, and Michael Marchetti (Blackwell, 2007)

In most cases, the genetic bases of reproductive isolation and speciation are unclear. One genus of freshwater fish, *Xiphophorus*, has provided some insights. When two species from the genus (the platyfish and swordtail species) are crossed, subsequent backcross hybrids develop cancer and die. The cause has been tracked down to a platyfish gene, *Xmrk-2*, which is misexpressed in the hybrids. In *Drosophila* fruit flies, several **speciation genes** have been identified that prevent viable or fertile offspring from developing after inter-species fertilization. Curiously, these correspond to ordinary genes with normal functions: to date their only distinctive feature tends to be that they are rapidly evolving. In addition, the jumping of a gene from one chromosome to another without any changes in sequence appears to be sufficient to account for hybrid sterility between two *Drosophila* species.

There is an additional route to the formation of a new species – hybridization between two existing species. **Allopolyploidy** is a form of polyploidy, common in plants, where chromosomes are derived from different species. More rarely, birth of new species by inter-species hybridization can occur between without change in chromosome number. Examples of this phenomenon of homoploid hybrid speciation include *Heliconius* butterflies, sunflowers, dill daisies and mallard ducks in New Zealand and the Marianas islands.

When considering speciation in the fossil record, additional concepts are required. Within a lineage one form can evolve over time into a

Furry pest: a native of Australia, the Common Brushtail Possum was introduced to New Zealand where it has devastated the country's flora and fauna.

morphologically distinct form, which would rank as a separate species had the two forms existed at the same time. The term **anagenesis** is applied to this form of non-branching speciation over time and the term **chrono-species** used to describe each distinct species within such a lineage (which may be difficult to demarcate if the change is gradual). **Cladogenesis** occurs when one lineage splits in two branches; the clearest example in the fossil record occurred around three million years ago and involved the splitting of one species of the planktonic diatom *Rhizosolenia* into two.

In the early 1970s, two American palaeontologists, **Niles Eldredge** and **Stephen Jay Gould** caused controversy by proposing a new model of speciation, **punctuated equilibrium** (or "punc eq"), drawn from fresh interpretations of the fossil record (particularly of Devonian trilobites and Pleistocene land snails) and the mechanisms of speciation. They proposed that most evolutionary lineages experience long periods of equilibrium or stasis, when the organism's appearance remains unaltered, punctuated by sudden, much shorter periods of intensive change, during which speciation occurs. They contrasted this with what they termed **phyletic gradualism**, where evolution occurs at a slow constant rate, with no clear lines of demarcation between one species and the next.

Gould's rhetorical skills helped popularize his theory among non-scientists. However, his ideas drew fire from many other evolutionary thinkers, particularly Richard Dawkins and Daniel Dennett. As a measure of the animosity between the camps, the two alternate views have been personalized as "**evolution by jerks**" (punctuated equilibrium) and "evolution by creeps" (gradualism). For many, Gould and his supporters exaggerated the importance of his ideas, particularly when these were portrayed as contradicting or superseding conventional Darwinian evolution. One objection was that Gould posited a "straw man" argument in pretending that Darwinism entailed a commitment to uniform rates of evolutionary change, when even Darwin didn't accept this, stating in later editions of *The Origin*: "the periods, during which species have undergone modification, though long as measured by years, have probably been short in comparison with the periods during which they have retained the same form". Another objection was that even the most rapid speciation events seen in the fossil record are, by any human time scale, extremely gradual.

Despite these criticisms, there is no doubt that Eldridge and Gould brought new excitement to the dry discipline of palaeontology and, with mantras like "stasis is data", forced researchers to look at fossil series previously considered too boring to bother with. With the passage of time, it has become clear that not only can both gradualism and punctuated equilibrium be seen in the fossil record depending where one looks, but that we remain some way from a "grand unified theory" of speciation.

Life drives life

One of Darwin's key insights was that evolution was driven not simply, or even primarily, in response to changes to the inanimate environment. Instead, he stressed that life itself drives the evolution of life, that the interactions of living organisms with other individuals within their own species, or from other species, can be more important than the impact of temperature, rainfall or the acidity of the soil. The emergence of **ecology** as a discipline, complete with its own nomenclature, has helped put flesh on the bones of Darwin's idea of an economy of nature. Darwin's contemporary, geologist Edward Suess (1831–1914), invented the term "**biosphere**" to cover that the parts of our planet that sustain

Darwin as early ecologist

The term "ecology" was coined in 1866 by Darwin's German admirer Ernst Haeckel. However, in the third and fourth chapters of *The Origin*, Darwin's emphasis on the "complex relations of all animals and plants throughout nature" guarantees him the status of an early ecologist. In Chapter 3, he puzzles out how enclosure of land can lead to a proliferation of fir trees and hits on the answer: the exclusion of cattle that graze on seedlings. He speculates on the cascade of ecological changes that might result from an increase in insectivorous birds in Paraguay. And he produces a beautiful English example of "how plants and animals, most remote in the scale of nature, are bound together by a web of complex relations", by pointing out that (1) humble bees are needed for fertilization of some plants; (2) the number of bees in an area depends on the number of mice; (3) the number of mice depends on the number of cats, so that the number of cats can determine the frequency of flowers in a district!

In Chapter 4, Darwin writes: "It has been experimentally proved that if a plot of ground be sown with one species of grass, and a similar plot be sown with several distinct genera of grasses, a greater number of plants and a greater weight of dry herbage can thus be raised." Many subsequent studies have confirmed this principle, which lies at the heart of organic farming methods. In his haste, Darwin unfortunately did not say when or where the pioneering study had been carried out. Recently, two British ecologists, Andy Hector and Rowan Hooper, have tracked the source of the information to George Sinclair, head gardener to the Duke of Bedford, who, in an 1826 article, described what must count as the first ecological experiments, conducted at Woburn Abbey in England, a few years earlier.

and are influenced by life. English botanist Roy Clapham coined the term "**ecosystem**" in 1930 to denote the physical and biological components of an environment considered in relation to each other as a unit. Around the same time, the word "**habitat**" emerged as a description of the ecological area inhabited by a single species.

Darwin used violent language to capture the idea of nature crammed full of species: "The face of Nature may be compared to a yielding surface, with ten thousand sharp wedges packed close together and driven inwards by incessant blows, sometimes one wedge being struck, and then another with greater force." The language of ecology now equips us with the gentler term "**niche**", drawn from architecture, to capture the notion that each species has its own way of making a living in the economy of nature. On an old continental landmass, every niche may be filled, but, as we have seen, in the simpler ecosystems of isolated islands, the availability of "vacant niches" may prime adaptive radiations.

Evolutionary arms races

The most obvious link between the individuals of one species and those from another is the **predator-prey relationship**. The population dynamics of the predator-prey relationship is complex, with numbers cycling up and down (experts capture these patterns with a bit of calculus known as the **Lotka-Volterra equations**). Natural selection will drive predators to optimize their ability to obtain food and prey species to avoid being eaten, resulting in an astonishing array of adaptations in both groups. However, in many situations, the outcome is an evolutionary arms race, in which the two populations, predator and prey, evolve continuously and progressively, each attempting to outdo the other. Prey develop a better defence against predators (gazelles run faster); predators in compensation develop a better offense (cheetahs run faster) and so on until repeated changes in relative fitness drive the organisms to astonishing heights of absolute fitness (a cheetah running at 70 miles per hour).

Just as the Cold War arms race led to an arsenal of nuclear weapons capable of destroying every city on the planet several times over, evolutionary arms races can drive plants, animals and bacteria to produce poisons that can kill a man many times over. For example, one-thirtieth of the toxin normally found in a rough-skinned newt (a species found in western North America) is enough to kill a human being. Why this overkill? The explanation is an arms race between the only predator capable of eating the newt, the common garter snake. Resistance to the toxin in the snake has driven an escalation in toxin production in the newt. In response, the garter snake has developed higher and higher levels of resistance, through mutations in a sodium channel targeted by the toxin.

American evolutionary biologist **Leigh Van Valen** (b.1935) coined the evocative phrase the **Red Queen effect** to capture the idea that in an evolutionary arms race "it takes all the running you can do, to keep in the same place" (as the Red Queen states in Lewis Carroll's *Through the Looking Glass*). Through this effect, evolutionary arms races provide a powerful explanation for progress or improvement during the course of evolution.

Mutual benefits

Arms races provide an example of a more general phenomenon: **co-evolution**, where one species exerts an influence on the evolution of another species and vice versa. This need not involve competition. Instead, it might lead to **mutualism**, a kind of biological barter, in which partners trade resources or services from which both species benefit. For example, nitrogen-fixing bacteria provide plants with nitrogenous compounds in exchange for carbohydrates.

The exuberant beauty of flowers derives from another kind of mutualism, where insects aid pollination in exchange for nectar. To prevent competition between different species of insects for nectar and to avoid wastage of pollen delivered to the wrong plant, there is an impetus towards specialized one-to-one relationships, with plants and insects often co-evolving elaborate contrivances to ensure fidelity. For example, a species of orchid from Madagascar, sometimes called **Darwin's orchid**, has an 18-inch long tube leading to its nectar. On seeing it, Darwin predicted the existence of a moth with an 18-inch long tongue that would pollinate the orchid. More than two decades after Darwin's death, the predicted moth, *Xanthopan morganii praedicta*, was discovered, complete with an 18-inch-long tongue.

Mimicry

Flowers do not always engage in mutually beneficial relationships with pollinators. Many species engage in **mimicry** to attract insect pollinators. Some species use sex, luring pollinators with a mimic of the female, twinned with sneaky chemical attractants. Swiss botanist **Florian Schiestl** has shown, in an interaction between an orchid and a wasp, that the mimics are not perfect replicas of the females. Instead, the orchids go for a "Jessica Rabbit effect", exaggerating the form and smell of the replica, so that males prefer the mimic to the real thing. Other plants opt for the smell of death. In Corsica, the aptly named **dead horse arum** mimics the shape and smell of a dead horse's rear end, thereby luring flies into its floral chamber to deposit and receive pollen.

Mimicry has evolved as an adaptation whenever an organism (the mimic) can gain a selective advantage by fooling another organism (the dupe) into thinking that the mimics belong to a different group (the model) than they actually do. English naturalist **Henry Walter Bates** (1825–92), who performed pioneering studies on butterflies of the Amazon, gave his name to what is now described as **Batesian mimicry**, where a harmless mimic poses as a harmful species to avoid predation.

Games of life

Further proof that life drives the evolution of life has come from the discipline of **evolutionary game theory**, first developed by Ronald Fisher in the 1930s in an attempt to solve the problem of equal sex ratios. The subject came of age in 1973, when **John Maynard Smith** published a paper with his troubled American collaborator **George Price** (who later gave all his possessions to the poor, before committing suicide) in which they outlined the concept of the evolutionary stable strategy. Maynard Smith's seminal text, *Evolution and the Theory of Games*, followed in 1982.

John Maynard Smith.

At the heart of evolutionary game theory is the need to understand the evolution of interactions between individuals. The inanimate environment generally presents a simple fixed target that evolution can respond to (for example, a snow-covered landscape favours the evolution of a white coat as camouflage).

JMS: vivacious voice of neo-Darwinian orthodoxy

John Maynard Smith (1920–2004) – JMS to his friends and associates – was born in London, educated at Eton, but spent much of his youth in and around Exmoor. His interest in evolution was awakened by borrowing a copy of Olaf Stapledon's *First and Last Men* from Minehead public library (the very same copy that launched Arthur C. Clarke's career in science fiction). He read engineering at Trinity College, Cambridge, where he joined the Communist Party. As war broke out, JMS attempted to enlist, but was rejected. He later joked that "under the circumstances, my poor eyesight was a selective advantage – it stopped me getting shot". Instead, he contributed to the war effort through military aircraft design. After the war he switched to zoology at University College London, studying under fellow Etonian and Marxist, J.B.S. Haldane, before taking up a lectureship in 1952. The Soviet invasion of Hungary in 1956 prompted him to leave the Communist Party. Two years later, JMS's career as a popularizer of science began with his book *The Theory Of Evolution*.

In the early 1960s, JMS joined the newly founded Sussex University, where he served as Dean of its School of Biological Sciences for many years. During the 1970s and 80s, he made his seminal contributions to evolutionary game theory. In 1985 he officially retired, coincident with a brush with colon cancer. But, as professor emeritus, he continued to outperform most of his younger colleagues. With Hungarian Eörs Szathmáry, he wrote the influential 1995 book, *The Major Transitions in Evolution*.

Despite his eminence as a scientist, JMS remained the quintessential English academic – eloquent but unpretentious, enjoying a pint in the pub, a keen gardener, always willing to argue the point but completely lacking in personal vanity. He was adept at summarizing complex ideas in a single phrase: the haystack model, the hawk-dove game, protein space, the evolutionary stable strategy or the cost of meiosis were all terms that he coined. In later life, he was evolution personified: in response to a misguided remark about what neo-Darwinian orthodoxy meant, he replied, "It doesn't, and I know this because I am the voice of neo-Darwinian orthodoxy". With impeccable timing, he managed to die on the very same day (19 April) as his hero, Darwin. Despite his age, JMS's death came as a shock: as Marek Kohn put it, JMS's death felt unfair, because, despite his age, it felt as if he was dying young.

However, other individuals represent a moving target: in many cases, the success or failure of a strategy open to an individual (whether to fight or to co-operate, to tell the truth or to cheat) depends on the strategies adopted by others. In this sense, the interaction resembles a game. However, unlike conventional human games, the evolutionary context assumes that strategies are not arrived at by conscious reflection but are hardwired by genes subject to natural selection.

Hawks and doves

Maynard Smith developed **the hawk-dove game** to predict whether an animal should fight to defend a resource in a one-on-one confrontation. In this game, **hawks** always fight for a resource and will injure or kill an opponent. The advantage of being a hawk is that if you win, you get all the resource. The disadvantage is that you may be injured or killed. **Doves**, on the other hand, never fight; they put on an initial display, but when challenged, they just fly away. The good news is that doves don't get killed; the bad news that they may not get a resource. While it may be risky to defend a territory, it is even more costly in reproductive fitness to have a low quality territory, or none at all.

To identify the strategies favoured by evolution, Maynard Smith formulated the question in mathematical terms, evaluating strategies in terms of **payoff** (an increase in reproductive fitness) and **expense**

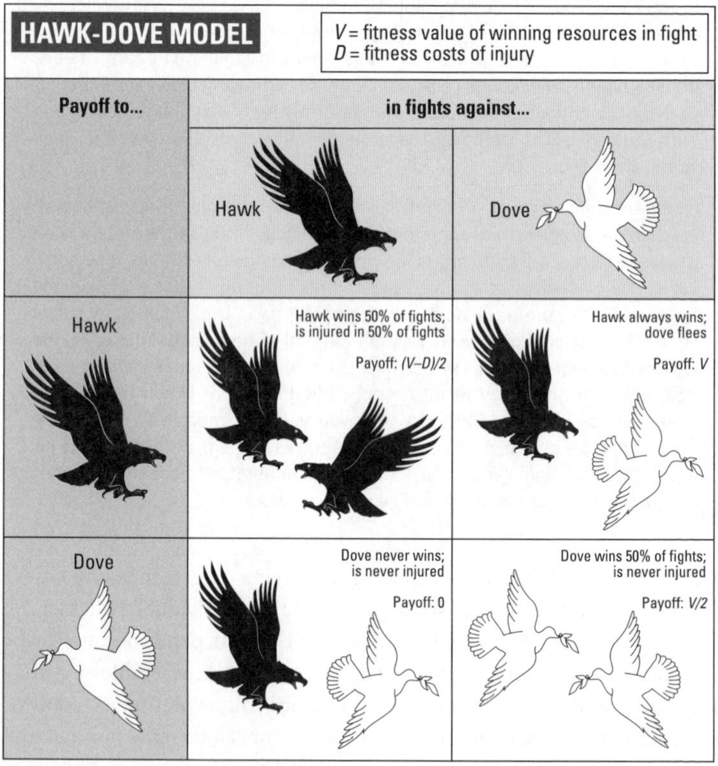

HAWK-DOVE MODEL

V = fitness value of winning resources in fight
D = fitness costs of injury

Payoff to...	in fights against...	
	Hawk	Dove
Hawk	Hawk wins 50% of fights; is injured in 50% of fights Payoff: *(V–D)/2*	Hawk always wins; dove flees Payoff: *V*
Dove	Dove never wins; is never injured Payoff: 0	Dove wins 50% of fights; is never injured Payoff: *V/2*

(the potential loss of fitness through injury, cost of display, etc). He then calculated the scores associated with each type of conflict and the average scores in populations comprised of various mixes of hawks and doves. Based on these analyses he came up with the idea of an **evolutionary stable strategy (ESS)**: that is a strategy, which, if followed by the whole population, would be immune to infiltration by mutants following a different strategy. To cut a long story short, the calculations show that an all-dove population is not an ESS, as hawks can invade, because the pay-off from a hawk-dove conflict is higher than from a dove-dove conflict. Similarly, an all-hawk population is not stable either, because the pay-off from a hawk-dove conflict is higher than from a hawk-hawk conflict. Instead, Maynard-Smith's calculations showed, counter-intuitively, that the only stable outcome is a population that includes a high proportion of hawks, but not as much as 100 percent.

Explorations of many other strategies and games by evolutionary game theorists have shed light on the distribution of behaviours and signals in nature. Colourful names for strategies abound: Bourgeois, Scrounger, Sneaky, Satellite, Transvestite, Sex-change. Maynard Smith even formalized the term "Sneaky Fucker" to describe a reproductive strategy in which some males, despite their subordinate status, manage to get to have sex! Games relevant to evolution have equally odd names: Rock-Paper-Scissors, War of Attrition, Prisoner's Dilemma. In recent decades, evolutionary game theory has outgrown its roots in biology and is now a vigorous research field in economics, computer science, ethics and sociology.

Kin selection

Aside from understanding conflicts, game theory has been applied to another key problem in evolution: the origins of **co-operation** and particularly apparently selfless **altruism**. Also relevant is the idea of **kin selection** (another term coined by Maynard Smith). J.B.S. Haldane formulated the concept but credit for its elaboration goes to British evolutionary biologist **William Hamilton** (1936–2000). The fundamental principle behind kin selection is that a gene that elicits behaviour enhancing the fitness of relatives, but lowering that of its owner, may nonetheless increase in frequency, because relatives often carry the same gene. In other words, the enhanced fitness of the many can compensate for the loss of fitness in the one. Haldane is credited with a pithy

Why do we have sex?

Much of human life – from soap operas to highbrow novels, from pornography to romance – is centred on sex, and the emotions (love and lust) that accompany it. But why do we bother? Why not just reproduce asexually, like a bacterium or a yeast cell? Sexual reproduction is a problem, which John Maynard Smith outlined as the **two-fold cost of sex**. Consider two populations with equal numbers of individuals: one population sexual, the other asexual. Suppose that each reproductively capable individual produces two offspring. In the asexual case, the population will double every generation. But in the sexual case, the population will remain constant, since half of all offspring are males, and they contribute to the next generation only by fertilizing the females. Thus, sex comes at a cost – the cost of males!

Numerous explanations have been put forward to explain why sex evolved and why it has been maintained, none of them universally accepted. Most centre on the fact that sex creates genetic variation among siblings, but differ in why this is important. According to the **tangled bank hypothesis**, each of the siblings exploits a slightly different niche, so that they can together extract more resources from the environment than could a group of identical organisms. Another suggestion is that sex, in shuffling genes every generation, will deliver new favourable combinations of genes more quickly than in an asexual population (but this, heretically, implies a benefit to the species at the expense of the individual).

Other potential explanations focus on the ability of sex to purge the population of deleterious genes. In a small asexual population, individuals with fitter genotypes may be lost randomly, so that the population slowly accumulates deleterious changes, a process known as **Müller's ratchet**. In a sexual population, recombination allows mutation-free genotypes to be restored.

summary of the principle, stating that he would not save a drowning brother, but would lay down his life to save two brothers or eight cousins. What he was describing was the break-even point at which you save the equivalent of your own genome (you share half your genes with a brother and one-eighth with a cousin). **Hamilton's rule** is a mathematical formulation of the same principle: $rB>C$ (where r = genetic relatedness of recipient to actor; B = additional reproductive benefit gained by recipient of the altruistic act; and C = the reproductive cost to the individual of performing the act). Kin selection clearly explains many facts of biology, for example, the existence of sterile workers in insect colonies, of risky alarm calls in social animals and subordinate males helping dominant males gain a mate. More controversially, it might also explain some facets of human behaviour.

Perhaps the most widely accepted explanation is that sex is a way of ensuring that organisms are able to stay ahead in an evolutionary arms race with their parasites. In a population of identical individuals, a parasite can latch on to conserved features present in all members of the host population, and quickly decimate the population. By contrast, a sexual population, in shuffling genes, presents a moving target much harder for the parasite to get to grips with. The extreme variation seen in genes involved in the immune response to pathogens provides evidence to support this idea. Why pornography? Blame the parasites!

Burt Lancaster and Deborah Kerr give in to their urges in *From Here to Eternity*.

From homology to cladistics

Evolution provides an explanatory framework for comparative anatomy. However, the most important concept in this field, **homology**, pre-dates evolution. In 1843, English comparative anatomist **Richard Owen** (1804–92) defined homology as "the same organ in different animals under every variety of form and function". He used the term to capture the deep similarities that were apparent to the informed eye, despite superficial differences in form and even radical differences in function (for example, the bones in a bird's wing compared to those in man's arm). To explain these homologies, Owen suggested that each group of organisms was constructed according to a master plan in the mind of God, which he called the "archetype".

Darwin's theory of evolution provided a simple explanation for these similarities, with a new definition; homology is **similarity** that arises because of descent from a **common ancestor**. Homology is often contrasted with analogy – similarity that arises when similar lifestyles favour the independent evolution of similar sets of morphological characters. Such analogous characters are often termed "**homoplasies**", a term which can encompass loss as well as gain of features (for example, independent loss of legs in snakes and whales). The process that drives the evolution of

Bill Hamilton: evolution's mathematical adventurer

William Donald Hamilton (1936–2000) was born in Cairo, brought up in Kent and educated at Cambridge. As a child his taste for adventure almost killed him, when some explosives he was playing with blew up, scarring him for life. Drawn to evolutionary biology after reading the work of Ronald Fisher as a 28-year-old PhD student, Hamilton precociously published a paper outlining the concept of kin selection. Largely ignored for a decade, it eventually became one of the most highly cited papers in biology.

From 1964 to 1978, Hamilton was a lecturer at Imperial College, but unlike his predecessor Thomas Huxley, he turned out to be hopeless at lecturing. After a few years in America, in 1984 Hamilton settled into an academic position in zoology at Oxford, where he stayed for the rest of his life. His intellectual interests ranged widely, covering such topics as the evolution of sex ratios, the influence of natural selection on aging, the evolution of the social insects and the evolution of human behaviour. He was a key proponent of the idea that sex evolved as part of an evolutionary arms race with parasites, proclaiming that species are "guilds of genotypes committed to free, fair exchange of biochemical technology for parasite exclusion". With Robert Axelrod, he exploited game theory to reveal the foundations of reciprocal altruism and helped lay the foundations of sociobiology.

Although best remembered for his theoretical contributions to the study of evolution, Hamilton was no couch potato; instead, his life was laced with risk and adventure. In Oxford, he insisted on cycling rather than driving, despite repeated collisions with other road users. Looking for ants, he hiked through Rwanda while the civil war was at its height, only to be mistaken for a spy, because no one believed the true explanation for his visit. In Brazil, he fought a mugger who was carrying a knife and ended up with some nasty wounds. In the last years of his life, Hamilton became fascinated with the hypothesis that the AIDS virus had been transmitted to humans through oral polio vaccines. In search of evidence to support this idea, Hamilton embarked on a field trip to the war-torn Democratic Republic of the Congo to collect chimpanzee faeces. He was flown back to London with malaria, but died from cerebral complications a few weeks later. He is buried near Oxford in Wytham Woods, his favourite local biodiversity hotspot.

similarity rather than divergence in form is called **convergent evolution**. German zoologist **Adolf Remane** (1898–1976) outlined two rules of thumb for distinguishing homology from analogy: structural and developmental similarities have to outweigh functional necessity and the similarities have to make sense in the context of the organism's phylogeny.

Homology has turned out to be one of evolution's most powerful concepts, underpinning advances in the evolutionary classification of organisms through morphological comparisons and through molecular-biological approaches. The concept of **serial homology**, which captures the similarities between different features in the same organism, for example, the anatomical arrangements common to your hands and feet, has also primed the development of evo-devo studies (see box, p.140).

The identification of homologous features is the starting point for an evolutionary classification of organisms. However, by the middle of the twentieth century, traditional Linnaean approaches to classification were widely criticized for relying on subjective judgements and for assuming a fixed hierarchy of levels (kingdom, phylum, class, order, family) across the whole of biology (see box, p.62). To highlight the absurdity of this one-size-fits-all approach, contrast the family that we humans belong to, Hominidae (which encompasses just seven existing species, with a common ancestor that lived less than 20 million years ago) with the orchid family, Orchidaceae, which contains over 22,000 species, sharing a common ancestor over 75 million years ago.

A revolution in taxonomy erupted in the 1960s and 70s in the form of **cladistics**, an approach that originated largely through the work of German entomologist **Willi Hennig** (1913–76). Cladistics placed evolutionary theory and quantitative objective working methods centre-stage in taxonomy. Hennig referred to his approach as **phylogenetic systematics** and the term **phylogenetics** is often used interchangeably with cladistics. Although cladistics met with hostility at its inception, it is routinely used by today's taxonomists on both living and extinct organisms. In fact, some now complain that cladistics has become the new orthodoxy, with the risk of jettisoning its stable, established and more pragmatic predecessor, Linnaean taxonomy.

Unlike the Linnaean system, cladistics is based not on anatomical similarity but on evolutionary relationships (phylogeny). Cladistics looks only at the topology of the tree of life and ignores distance (a loose analogy can be made with the way stations are linked into lines on the map of the London Underground without worrying about the true distances between them). According to the cladists, taxonomy should capture the branching

Willi Hennig: quiet revolutionary of taxonomy

Emil Hans Willi Hennig (1913–76) was an unassuming expert on the classification of flies who unleashed a revolution in taxonomy. Hennig was born in Saxony and attended school in Dresden. His precocious interest in taxonomy is evident from an essay entitled "The position of systematics in zoology", written when only eighteen. Hennig studied natural sciences at the University of Leipzig and in 1936 completed his doctoral thesis on the unlikely subject of the genitals of flies. The following year he started work at the German Entomological Institute in Berlin. By 1939, Hennig's prolific research output had reached 41 publications totalling over a thousand pages.

Although never a member of the Nazi party, Hennig was conscripted when war broke out. He served as an infantryman in Poland, France, Denmark and Russia, where he was wounded in 1942. Later in the war, he contributed to malaria control in Greece and Northern Italy and served a similar role for the allies while a British prisoner of war in 1945. His wartime brushes with mortality (plus the fact that one of his brothers never returned from Stalingrad) drove Hennig to make the best of the time available to him during the difficult years after the war. After a brief spell in Leipzig, he returned to his old position in Berlin. In 1950, Hennig published (in German) his groundbreaking *Principles of a Theory of Phylogenetic Systematics*. In 1961, with his home cut off from his work place by the erection of the Berlin Wall, he was forced out of work. Fortunately, within a couple of years, Hennig managed to secure a position at the State Museum of Natural History near Stuttgart where he stayed until his death, carving out a new research niche studying insects trapped in amber. In 1966 a revised English-language version of Hennig's magnum opus from 1950 brought his ideas to a wider audience.

Despite his immense research output, Hennig retained an interest in high culture: he loved the music of Mozart and Handel, enjoyed vacations in Rome and Florence and, like Darwin, reveled in the prose of his hero Humboldt. Though famous by the end of his life, Hennig modestly disavowed any mention of "Hennigian systematics", arguing that that the term "phylogenetic systematics" should suffice. Cut down by a heart attack at 63, Hennig is buried in a mountain cemetery in Tübingen.

topology of the tree of life – classifications should reflect **common ancestry** and nothing else. By contrast, traditional taxonomists wish to include considerations of how far and how fast evolution has proceeded and whether evolutionary innovations or transitions have occurred. Cladists object to this on the grounds that deciding whether a feature counts as an innovation relies on subjectivity.

Frustratingly, cladists seem to have swallowed a Greek dictionary in devising their rather indigestible terminology. The basic taxonomic unit for the cladist is the **clade** (from the Greek for "branch"), which includes the common ancestor of species X and Y and all the descendants of that ancestor. In place of a universal hierarchy of equivalent ranks, the cladist views all taxonomic groups above species simply as clades, generated through the process of evolutionary branching (**cladogenesis**). Cladists generate tree-like diagrams called **cladograms** to represent the evolutionary relationships between clades, each node defining the ancestor of two branching lines of descendants. For the cladists, all taxonomic units should be **monophyletic**, i.e. consist of species that share a unique common ancestor. Thus, mammals all belong together in one clade, birds in another. However, crocodiles, dinosaurs and birds are all placed together in a clade (**Archosauria**) with no equivalent in traditional taxonomy, because they all share a unique common ancestor that is not shared by mammals or lizards.

The term **paraphyletic** is used to denote a group linked by common ancestry but which does not contain *all* the descendants of the last common ancestor. For the cladist, there is no such thing as "reptiles", as the taxonomic group Reptilia (reptiles) is paraphyletic (and so frowned on by cladists) in including crocodiles and snakes but excluding birds, even though birds number among the descendants of the most recent common ancestor shared by crocodiles and snakes. The term **polyphyletic** is used to describe a set of organisms linked by traits that

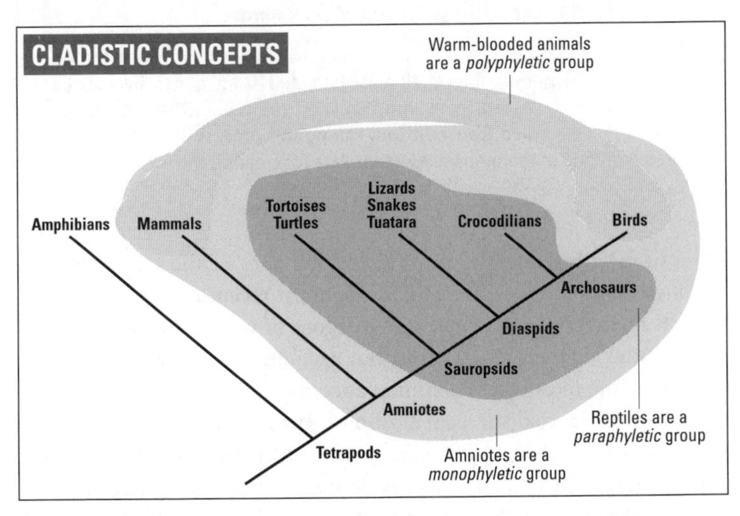

Monophyly, paraphyly and polyphyly are central concepts in cladistic classification.

Molecular biology: a brief primer

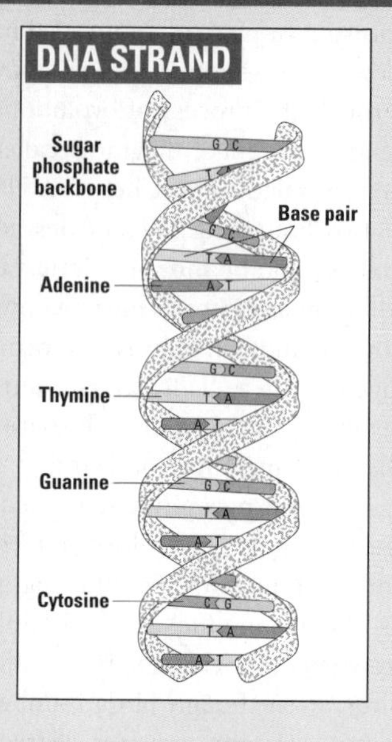

In all living organisms (except for a few viruses), the hereditary material is **DNA** (deoxyribonucleic acid). The information encoded in the DNA is used to construct all the other components of the cell, and, ultimately, the whole organism. DNA is a long double-stranded molecule with a helter-skelter twist. Each strand is a long chain of simple units called **nucleotides**, with a backbone made of sugars and phosphate groups. Attached to each sugar is one of four types of molecules called bases. Information is encoded in the **DNA sequence**, the sequence of these four bases along the DNA molecule. These bases, usually represented by the letters **ATGC**, supply the alphabet of life. In DNA's double helix, each type of base on one strand forms a bond with just one type of base on the other strand: A always pairs with T, and C with G: a phenomenon known as **complementary base pairing**. The double-stranded structure of DNA provides a simple mechanism for **DNA replication**, in which the two strands separate and two new complementary strands are synthesized by the enzyme **DNA polymerase**.

DNA segments that carry genetic information to make other cellular components are called **genes**. Information is extracted from genes by copying stretches of the DNA sequence into a related molecule, **RNA** (ribonucleic acid), in a process called **transcription.** During this process, the DNA strands separate and one of them serves as a template for creating the RNA molecules

have arisen independently on different branches of the tree of life. A polyphyletic group does not contain the most recent common ancestor of all its members: for example, the category of warm-blooded animals, which unites mammals and birds, but excludes many groups that also share the same most recent common ancestor.

An exercise in cladistic classification begins with the cataloguing of character states (get ready for some more Greek). These can be classified as **plesiomorphies** (ancestral character states) or **apomorphies** (derived character states). Only the distributions of shared derived character states (**synapomorphies**) are taken as evidence in the identification of clades

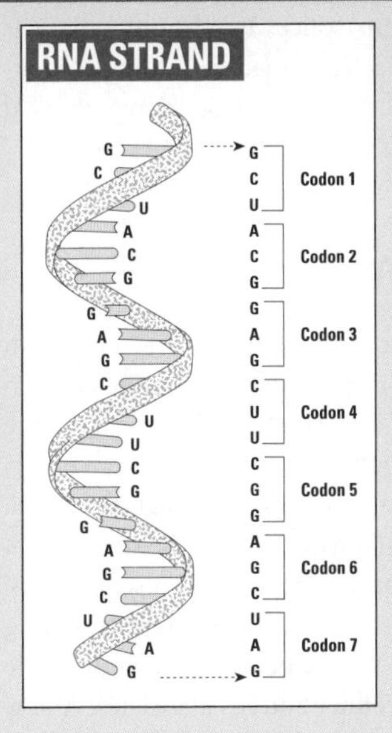

through the action of a protein called RNA polymerase. Some RNA molecules have functions in their own right but more commonly, as **messenger RNA** (or mRNA), they act as intermediaries in supplying information for the synthesis of proteins.

Proteins consist of strings of subunits, known as **amino acids**. Most of the time, the subunits used to synthesize proteins are drawn from a repertoire of twenty amino acids, although in rare cases, two additional amino acids are used. Amino acids are joined together in proteins like pearls on a string. The order of amino acids in a protein molecule – its **protein sequence** – determines its structure and function. Protein synthesis is carried out by a complex molecular machine called the **ribosome**. Through a process called **translation**, information in the mRNA specifies the sequence of the amino acids within proteins, using the genetic code to convert the sequence of bases into a protein sequence. The position of each amino acid in the protein molecule is determined by a stretch of three bases in the mRNA (a **codon**). The final protein molecule typically consists of several hundred amino acids linked together according to the instructions encoded in the mRNA. Proteins perform many functions in the cell, including catalyzing reactions between molecules, carrying messages, and acting as structural components.

The Rough Guide to Genes and Cloning Jess Buxton, Jon Turney (Rough Guides, 2007)

– shared ancestral character states (**symplesiomorphies**) are ignored. Thus, the presence of legs or hair would be useless in classifying groups of mammals, but the presence of a placenta might be useful in defining the clade of eutherian mammals.

In an ideal world, any reconstruction of phylogeny based on any subset of characters would agree with any other. In reality, it is often impossible to distinguish homologies from homoplasies or ancestral from derived characters, so that the taxonomist is faced with conflicting evidence. Several tricks of the trade are used to overcome these limitations. Large datasets are collected. A distantly related species outside the group one is studying (the outgroup) can be used to identify ancestral character states (for example, information from a rhesus macaque could be used to resolve the human-chimpanzee-gorilla three-way split). **Parsimony analysis** involves a series of computational approaches aimed at finding cladograms that depend on the smaller number of changes from a hypothetical ancestral state.

Molecular evolution

The cladistics revolution has been matched by another revolution in evolutionary biology: caused by the application of molecular biology to evolutionary problems, it was kick-started by several key discoveries in the mid-twentieth century. In the 1940s, through the work of American bacteriologist **Oswald Avery**, it became clear that DNA was the genetic material. In the 1950s, **Francis Crick and James Watson** deciphered the structure of DNA and revealed that genetic information was encoded in a sequence of bases along strings of DNA. In the 1960s, the mechanisms for translating DNA sequences into protein sequences were elucidated, including the cracking of the genetic code by **Sydney Brenner** and his associates. A new definition of a gene emerged: a sequence of DNA characters that encoded an "action molecule" like a protein or, less commonly, one of various kinds of RNA. The term genome was increasingly used to describe the set of all the genes in any given organism. The molecular basis of mutations and of genetic variation became clear.

For the non-biologist, the easiest way to envisage how genetics works at the molecular level is to draw on an analogy between a genome and a handwritten manuscript carefully copied by generation after generation of monks. Just as a manuscript is written in the letters of an alphabet, the genome is built from four DNA characters, A, C, T and G. In the manuscript, strings of letters make up larger units (words, paragraphs, chapters) that, when read through the human eye and brain, produce meaning. Similarly, sequences of bases in DNA make up genes that, when read through a cell's molecular machinery, elicit functions, usually by providing recipes for proteins. Changes arise in the manuscript and the genome during the process of copying. We see similar changes in both contexts: one character can change into another; one or more character can be inserted or deleted; strings of characters can be moved from one part of a manuscript/genome to another or to an entirely different document/cell.

Homology and model organisms

At first sight, nothing seems more absurd than imagining that one could gain a better understanding of cancer by investigating yeast cells or gain insights into immunity or aging from studying microscopic worms. And yet in 2001, two scientists (Leland Hartwell and Paul Nurse) won the Nobel Prize in Medicine for studies on genes that control cell division in yeast, while the following year, the same prize went to three scientists (Robert Horovitz, John Sulston and Sydney Brenner) who pioneered scientific scrutiny of a one-millimetre-long nematode worm, *Caenorhabditis elegans*. These advances rely on the fact that **model organisms** quite distant from humans rely on similar molecular pathways and even on homologous genes and proteins in their day-to-day lives.

Homology at the molecular level is most commonly detected by looking for similarities between the sequences of amino acids in different proteins. Computer programs allow scientists to search a protein sequence against a database, representing the accumulated knowledge of humankind, in just a few minutes. In fact, in the genome-sequencing era, more genes have been assigned a function using these **bioinformatics** approaches than through experimentation. The human genome sequence would be nearly worthless without the value added by evolutionary comparisons with sequences from other organisms. Although assignments of function based on sequence homology are initially only tentative, they provide a valuable framework for framing hypotheses that may then be tested in the laboratory. For example, Grant Bitter and his associates have been investigating the effects of cancer-causing mutations in human DNA repair genes by studying the effects of the same mutations at homologous positions in equivalent yeast genes.

The molecular bases of many human diseases, particularly those with a known genetic component, are now routinely studied in animal models, from yeast to laboratory mice. The ability to make **knockout mice**, which lack a gene conserved in humans, allows sophisticated investigations of gene function. Perhaps the most bizarre use of model organisms to investigate human disease relies not on mice but on fruit flies. Flies, like humans, get intoxicated, build up tolerance and develop a liking for alcohol. Molecular analysis of such alcohol-induced behaviours in flies has already revealed some surprising parallels with mammals.

Neutral theory

Some copying errors in a manuscript will affect meaning (for example, turn one word into another; turn a meaningful phrase into nonsense), but many will not (say putting a slightly larger gap between two words; ending a line at a different place in a sentence; replacing one valid spelling with another, say changing "synthesize" to "synthesise"). Surprisingly, most of the analogous changes in genomes have no effect on function – an idea forcefully articulated in the late 1960s and early 70s by Japanese biologist

The great tree of life

In *The Origin*, Darwin wrote poetically about a great tree of life and speculated about a common origin. In the century that followed his death, biologists and palaeontologists made progress in sketching out branching patterns close to the crown of the tree, but were stumped when it came to making any sense of the deep branches. Early in the twentieth century, biologists recognized a key distinction between **eukaryotes**, organisms such as animals, plants, fungi and protozoa that partitioned their DNA away within a membrane-bound nucleus, and **prokaryotes** that lacked a nucleus. In the 1970s, the pace of discovery speeded up with the arrival of molecular phylogenetics. American microbiologist **Carl Woese** (b.1928) proposed that, at the deepest level, life was divided into three domains: eukaryotes, bacteria and **archaea** (a group of single-celled microorganisms as distant evolutionarily from bacteria as we are). Although archaea were first recognized only in extreme environments, they are now known to be common elsewhere, perhaps accounting for a fifth of our planet's biomass. Working out how these three branches should be joined – rooting the tree of life – has been a cause of contention. In recent years, recognition of the potential for genomic mergers or large-scale exchange of genes has led some to reformulate the connections between the earliest organisms as a **web of life** or **ring of life**. However, branching evolution is still an accepted model for most of life's history.

Molecular surveys continue to highlight life's unexpected diversity. Curiously, multicellularity appears to have evolved several times: in animals, fungi, plants, as well as in amoebozoans (a group that includes slime moulds) and heterokonts (a group that includes common seaweeds). Most eukaryotes can now be slotted into one of eight major groups: animals and fungi, for example, belong together in a branch called the **opisthokonts**. But molecular surveys have revealed huge diversity among extremely small eukaryotes, which populate new deep branches scattered across the tree. The **Tree of Life Web Project** (www.tolweb.org/) is an ambitious collaborative project aimed at providing information about the diversity and phylogeny of all life on Earth, arranged in branching hierarchy of interlinked web pages. In addition, German bioinformatician **Peer Bork** (b.1963) has created the **Interactive Tree Of Life** (http://itol.embl.de), a web based tool for the display and manipulation of life's deepest branches.

The Tree of Life: A Phylogenetic Classification Guillaume Lecointre Hervé and Le Guyader (Belknap Press, 2007)

The Ancestor's Tale: A Pilgrimage to the Dawn of Evolution Richard Dawkins (Houghton Mifflin, 2004)

Motoo Kimura (1924–94), under the title of the **neutral theory** of molecular evolution. Inherent in Kimura's theory was the idea that most evolutionary change is due to genetic drift acting on neutral molecular variants. Often one variant will increase in frequency and even take over

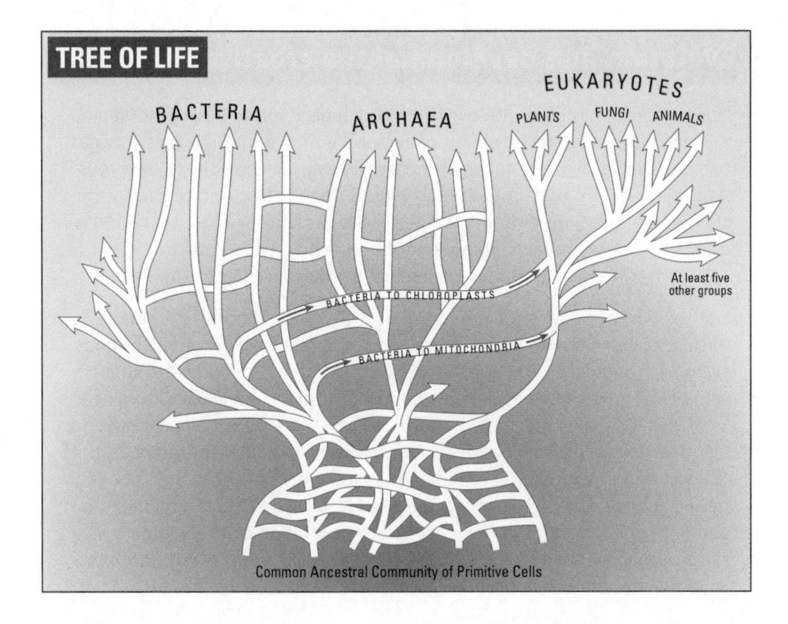

a population, just by chance. When it was first introduced the neutral theory was seen as a controversial challenge to Darwin's evolution by natural selection, but it is now accepted as the orthodox view of evolution at the molecular level.

The neutral theory still allows for natural selection to act on non-synonymous mutations that do affect protein function. In fact, it has provided a handy method to identify proteins that are evolving rapidly under the influence of natural selection. The reasoning goes like this. Neutral change is taken to be the default option in protein evolution, with the proportion of synonymous changes (dS) outnumbering the proportion of non-synonymous mutations (dN) by a considerable margin – **the dN/dS ratio** is typically about 1:5 or 0.2. However, when natural selection is active, the rate of non-synonymous change goes up. Once the dN/dS ratio exceeds 1.0, scientists take this as good evidence that natural selection has been acting to change the protein sequence. Using such approaches, hundreds of proteins have been identified in the human genome that have been evolving rapidly, particularly since our divergence from our common ancestor with our nearest relatives, the chimpanzees.

Another consequence of the neutral theory is that we can use the number of changes between homologous sequences to calculate the divergence times. **Emile Zuckerkandl** (b.1922) and **Linus Pauling** (1901–94) noticed

Evolution and development: a creative clash

Biologists are confronted with two kinds of change: changes during the lifetime of an individual (the development of an adult from a single cell) and changes that have unfolded across an organism's evolutionary lineage. The claim of "evo-devo", a movement that arose in the 1990s, is that the first kind of change can afford insights into the second. The roots of evo-devo go way back to the 1890s, to Bateson's description of homoeotic abnormalities, where one body part is replaced by another (for example, a leg grows where an antenna should be).

With the advent of genetics, it became clear that such abnormalities in the fly were associated with ***Hox* genes**, expressed at different positions along the body, telling each segment what appendage it should grow. Curiously, the arrangement of *Hox* genes in the fly genome mirrors the order of body parts, from head to tail, that the genes specify. A big surprise came in the 1980s, when it became clear that *Hox* genes were not restricted to insects and other invertebrates but also occurred in vertebrates, including humans, where they did the same job, specifying what goes where along the head-to-tail axis. One recent study has even implicated changes in *Hox* genes to the **evolution of the vagina** (a late mammalian innovation – reptiles, birds and platypuses have to make do with a single channel, the cloaca, for pooing, peeing and procreating).

These discoveries imply that animals use conserved developmental toolkits, homologous genes and pathways, that must have evolved over half a billion years ago, before the emergence of animal body plans in the Cambrian explosion. These findings have breathed new life into a favourite adage of Darwin's: "Nature is profligate in variety, though niggard in innovation". Beneath the ebb and flow of morphological diversity lies a bedrock of molecular conservation.

Several other themes govern the evo-devo outlook. Body plans and developmental processes are **modular**, with module duplication and adaptation acting as drivers of evolution. Rather than radical changes in gene sequences or repertoire, most morphological evolution derives from changes in the strength, timing and spatial distribution of the expression of genes or chemical signals that influence the fate of cells. In other words, evolution is primarily regulatory, not structural. For example, when sticklebacks colonize a lake lacking predators, evolution drives loss of their pelvic spines through changes affecting a single gene, *Pitx1*. However, this gene produces the same protein in both kinds of fish. The crucial difference is that in the ancestral fish, *Pitx1* is expressed in the head, trunk, pelvis, tail, but in the lake fish the pelvis has stopped expressing it.

Endless Forms Most Beautiful: The New Science of Evo Devo Sean B. Caroll (Norton, 2005)

in the early 1960s that the number of amino acid differences in haemoglobin molecules from various species roughly reflected their divergence times according to the fossil record. The term **molecular clock** was popularized

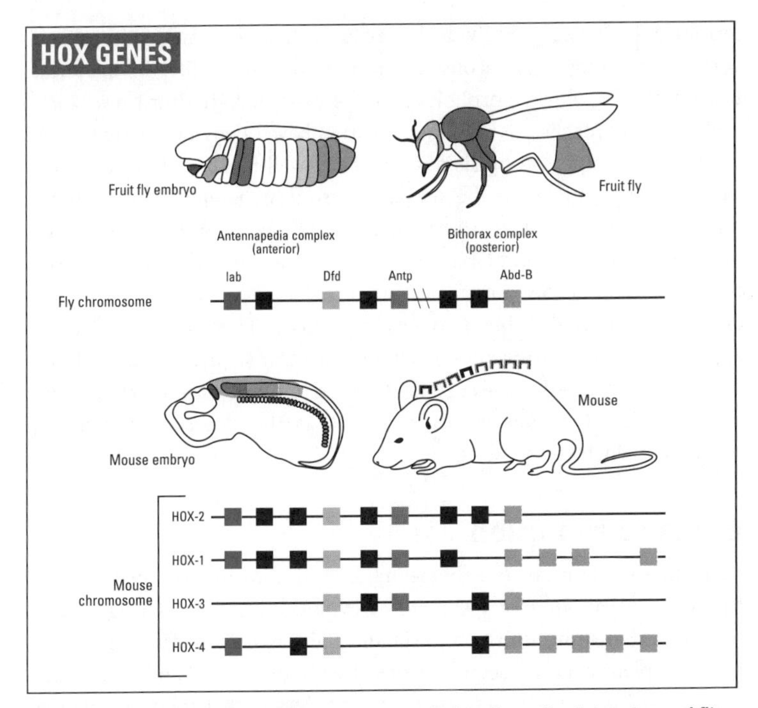

Similar Hox genes control development along the head to tail axis in mice and flies.

later in the sixties by New Zealander **Allan Wilson** (1934–91) who, along with his student **Vincent Sarich** (b.1934), used this approach to show that humans and chimpanzees shared a most recent common ancestor more recently than anyone at that time suspected.

Molecular systematics

Molecular biology has had its greatest impact on evolution through the development of **molecular systematics**, that is the use of similarities and differences in molecular sequences to classify living organisms and reconstruct their lines of evolutionary descent. In short, this approach, a molecular version of cladistics, relies on determining the sequences of homologous genes or proteins from different organisms, aligning those sequences to identify equivalent positions in each of them, and then exploit computational and mathematical approaches to build a **phylogenetic tree** based on patterns of sequence similarity. Molecular approaches have transformed taxonomy at every level, from unravelling the detailed relationships between individual species such as humans, chimps and gorillas to attempts to reconstruct the entire tree of life. In **bacterial taxonomy**, for which morphology or the fossil record afford few clues, molecular approaches to bacterial evolution have opened up entirely new vistas. For example, mycoplasmas, small cell-wall-less bacteria, were once thought to be similar to the simple ancestral archetype of all bacteria, but now, through molecular systematics, we know them to be degenerate relatives of more complex Gram-positive bacteria. Similarly, mitochondria, the small cigar-shaped organelles that provide power to our own cells, are now recognized as highly divergent bacteria. Even where traditional approaches work, phylogenies based on molecular sequences have refined or even overturned earlier conclusions. Examples have already been discussed: giant pandas as bears; hippos as the closest land-dwelling relatives of whales.

Genome evolution

In the new millennium, one theme in the application of molecular biology to evolution has come of age: **genome evolution**. In 1995, **Craig Venter** (b.1946), **Hamilton Smith** (b.1931) and their associates determined the first complete genome sequence of a free-living organism, that of the bacterium *Haemophilus influenzae*. In the decade that followed, at first dozens, but then soon thousands of bacterial genomes were sequenced. Comparisons between these genomes revealed the forces that have shaped their evolution. One surprise to arise from this work was the high level of horizontal gene transfer in bacteria, that is the exchange of genes between one cell and another outside the usual route from parent to offspring. Indeed, in some bacteria like *E. coli*, there is such rampant **gene exchange** that one is driven to a vision of bacteria promiscuously engaged in an orgy of global group sex. However, equally surprising was the discovery that some important bacteria have evolved by losing genes rather than gaining them, a process known as **reductive evolution**.

While determining the genome sequences of multicellular organisms (such as plants and animals) is much slower and more laborious than sequencing a bacterial genome, the last decade has seen notable successes here too. We now have genome sequences from worms, flies, fish, rodents, humans, apes, dogs, cats, rice and maize. Comparisons between these genomes are providing us with valuable insights into the molecular changes that led to the evolution of key groups. For example, genome sequences have provided fresh evidence for the **2R hypothesis** first proposed in the 1970s by

J. Craig Venter, whose pioneering work on genome sequencing has opened a new window on the diversity and evolution of life.

Asian-American geneticist **Susumu Ohno** (1928–2000) that the emergence of the vertebrates was marked by two separate doublings of the entire genome. Comparisons between the human, chimp and macaque genomes are even helping provide insights into the changes that make us human.

The target of natural selection: gene, organism or group?

One of the liveliest debates in evolutionary biology in recent years has centred on defining the level at which natural selection acts. In his discussions of the struggle for existence, Darwin made it clear that he saw the struggle between individuals as the key driver of evolution: "But the struggle almost invariably will be most severe between the individuals of the same species, for they frequent the same districts, require the same food, and are exposed to the same dangers". Despite the shortsighted nature of selection acting at the level of the individual, popular interpretations of Darwinism often misleadingly evoke evolution acting "for the good of the species" – a phrase guaranteed to make orthodox Darwinians wince.

But as the individual sits half way up a hierarchy stretching from genes to the whole biosphere, many evolutionary biologists have been tempted to view evolution acting at higher or lower levels. American evolutionist **George Williams** proposed a **gene-centred** view of evolution in the 1960s. However, it was only after the publication of the popular and controversial book *The Selfish Gene* (1976) by Oxford zoologist **Richard Dawkins** (b.1941) that this viewpoint gained widespread attention. The book's provocative title encapsulated Dawkins' idea that evolution is best envisaged as acting on genes. Dawkins classed individual organisms as disposable **survival machines** at the service of genes as long-lived **replicators**. He also made an analogy between the propagation of ideas and genes, coining the term **meme** to describe ideas that spread from mind to mind just as genes move from body to body. Dawkins elaborated his gene-centric view in later publications, particularly his book, *River Out of Eden* (1995). In the mid-1990s, American philosopher Daniel Dennett effectively became Dawkins' bulldog, defending and extending the selfish-gene concept in his 1995 book *Darwin's Dangerous Idea*.

Opponents of Dawkins' view have argued that it is too narrow and reductionist; that it relies on a poor anthropomorphic analogy (genes cannot really be selfish); that it scarcely qualifies as scientific (how could

Spandrels, squinches and squabbles

A lively dispute in evolutionary biology centres on whether every feature of an organism can be explained as an adaptation, or whether some features are simply by-products of other processes. In *The Origin,* Darwin recognized "correlation of growth" as a cause of evolutionary change independent of natural selection (for example, if selection favours increased strength in one arm, then the other arm might also grow stronger, given a common underlying developmental programme for the two limbs). In modern times, this topic was thrust centre stage by a paper from 1979 by Stephen Jay Gould and Richard Lewontin, *The Spandrels of San Marco and the Panglossian Paradigm: A Critique of the Adaptationist Programme*. They borrowed the term "spandrel" from architecture to describe a characteristic that has arisen during evolution as a side effect of a true adaptation. In architecture, spandrels are roughly triangular areas that result from the design of fan-vaulted ceilings. Although the spandrels of San Marco frame beautiful mosaics, Gould and Lewontin argued that spandrels were not originally designed for that purpose – they were just by-products of other architectural decisions, and then architects were forced to make best use of them. Gould subsequently elaborated on his analogy in his magnum opus, *The Structure of Evolutionary Theory* (2002). In response to Gould's analogy, critics such as philosopher Daniel Dennett have argued that Gould has not just his biology wrong, but also his architecture. Dennett argues that architectural spandrels (members of a class of features known as pendentives) are not un-designed accidents, but have been deliberately chosen as one of several possible solutions (alternatives include a simpler structure known as a squinch) to an architectural problem.

The domes and spandrels of St Mark's Basilica in Venice, Italy.

it be falsified?) and presupposes a clear-cut definition of what a gene is (the concept is decidedly fuzzy) or how long a gene survives before mutating beyond recognition. Nonetheless, Dawkins' selfish gene has provided biologists with a new conceptual viewpoint and has helped focus the popular imagination on to evolutionary biology. It also helps explain the proliferation within bacterial and human genomes of **mobile genetic elements**, whose only action is a selfish one, the creation of multiple copies of themselves.

At the opposite extreme to the supporters of the selfish gene are those that advocate the phenomenon of **group selection**, i.e. that evolution can favour the survival of the group over and above any benefits for the individual. This view was championed by British zoologist **Vero Copner Wynne-Edwards** (1906–97) in his book *Animal Dispersion in Relation to Social Behaviour* (1962), but largely discredited after critiques by George Williams, John Maynard Smith and others. In recent years, interest has re-focused on the possibility selection at multiple levels (for example, when viewing genomes as ecosystems of genes), but the orthodox Darwinian view still limits selection to the level of the gene or individual.

Part 2

The greatest story ever told

6 A brief history of life

Darwin ducked the origin-of-life question in *The Origin of Species*, but in a later letter to his friend Joseph Hooker he wrote: "It is often said that all the conditions for the first production of a living organism are now present … But if (and oh, what a big if!) we could conceive in some warm little pond, with all sorts of ammonia and phosphoric salts, light, heat, electricity, &c., present, that a proteine compound was chemically formed ready to undergo still more complex changes, at the present day such matter would be instantly devoured or absorbed, which would not have been the case before living creatures were formed."

The origin of life

Darwin's letter to Hooker raises two key points. Firstly, that there is no need for a supernaturalist explanation for the **origin of life** – life could have arisen through natural chemical processes. Secondly, to avoid confusion with the discredited phenomenon of **spontaneous generation** (the idea that living organisms are continually springing into existence), we have to assume that the first life forms originated under conditions that no longer exist.

Although the origin-of-life question remains unresolved, there has been considerable progress since Darwin's time. Instead of a blank sheet of paper, we now have the equivalent of a half-completed sudoku puzzle, with some answers firmly written in ink, whereas in other cases several alternatives have been outlined with a pencil. Life as we know it is **carbon-based** and depends on **water**. The nuclei of the heavy elements needed for life (carbon, nitrogen, oxygen, phosphorus, sulfur and iron) were formed during the explosive destruction of the first generations of stars as they became supernovae. In a quite literal sense, we are **stardust**. Whether exotic forms of life built from entirely different elements could ever exist remains the stuff of science fiction.

Extraterrestrial origins?

A key question is where did life originate? Most investigators make the parsimonious assumption that life originated here on Earth (**geogenesis**). However, the idea that life originated elsewhere and was then seeded here (**panspermia** or **exogenesis**) has a long and distinguished pedigree. Proponents have included Swedish chemist **Svante Arrhenius** (1859–1927), Irish physicist Lord Kelvin and co-discoverer of the structure of DNA, Francis Crick. Crick even went so far as to suggest **directed panspermia**, where the seeds of life are purposely spread by an advanced extraterrestrial civilization. Beyond the fringe of scientific acceptability sit claims made by astronomers Fred Hoyle (1915–2001) and Chandra Wickramasinghe (b.1939) that living

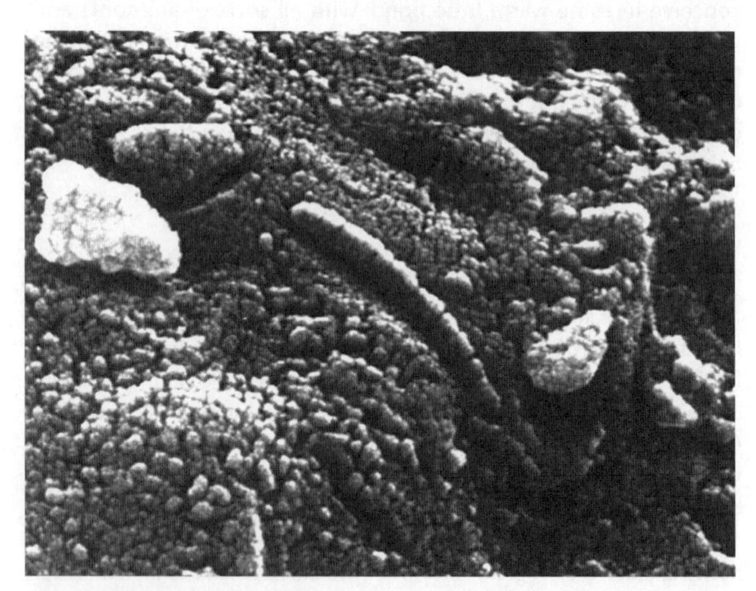

It has been controversially suggested that the microscopic structures visible in this view of meteorite ALH84001 could represent fossilized Martian microbes.

organisms from extraterrestrial environments continually enter the Earth's atmosphere, priming new epidemics and driving evolutionary innovations. Similarly, claims that bacteria on the Surveyor 3 camera survived for over two years on the Moon have now been discredited.

The discovery in the mid-1990s of what appeared to be fossilized bacteria in **ALH84001**, a meteorite of Martian origin, stimulated interest in a limited form of panspermia, whereby life might have originated elsewhere in our solar system. This idea has gained support from the realization that ice occurs on the Moon and Mercury and that liquid water may have existed on early Mars and Venus and probably still exists on some satellites of large planets (Europa, Enceladus and Triton).

The fledgling field of **astrobiology** brings an inter-disciplinary slant to the potential for life in the universe. There is as yet no direct evidence for the existence of life outside of Earth. However, at the very least, investigations of meteorites and samples from other planetary bodies are likely to afford new insights into the conditions that give rise to the chemical prerequisites for life. At best, scientists will one day hit the jackpot by finding the first evidence of extraterrestrial life. Similarly, the discovery of increasing numbers of extra-solar planets increases the likelihood that life exists elsewhere in the universe. Some of these planets might even occur in the **habitable zones** around their associated stars (the region in which liquid water could exist), with the planet **Gliese 581**, twenty light years from Earth, the best candidate so far. Spectroscopic detection of oxygen-rich Earth-like atmospheres from extra-solar planets might one day also provide evidence of extraterrestrial life.

Assuming life did originate on Earth, one thing is clear: conditions on the early Earth during the **Hadean eon** (4.6–3.8 billion years ago) were quite different from what we see today. At first, there was a series of cataclysms: the Earth formed from the accretion of particles within a disk around the Sun; collision with a Mars-sized planet spawned the Moon; iron and other heavy elements sank into the planet's core; heavy bombardments kept the planet's surface molten. Even after the crust solidified and oceans formed, the Sun was weaker, the day much shorter, the Moon much closer (and so tides much stronger) and the atmosphere completely unlike that of today (for a start, there was no oxygen).

Life's building blocks

The first real progress in addressing the chemical basis for the origin of life came in the 1920s when Russian biochemist **Alexander Oparin**

(1894–1980) and English evolutionist J.B.S. Haldane independently suggested that life must have originated in a **primeval soup** of organic molecules. The discovery of methane in the atmosphere of gas-giant planets like Jupiter led Oparin to suggest that the primaeval Earth possessed a strongly reducing atmosphere of methane, ammonia, hydrogen and water vapour. In 1953, based on Oparin's predictions, **Stanley Miller** (1930–2007) and Nobel-prize-winning chemist **Harold Urey** (1893–1981) performed the classic **Miller-Urey experiment** at the University of Chicago. They sealed water, ammonia, methane and hydrogen together in a glass apparatus, supplying heat, electrical sparks and a condenser, and left the experiment running for a week. Within that time, over ten percent of the carbon in the mixture had become incorporated into organic compounds, with two percent forming amino acids. The experimental conditions led to the synthesis of thirteen of the twenty amino acids found in proteins and proved that most of the building blocks of life could arise through non-biological processes. A few years later, American biochemist **Sidney Fox** (1912–98) built on the Miller-Urey experiment, showing that amino acids could link up under similar conditions into small protein fragments.

Although it now seems less likely that the early atmosphere contained precisely the components Urey and Miller used, analogous experiments using other plausible oxygen-free atmospheres have afforded similar results. The discovery of amino acids and other biologically relevant molecules in the **Murchison meteorite** (which fell on Australia in 1969), and in other similar carbonaceous chondrites, has shown that the building blocks of life can be synthesized even in extraterrestrial environments, and may have been supplied to the early Earth through meteorite bombardments.

From chemistry to biology

But chemicals alone do not make life. Somehow, simple chemicals had to be assembled into something more complex. The precise kind of environment in which this happened remains unclear for the time being. Darwin's warm little pond or Oparin's primaeval soup implies that it happened to molecules floating free in a watery liquid. But for many scientists a solid surface seems more likely. Glasgow chemist **Alexander Cairns-Smith** (b.1931) has championed a controversial **clay theory**, where life originated in the self-replication of clay crystals. Munich chemist **Günter Wächtershaüser** has suggested mineral surfaces such as iron pyrites as a matrix for linking amino acids together. Wächtershaüser has also proposed hydrothermal vents (cracks in the Earth's surface from which heated water escapes) as a likely starting place for life, free of the need for energy from photosynthesis. One kind of vent discovered in the 1970s, the **black smoker**, supports a rich ecosystem based on organisms that rely on chemicals such as hydrogen sulfide issuing from the vent as a source of energy, rather than light from the Sun. In a quite different setting, the mid-Atlantic **Lost City vents** produce no sulfides, but through **serpentinization** (a reaction between rocks from deep in the Earth's mantle and water) liberate molecular hydrogen and methane, two other potential fuels for early life.

Beaches provide another plausible venue for the emergence of life. Drawing on the example of a natural nuclear reactor from **Oklo** in Africa, Seattle astrobiologist **Zachary Adam** has suggested that stronger tidal forces on the early Earth might have concentrated uranium at high tide mark, producing **radioactive beaches** that energized the assembly of life's building blocks. Tidal pools also provide a mechanism for concentrating chemicals and for chemical and thermal cycling as they fill, evaporate and re-fill or warm up, cool down, then warm up again. Such cycles might have driven the replication of the progenitors of DNA by allowing template-driven synthesis then dissociation of new strands.

Any theory of the origin of life has to account for the three cardinal properties of life-as-we-know-it: **genetics**, **metabolism** and **cellularity**. One lively chicken-or-egg debate concerns whether metabolism (the conversion of one kind of small molecule into another) was invented before genetics (maintenance of information as a string of subunits in a large molecule), or vice versa. Wächtershaüser is a keen proponent of the **metabolism-first** view, proposing the existence of an **iron-sulfur world**, which harboured ancestral pre-living reactions that powered the synthesis of organic building blocks from simple gaseous compounds. American theoretical biologist **Stuart Kauffman** (b.1939) also favours the metabolism-first view. However, Kauffman bases his arguments on the notion of **autocatalytic sets** of reactions, where complex molecule A helps catalyze formation of B, which helps catalyze the synthesis of C, and so on until the loop is closed with Z facilitating the synthesis of A.

Primordial genes

Genetics relies on the replication and copying of information based on a template molecule, together with a pathway to translate information into action. A big problem with life's current DNA-makes-RNA-makes-protein approach is its circularity: DNA needs proteins for its own replication. In the late 1960s, Carl Woese, Francis Crick and British evolutionary theorist **Leslie Orgel** (1927–2007) all independently suggested that the middleman, RNA, might provide an escape from the causal loop. American biochemist **Tom Cech** (b.1947) subsequently showed how the twin attributes of information and action could co-exist within RNA molecules that could act as enzymes (**ribozymes**). Drawing on this work, Nobel-laureate **Walter Gilbert** (b.1932) coined the term **RNA world** to describe a primordial biology that used RNA for information storage and catalysis. Catalytic RNA molecules at the heart of today's ribosomes (the cell's protein synthesis factories) are probably remnants of this RNA world. Cell-free reproduction and even evolution of RNA molecules has been replicated in the laboratory. But some scientists are keen to reconstruct a pre-RNA world, made from even simpler macromolecules. Amsterdam astrobiologist **Pascale Ehrenfreund** has suggested polycyclic aromatic hydrocarbons (organic molecules found in meteorites and elsewhere in space) as the original replicators (the PAH world).

Another thing that has to be explained is how the chemistry of life got to be sealed off within cells from the outside world. Globules form in numerous settings: Stan Fox found them in his experiments with peptides; bubbles in sea foam are another suggested source. Wächtershaüser has proposed a **primordial sandwich** model in which membranes form first on a mineral surface before bubbling off primordial cells. Oxford evolutionary biologist **Tom Cavalier-Smith** (b.1942) has promoted a controversial evolutionary scheme in which the genetic material was at first attached to the outside of an inside-out cell (the **obcell**) before an involution of membranes brought it inside.

Another puzzling difference between lifeless chemistry and the world of biology is a preference for one mirror-image form over another in the building blocks of life – a phenomenon known as chirality. The forces that have driven the **origin of chirality** in living systems are unclear. An excess of one mirror-image form over another has been detected in non-biological samples, such as the Murchison meteorite. **Donna Blackmond**, a chemist at London's Imperial College, has described a physicochemical process that generates a strong bias towards one form from a small initial imbalance. Researchers in Sweden have shown that biased mixtures of amino acids can in turn catalyze the asymmetric synthesis of carbohydrates.

LUCA: the elusive progenitor of all life on Earth

Although one cannot rule out the possibility that life might have arisen multiple times in Earth's early history, molecular studies have confirmed that all living organisms trace their ancestry back to a single common ancestor, poetically named **LUCA** (for last universal common ancestor). However, LUCA should not be confused with the first-ever living organism, nor with the most primitive possible life form. Disputes rage over how far one can reconstruct LUCA, given the potential for early genomic mergers and gene transfers. Some have even suggested that LUCA was not a discrete cellular entity, but a community.

But as an early advert for communal living, LUCA appears to have run a sophisticated show. She apparently possessed all the enzymes necessary to make DNA, to transcribe it into RNA and to translate the information in RNA into proteins. She probably also ran a versatile metabolism that worked without oxygen, scavenged energy from chemicals yet used the same chemical currency as all modern cells (the molecule ATP).

If, as is commonly assumed, life originated on the early Earth in conditions similar to those seen in hydrothermal vents, then one would expect LUCA to be a **thermophile** – capable of growth at high temperatures. A team headed by **Eric Gaucher** in Florida recently provided a striking confirmation of this prediction by resuscitating an ancient protein from LUCA. They compared over two dozen versions of an ancient conserved protein (EF-Tu) drawn from a menagerie of organisms, reconstructed various ancestral forms and then tested their abilities to function at high temperatures. They found a steady increase in heat resistance as they progressed back towards the ancestral form found in LUCA.

Similar sequence comparisons allow us to gain a glimpse of LUCA's simpler ancestors. For example, the enzymes that build the transfer RNA molecules used to translate nucleotide sequences into protein sequences belong to two large families of homologous proteins, suggesting that the primordial genetic code might have encoded just two forms of amino acid. Francis Crick suggested that the current universal **genetic code** is a **frozen accident** – an arbitrary arrangement that has been conserved since before LUCA (just as the QWERTY keyboard has survived beyond the typewriter era). More recently, it has been suggested that some of aspects of the code might be adaptive, for example making it more

Our inner bacteria: the mitochondrial merger

Most people have heard of the "friendly bacteria" that live on our skin and in our guts. In fact, there are more of these bacterial cells associated with you than there are human cells. But the association with bacteria is much deeper and more ancient than you think. Most cells in your body are stuffed with hundreds or thousands of minute cigar-shaped structures called mitochondria – miniature powerhouses of the cell, which allow you to exploit oxygen as a fuel. The average adult human body contains 10 million billion of them, and they make up 10 percent of your weight! But it turns out that mitochondria are highly specialized, degenerate bacteria.

In the early twentieth century Frenchman Paul Portier and American Ivan Wallin suggested that mitochondria might have originated as beneficial bacteria that took up life inside our cells (endosymbionts). Decades later, American biologist Lynn Margulis (b.1938) breathed fresh life into the **endosymbiotic theory of mitochondrial origins**, particularly through publication of a landmark book *Origin of Eukaryotic Cells* (1970). In this scenario, chloroplasts were also seen as endosymbiotic descendants of photosynthetic bacteria. The discovery in the 1980s that mitochondria and chloroplasts have their own **genomes** placed the theory beyond doubt.

Possession of mitochondria is a hallmark of the eukaryotes (organisms that wrap their DNA in a membrane to form a nucleus). Eukaryotes form one of the three great domains of life (along with bacteria and archaea). Although some unicellular eukaryotes lack mitochondria and were once misclassified as primitive "archaezoa", it is now clear that all eukaryotes have evolved from a mitochondrion-bearing ancestor. In fact, many scientists now see the acquisition of mitochondria as the defining moment in the birth of **eukaryotes**. However, arguments rage over what kind of cell did the acquiring. Various scenarios have been proposed, many identifying the engulfing cell as an archaeon. However, molecular biologists Anthony Poole and David Penny have recently argued persuasively that the **proto-eukaryote** must have already diverged from the archaea by the time of the mitochondrial merger. Whatever the train of events, without mitochondria there would be no animals or humans, as these specialized bacteria power our muscles, brains and even our sperm; they also keep cancer in check by controlling whether our cells live or die. Rejoice in your inner bacteria!

Power, Sex, Suicide: Mitochondria and the Meaning of Life Nick Lane (Oxford, 2005)

robust to the effects of mutation and that there might be a non-arbitrary affinity between codons and amino acids. Evolutionary computational biologist **Eugene Koonin** has suggested that the genetic code arose from pre-existing amino-acid-RNA interactions that enhanced the catalytic activities of ribozymes.

Life in the Precambrian

The oldest known sedimentary rocks, which are at least 3.8 billion years old, are found on **Akilia Island**, off Greenland. These rocks formed at or near the end of a late heavy bombardment of the Earth's surface that would have precluded an earlier origin of life. Analysis of carbon in these rocks shows what is arguably the first evidence of life in the geological record: a change in isotopic composition (in the ratio of one form of carbon atom to another) that is commonly associated with biological processes. However, this evidence has been challenged, given that other non-biological explanations may account for this anomaly.

Among the most striking first features in the geological record are **stromatolites**, layered structures that form in shallow water by the capture and accretion of grains of sediment. The earliest stromatolites, dated at 3.4 billion years old, occur in the **Pilbara Craton** rock formation in Western Australia. Although doubts have been expressed as to whether the earliest stromatolites are biological in origin, modern stromatolites are clearly the products of living organisms – communities of bacteria that trap and cement the grains.

Modern stromatolites in Shark Bay, Western Australia. Fossilized stromatolites are thought to provide the earliest macroscopic evidence of life.

The first convincing microscopic fossils (microfossils), which resemble contemporary photosynthetic bacteria, also hail from the Pilbara Craton formations. Thereafter, there is a fairly solid record of bacterial life through the rest of the Precambrian. The existence of **banded-iron**

Geological time: eons, eras, periods and epochs

In the late seventeenth century, Danish geologist **Nicholas Steno** (1638–86) proposed that rocks were laid down in horizontal layers or strata. Crucially he recognized that younger upper strata were deposited after older lower strata, the key insight underlying the discipline of **stratigraphy**. Armed with Hutton's recognition of deep time and the identification of signature fossils in selected strata, geologists were able to build a sequence of strata that reflected the Earth's history and to divide geological time into units based on major geological events (mountain building) or biological incidents (mass extinctions). With the arrival of radiometric dating methods in the twentieth century (see box, p.12), absolute dates could be attached to each unit of geological time, typically measured in millions of years or **mega-annums** (abbreviated as Ma; this nomenclature is also used to describe past periods, where the term "ago" is taken as read, ie 65 Ma can mean 65 million years ago). Maintenance of a standard global geological time scale is now the responsibility of the International Commission on Stratigraphy (www.stratigraphy.org).

The largest unit of geological time is the **eon**. The first three eons (the **Hadean** (after the Greek for "hell"), the **Archaean** ("old") and the **Proterozoic** ("first life") are often shoved into one supereon called the **Precambrian**. Spanning from about 542 million years ago to the present, the **Phanerozoic** ("visible life") eon includes all the strata that bear abundant fossils of multi-cellular life. The next largest unit is the **era**. The first two eras of the Phanerozoic are the **Paleozoic** ("ancient life") and **Mesozoic** ("middle life"). The Paleozoic era is divided into **periods**, with names that reflect the predominence of British geologists in the early days of stratigraphy: Cambrian (after the Latin for "Welsh"), Ordovician (named after the Welsh tribe Ordovices by Birmingham geologist Charles Lapworth), Silurian (named after another Welsh tribe, the Silures), Devonian (after the English county of Devon); only the Carboniferous ("coal-bearing") and Permian (named after a Russian city in the Urals) bear names with no British links. The Mesozoic contains three periods: the **Triassic** (named after a triad of rock layers associated with the period), the **Jurassic** (after the Jura moutains) and **Cretaceous** ("chalk-bearing").

The most recent era is the **Cenozoic** ("recent life"), which is commonly finely sub-divided into seven epochs: the Palaeocene ("old recent"), Eocene ("new recent"), Oligocene ("little recent"), Miocene ("middle recent"), Pliocene ("more recent"), Pleistocene ("most recent") and Holocene ("entirely recent") or Recent. A handy mnemonic for remembering the order of the most important geological units is: **C**amels **o**ften **s**it **d**own **c**arefully. **P**erhaps **t**heir **j**oints **c**reak. **P**ossibly, **e**arly **o**iling **m**ight **p**revent **p**ainful **r**heumatism.

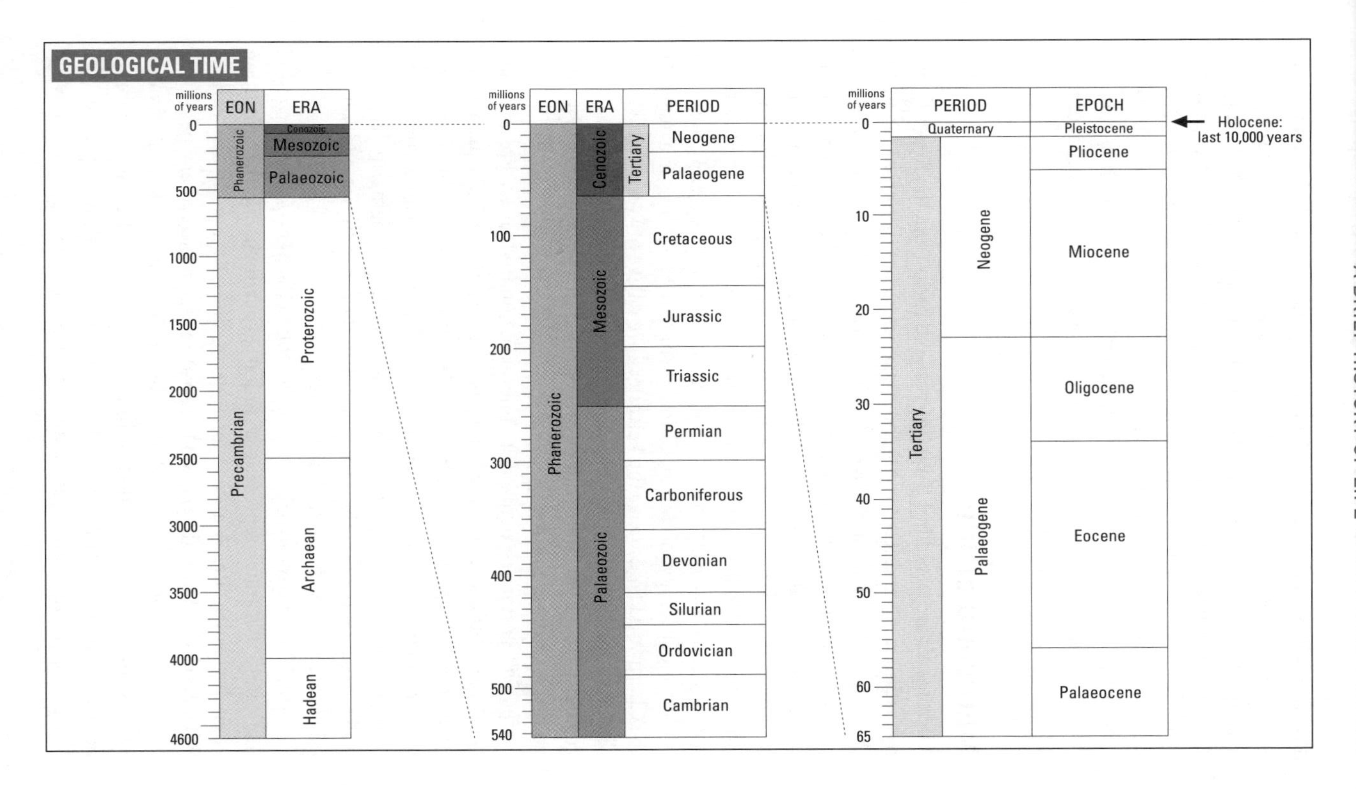
GEOLOGICAL TIME
millions of years
EON
ERA
Phanerozoic
Precambrian
Cenozoic
Mesozoic
Palaeozoic
Proterozoic
Archaean
Hadean
0
500
1000
1500
2000
2500
3000
3500
4000
4600
PERIOD
Tertiary
Neogene
Palaeogene
Cretaceous
Jurassic
Triassic
Permian
Carboniferous
Devonian
Silurian
Ordovician
Cambrian
100
200
300
400
500
540
EPOCH
Quaternary
Pleistocene
Pliocene
Miocene
Oligocene
Eocene
Palaeocene
10
20
30
40
50
60
65
Holocene: last 10,000 years

formations in many Precambrian rocks is generally accepted as evidence that oxygen released into seawater through primordial photosynthesis was mopped up by oxidation of iron. Eventually so much oxygen was released that mop-up mechanisms were overwhelmed, so, from around 2.4 billion years ago, the oxygen level in the atmosphere rose dramatically. This is likely to have had widespread ecological effects (it has been colourfully named the **Oxygen Catastrophe**), triggering the extinction of many anaerobic bacteria, but paving the way for the evolution of more complex eukaryotic cells.

Multicellular life

Microfossils preserved in 1.5-billion-year-old rocks from the Roper Group formation in northern Australia show characteristics that probably qualify them as early eukaryotes. However, the most exciting event in the fossil record of the Precambrian occurs much later, just before it ends in the **Ediacaran Period** (635–542Ma): the appearance of the **first multicellular organisms**. Until the mid-twentieth century it was widely believed that the Precambrian was devoid of complex life forms. However, all that changed with a chance discovery in 1957 by English schoolboy **Roger Mason** (now professor of geology in Wuhan, China). While rock-climbing in a quarry in the Charnwood Forest near the Leicestershire village of Woodhouse Eaves, Mason spotted what he thought was a leaf embedded in the rock. Leicester geologist Trevor Ford subsequently recognized it as Precambrian fossil and named it ***Charnia masoni***.

Charnia is now recognized as one of a hundred or so Precambrian genera of multicellular organisms that make up the **Ediacaran biota** (named after a range of hills in South Australia). Appearing 575 million years ago and persisting for about 33 million, most of these organisms look like nothing else seen in the history of life. Ranging in size from millimetres to metres, they resemble, among other things, discs, mud-filled sacks, or exotic quilted mattresses.

In the 1980s, German palaeontologist **Dolf Seilacher** (b.1929) suggested that the Ediacarans represent a **failed experiment** in the evolution of multicellular life. A recent analysis of diversity among the earliest Ediacarans (romantically named the **Avalon assemblage** after a peninsula in Newfoundland) led geobiologist **Shuhai Xiao** to speak of an **Avalon Explosion**, rivalling the burst of innovation seen in the later Cambrian period. Others suggested that the Ediacaran organisms represent the "long fuse" of that later Cambrian Explosion, envisaging links between late-Ediacaran organisms and subsequent more conventional animals. As the fossil record is poor for the final few million years of the Ediacaran period, a definitive verdict on the link between the Ediacarans and subsequent organisms remains elusive.

The Cambrian Explosion

Like the low rumbling chord that presages the explosive start of Strauss's tone poem *Also Sprach Zarathustra*, numerous small shelly fossils appear early in the Cambrian period. However, the brass fanfare, cymbals and drums of the **Cambrian Explosion** soon follow – the name given to the apparently sudden appearance of many large and diverse animals in the fossil record, including representatives of all except two of today's animal phyla that possess hard, easily fossilizable skeletons. The Cambrian Explosion was brought to the attention of a wide readership by Stephen Jay Gould's 1989 award-winning book, *Wonderful Life*. Here, Gould discussed studies by palaeontologists **George Walcott** (1850–1927), **Derek Briggs** and **Simon Conway Morris** of the Burgess Shale, a Middle-Cambrian rock formation from the Canadian Rockies (such fossil-rich sediments are often referred to by the unwieldy German word **lagerstätten,** particularly when soft parts are preserved).

Gould characterized the Cambrian period as one of riotous innovation, arguing that many of the fossils from the Burgess Shale could not be classified into any known animal group – such oddities included ***Opabinia***, with five stalked eyes and a hose-like snout that ended in a spiny claw, and ***Hallucigenia***, a wormlike animal that, Gould supposed, walked on two arrays of symmetrical spines. Gould went on to argue that mere serendipity decided which forms from the Burgess Shale community survived to populate subsequent eras, implying that there was no inherent predictability to the history of life. Gould's interpretation of the Burgess Shale assemblage was soon challenged – even by Conway Morris. It is now generally accepted that even bizarre Cambrian fossils, when scrutinized in detail, often show affiliations to some existing animal group. For example, in a 1991 *Nature* article, Swedish dentist-turned-palaeontologist **Lars Ramsköld** and Chinese geologist **Hou Xianguang** flipped *Hallucigenia* over, re-interpreting its tentacles as legs for walking and its spines as protective structures; they then re-classified it as a relative of modern velvet worms.

The Cambrian fossil collections are dominated by **arthropods**, a grouping that still ranks as the largest phylum of animals, containing insects, spiders, crustaceans and similar organisms that possess a segmented body and a hard external skeleton. **Trilobites** are the most famous group of extinct arthropods, representing the majority of known fossils from the Cambrian period and surviving for over 250 million years until the end of the Permian. As their name suggests, trilobites possessed an oval flat segmented body, divided lengthwise by furrows into three lobes: a raised middle lobe with a lower lobe on each side and split crosswise into three roughly equal sections (head, thorax and tail). Most species had two compound eyes, with multiple calcium carbonate lenses. Trilobite fossils are found worldwide: in England they are found in large numbers in the West Midlands town of Dudley, where they have even been incorporated into the local coat of arms.

The largest Cambrian animals were the **anomalocarids**, swimming segmented animals up to two metres in length. Equipped with strange pineapple-slice mouths flanked by two shrimp-shaped feelers, they were the Cambrian's fiercest predators. From the perspective of human evolution, the most exciting finds among the fossils of the Cambrian explosion are

Trilobites existed from the Cambrian period until the end of the Permian. Some developed bizarre features, such as this blind, bumpy-headed *Encrinurus*.

the first members of our own phylum, **the chordates**. From Chinese **Chengjiang lagerstätten** come ***Haikouichthys***, a fish-like animal, complete with fins and skulls (but lacking jaws). From the Burgess Shale comes ***Pikaia***, resembling lancelets (a group of primitive chordates alive today).

The Late Cambrian also saw the appearance of the first **conodonts**. Until recently, these organisms were known only from small mysterious tooth-like structures (**conodont elements**), which are so abundant in the fossil record for so long, that they can be used to date rocks from the Cambrian through to the Late Triassic. However, in the early 1980s, a few well-preserved Scottish fossils revealed the conodont animal in its full glory, including the soft-bodied parts that would not normally fossilize. It is now clear that conodonts were early vertebrates, with an eel-like body complete with eyes and fins.

The Cambrian enigma?

Before leaving this period, one has to ask: how can we explain the Cambrian Explosion? The sudden appearance of so many phyla seems puzzling. However, fresh interpretations have chipped away at the taxonomic uniqueness of many Cambrian species, new fossil finds have closed the gap between the Cambrian and Ediacaran worlds and several molecular phylogenetic studies suggest that the diversification of animal phyla pre-dates the Cambrian. So, perhaps the explosion was merely an apparent explosion, simply reflecting the emergence of animals large enough and hard enough to be preserved in the fossil record?

Three other kinds of explanation have been put forward: environmental, ecological and developmental. One potential environmental trigger was a fresh rise in **oxygen levels** in the atmosphere at the end of the Precambrian, providing an improved source of energy, plus an ozone layer to shield the planet. In addition, it is possible that the chemistry of the oceans changed so as to favour the formation of mineralized body parts. Ecological explanations posit a life-driving-life scenario, where **evolutionary arms races** fuel diversification. Andrew Parker at the Natural History Museum in London has proposed the **Light Switch Theory**, suggesting that the evolution of eyes and vision changed the ecological landscape. He suggests that the ability of predators to see their prey triggered innovations such as hard body parts.

Proponents of **evo-devo** (see box, p.140) see the Cambrian Explosion as the crossing of a threshold in the complexity of the developmental tool kits available to multicellular life (particularly the evolution of *Hox* genes), an idea which can be tested once representatives of all animal phyla

Replaying life's tape: contingency or convergence?

Stephen Jay Gould started a lively argument with his book *Wonderful Life*. Drawing on his own analysis of the Cambrian Explosion, Gould proposed that **contingency**, the chance survival of one lineage rather than another, was a key force in the history of life, so that if the tape of life were replayed, we would not necessarily arrive at the same outcome. Gould's analysis is strongly contested in *The Crucible of Creation*, a 1998 book by Simon Conway Morris, the English expert on the Cambrian Explosion, upon whose finds Gould had based his arguments. In his more recent *Life's Solution: Inevitable Humans in a Lonely Universe* (2003) Conway Morris argues forcefully that the evolutionary history of life is generally predictable, citing multiple examples of **convergent evolution**, through which evolution repeatedly arrives at the same biological solution even in independent lineages coming from different starting points. The numerous textbook examples of convergent evolution include marsupial and placental wolves; three different lineages of sabre-toothed tigers (one marsupial, two carnivores); Old World and New World vultures and desert plants; the body shapes of modern crocodiles and extinct phytosaurs and champsosaurs or of ichthyosaurs and dolphins. Rather implausibly, Conway Morris argues that even the evolution of (something like) humans was inevitable (but if so, why no tool-using bipedal therapsids or dinosaurs?). He also unashamedly strays from biology to religion in his lectures and books.

So, which rules? Contingency or convergence, chance or necessity? The answer is obviously both. The marsupial wolf (had it not been hunted to extinction) would not look out of place in a European safari park. But when did you last see anything in the Old World that looks like a kangaroo?

have been genome-sequenced. In conclusion, the Cambrian Explosion is no longer an eternal inscrutable mystery, never to be solved. Instead, a combination of new theories, new fossils, new genomes, and new phylogenies promise to shed new light on this apparent enigma.

Life in the rest of the Palaeozoic

Partway through the second period of the Palaeozoic, the **Ordovician** (488–444Ma), life experienced a second remarkable increase in diversity termed the **Great Ordovician Biodiversity Event**. Trilobites branched out; some species evolved bizarre spines to ward off predators, others developed shovel-like muzzles for ploughing through the muddy sea floor. Some lost their eyes; other grew elaborate eyestalks. Both conodonts and graptolites (a sister group to the vertebrates) thrived, while the first corals appeared. Molluscs also became more common and varied; one group, the **nautiloid cephalopods**, diversified into the top predator niches, becoming the largest (up to ten metres long) and perhaps the most intelligent animals in the Ordovician oceans. Ordovician seas were home to **ostracoderms**, large armoured jawless fish, with slit-like mouths, and **thelodonts**, smaller jawless fish, covered in tooth-like scales. However, from a human perspective, the most dramatic occurrence during the Ordovician was the **colonization of the land**. Ordovician microfossils document the first land plants (resembling today's liverworts) and fungi. Tracks were left by the first unidentified terrestrial animals.

The Silurian and Devonian periods

During the subsequent **Silurian period** (444–416Ma), coral reefs make their appearance. **Brachiopods** (shelly invertebrates, superficially similar to clams, but distinct in details of anatomy) emerge; jawless fish reach new heights of diversity. The first jawed vertebrates (**Gnathostomata**) appear and split into two lineages (the **ray-finned** and the **lobe-finned fish**). Silurian rocks yield the first large fossils of land plants. During the latter half of the period, the **first vascular plants** (plants with transport systems for water and nutrients) appear. Examples include *Cooksonia,* a small simple branching plant devoid of leaves, flowers or roots and the more sophisticated *Baragwanathia*, which sprouted roots and leaves.

In the seas, rivers and lakes of the **Devonian period** (416–359Ma) lurked **huge sea scorpions** (not closely related to true scorpions; more accurately termed **eurypterids**). In 2007, Simon Braddy and Markus Poschmann described an eighteen-inch claw from the eurypterid ***Jaekelopterus rhenaniae***; by extrapolation, the whole animal measured over eight feet in length! The Devonian oceans were also home to massive barrier reefs and to an increasing variety of predatory fish, including the **first sharks** and the ten-metre-long monster *Dunkleosteus* (thought to have the most powerful bite of any fish that has ever existed).

The Devonian period saw the evolution of larger, more varied and more sophisticated land plants, culminating in the growth of the first forests and creation of the first soils. In 2007, a trans-Atlantic team of palaeobotanists showed that unexplained mid-Devonian tree stumps from Gilboa in New York state belonged to *Wattieza,* an eight-metre-tall relative of modern ferns, now heralded as the **first tree**. In 2004, Scottish bus driver and amateur fossil hunter **Mike Newman** discovered one of the oldest known body fossils of a land animal: a 428-million-year-old millipede now named in his honour ***Pneumodesmus newmani.***

In the late Devonian, one group of lobed fish evolved into tetrapods (vertebrates with four limbs). One genus of Devonian tetrapod, ***Acanthostega***, was first named from skull fragments discovered in the 1930s. At least eight other genera of tetrapods are now known from the period. In 1987, Cambridge palaeontologist **Jenny Clack** discovered a 365-million-year-old near-complete specimen of *Acanthostega*, equipped with eight fingers but no wrists. From detailed studies of its anatomy, Clack concluded that *Acanthostega* was primarily aquatic, overturning the earlier assumption that tetrapods evolved limbs to crawl on land; instead limbs developed in the water and only later proved useful for terrestrial life. Curiously, another group of lobed fish that evolved in the Devonian, the **coelacanths**, were thought to have become extinct in the Cretaceous, until they were re-discovered as **living fossils** in the 1930s.

The Carboniferous and Permian periods

During the **Carboniferous period** (359–299Ma), extensive and diverse plant populations evolved on land. The scale tree *Lepidodendron* reached heights of over thirty metres, with trunks more than a metre across. The horsetail *Calamites* reached similar heights. Extensive coal measures were laid down in lush equatorial forests. Burial of so much organic carbon led to a build-up of **oxygen** in the atmosphere, reaching a high of 35 percent (compared to 21 percent today). In fact, today's "inconvenient" rise in CO2 levels, largely derived from burning Carboniferous coal, is a mirror image of that earlier period's elevated oxygen. Towards the end of the Carboniferous, the **first conifer**, *Walchia*, appeared (preserved in a spectacular fossil forest from Brule, Nova Scotia). In 2007, British fossil forest expert **Howard Falcon-Lang** described a huge fossilized rainforest from the Carboniferous. Sprawling through the underground workings of an Illinois coalmine, it rivals a modern city in size.

In the Carboniferous oceans, trilobites became less common, while brachiopods, sea lilies (crinoids), sea scorpions and sharks were still abundant. A group of lobe-finned fish, the **rhizodonts**, became top-level predators in the lakes and rivers: one of them, *Rhizodus hibberti*, at seven metres long, ranks as the largest known freshwater fish. Freshwater eurypterids flourished, some even taking to the land. British palaeontologist Martin Whyte has described a complex terrestrial trackway in Scotland, sculpted by a giant lumbering eurypterid a third of a billion years ago.

Several other kinds of arthropod thrived on the land, many growing to massive sizes thanks to the oxygen-rich atmosphere. A group of stiff-beaked insects with the unwieldy name **Palaeodictyopteroidea** became the first major terrestrial herbivores. Giant predatory insects superficially similar to dragonflies took to the air: with a wingspan of over two and half feet and the body weight of a crow, ***Meganeura monyi*** was probably the largest insect ever to appear on the planet. ***Arthropleura***, a gigantic relative of millipedes and centipedes, crept across the forest floor, leaving fossilized tracks over a foot wide in what is now New Mexico and on the Scottish Island of Arran.

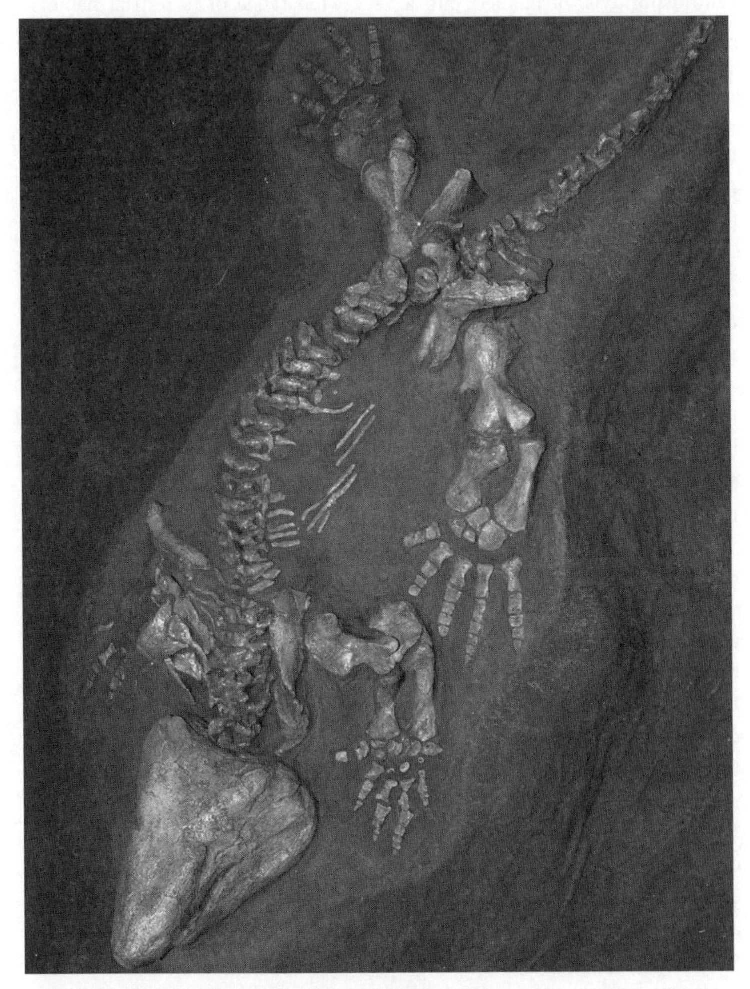

Limnoscelis, a five-foot-long reptile-like amphibian. During the Permian period, terrestrial tetrapods evolved into the first large herbivores and carnivores.

Pulmonoscorpius kirktonensis, a true scorpion living on land, reached lengths of two-thirds of a metre. In 2001, researchers from Ohio described a Carboniferous cockroach the size of a mouse.

During the Carboniferous, amphibians also grew in diversity and number. Some species stayed small, occupying niches similar to present-day newts, lizards and snakes, while others, such as *Proterogyrinus*, rivalled today's crocodiles in size and lifestyle. One group of primitive amphibians, the **temnospondyls**, diversified into a very wide range of terrestrial habitats despite having to lay their eggs in water. Another lineage, the amniotes, exploited a key innovation, the amniote egg (which can survive out of water) to expand into dry terrestrial environments. The amniotes split into two lineages: the **sauropsids** (which later gave rise to reptiles, dinosaurs and birds) and the **synapsids** (which later gave rise to the mammals). Late in the Carboniferous, the synapsid lineage spawned the **pelycosaurs**, which soon became the dominant land animals.

During the final period of the Palaeozoic, the **Permian** (299–251Ma), conifers took over as the dominant large land plants; ginkgos and cycads also emerged during this time. Cockroaches and beetles also made their mark on the Permian fossil record, but it is the Permian tetrapods – the amphibians and pelycosaurs – that capture the popular imagination as the first large herbivores and carnivores. Some **large pelycosaurs**, herbivores like *Edaphosaurus* and carnivores such as *Dimetrodon*, flaunted spectacular **sails** along their backs, which are thought to have helped them warm up more quickly in the sun.

Later in the Permian, an offshoot of the synapsids, the **therapsids**, emerge as the masters of the terrestrial environment. **Dinocephalians** were early therapsids that grew to enormous sizes: *Estemmenosuchus*, a massive omnivore, sported large horns atop its bull-sized head; giant bone-headed two-ton herbivores like *Tapinocephalus* turned head butting into a new art, while the carnivorous dinocephalians grew fearsome canines and incisors. In the Late Permian, one group of therapsids became top predators: the **gorgonopsians** (some reaching the size of modern rhinos), while another group, the **dicynodonts**, diversified as herbivores. Fossils of one common dicynodont, *Diictodon* (the Permian equivalent of a gopher) are often found in pairs, earning the status as probably the earliest known exponent of monogamy. In the second half of the Permian, a group of therapsids ancestral to mammals evolved, the **cynodonts**, which were probably covered in hair and warm-bloodied. The most primitive known cynodont, the tiny carnivore *Charassognathus gracilis* was described as recently as 2007.

A sister group to the cynodonts, the **therocephalians**, also flourished in the Late Permian; some have even used poison-bearing fangs to kill their prey. The oldest known **airborne tetrapod**, *Coelurosauravus jaekeli*, dates from the late Permian – it resembled a tree-dwelling gliding lizard and will be familiar to fans of the TV show *Primeval* as Rex. Sauropsids also diversified during the Permian, producing ungainly stocky herbivores such as the armour-plated *Scutosaurus* and, very late on, the primordial prototypes of the **archosaurs**, which later gave rise to dinosaurs, pterosaurs, birds and crocodiles.

Life in the Mesozoic

The **end-Permian mass extinction**, sometimes labelled the **Great Dying**, eliminated as much as 96 percent of marine species and three quarters of terrestrial organisms. Trilobites, eurypterids and graptolites all disappeared. Among fish, just a few families pulled through into the first period of the Mesozoic, **the Triassic** (251–199Ma). **Trematosaurids**

Skeleton of the saurischian dinosaur *Tyrannosaurus rex*. Dinosaurs were the dominant land animals of the late Mesozoic.

Mary Anning: fossil hunter extraordinaire

Mary Anning (1799–1847), from the English coastal town of **Lyme Regis**, was, according to British historian and geologist Hugh Torrens, "the greatest fossilist the world ever knew". Left destitute by the death of her father in 1810, Mary and her brother Joseph turned to collecting fossils from the local coastline (now styled the **Jurassic Coast**) in the hope of selling them to amateur collectors. At the age of twelve, Mary made a spectacular find that brought her to the attention of the scientific community: the first complete skeleton of an **ichthyosaur** ever found (her brother had found the skull earlier, but Mary was responsible for locating the rest of the fossil). Anning's subsequent discoveries included the first **plesiosaur** fossil, *Plesiosaurus dolichodeirus*, in 1821 and a remarkable specimen of an extinct ray-finned fish, *Dapedium politum*, in 1828. Anning also described the first complete skeleton of a **flying reptile**, the pterosaur *Dimorphodon macronyx*. Later in life, Anning's fame secured her financial support from the British Association for the Advancement of Science and honorary membership of the Geological Society of London – the only woman in an exclusively male club. The chief impact of Anning's work was that her fossils established beyond doubt the concept of extinction, proving that some extinct animals looked nothing like anything alive today. Anning died from breast cancer in her forties and is buried with her brother at St Michael's Church, Lyme Regis, where a stained-glass window is dedicated to her memory.

> **"...the extraordinary thing in this young woman is that she has made herself so thoroughly acquainted with the science that the moment she finds any bones she knows to what tribe they belong".**
>
> **Lady Harriet Silvester (1824)**

became the only amphibians ever to take to the sea. Some reptiles, the **sauropterygians**, also adjusted to aquatic environments, growing flippers on specially adapted shoulders; the most primitive sauropterygians, **pachypleurosaurs**, resembled elongated semi-aquatic lizards. Another early group, the **nothosaurs**, lived like seals, feeding in water and basking on beaches. Some aquatic reptiles, such as the **placodonts**, grew armour plating, while others, the **thalattosaurs**, developed sleek elongated bodies up to four metres long. Towards the end of the Triassic, the **first plesiosaurs** appeared: long-necked, carnivorous, fully aquatic reptiles that were to flourish later in the Mesozoic. Another star of the Mesozoic oceans, the **ichthyosaurs** (large dolphin-like reptiles), took to the sea early in the Triassic and then diversified. The largest of these, *Shonisaurus sikanniensis,* was equivalent in size to a sperm whale.

In what has been dubbed the **Triassic takeover**, various archosaurs steadily took over from synapsids as the dominant large land animals: these include the **rauisuchians** (large predators with a distinctive erect stance), the crocodile-like **phytosaurs**, the plant-eating armoured **aetosaurs** and the carnivorous **poposaurs**. **Rhynchosaurs**, stocky beak-snouted herbivorous sauropsids, also flourished in Triassic terrestrial environments. In the late Triassic, the first turtles appeared and one group of archosaurs, the **ornithodira**, split into two lineages whose descendents would characterize the rest of the Mesozoic: the **pterosaurs** and the **dinosaurs**. *Eoraptor*, a small two-legged carnivore discovered in the early 1990s is probably the best-known candidate for the first dinosaur. Also, in the Late Triassic, a new offshoot of the cynodont lineage appeared, which gave rise to the mammals (although what counts as a mammal and what counts as a "mammaliform" remains a matter of semantics). ***Morganucodon watsoni*** (named after Glamorgan in Wales), is the best example of an abundant, early proto-mammal from the Late Triassic.

Dinosaur world: the Jurassic and Cretaceous periods

As every schoolchild knows, during the **Jurassic** (199–145Ma) and **Cretaceous** (145–65Ma) periods the oceans swarmed with fearsome reptiles (plesiosaurs, ichthyosaurs, crocodiles), pterosaurs ruled the skies, while dinosaurs thundered across the land. But one mustn't forget the little creatures: the wonderfully named **rudists**, bizarrely shaped bivalve molluscs that built Cretaceous reefs, or the **diatoms**, algal cells encased in ornate glassy shells.

The dinosaurs fell into two lineages: the lizard-hipped saurischians and the bird-hipped ornithiscians. The **saurischians** in turn fell into two groups: the therapods (bipedal carnivores) and the sauropods (four-legged herbivores). The **therapods** include the largest predators ever seen on land. *Spinosaurus* at nearly ten tons was the largest of all known carnivorous dinosaurs. The legendary ***Tyrannosaurus rex*** (see box, p.172) was not much smaller at seven and a half tons. *T. rex* belonged to a group of therapods known as the **coelurosaurs**, which included the lineage that gave rise to birds. *Archaeopteryx* appeared during the Late Jurassic, while **birds diversified** during the Cretaceous. Noteworthy primitive birds include *Confuciusornis* (toothless, but clawed), the recently discovered *Sapeornis* and an extinct but diverse lineage, the Enantiornithes.

Sauropods included the largest animals ever to walk on earth – massive plant-eaters such as *Apatosaurus* (formerly known as *Brontosaurus*), *Diplodocus* and *Brachiosaurus*. The aptly named *Supersaurus* was up to

Tyrannosaurus rex: hot-bloodied paragon of a fast-moving field

For over a hundred years, *Tyrannosaurus rex* has occupied pride of place in the pantheon of ferocious dinosaurs. Like most dinosaurs, *T. rex* was first thought to be cold-blooded. But things changed in the **Dinosaur Renaissance**, a revolution in palaeontology that began in the 1960s. The upheaval started with the realization that birds were a group of dinosaurs that had survived previous mass extinctions (pedants now classify *T. rex* and its kin as "non-avian dinosaurs"). Then American palaeontologist **"Dinosaur Bob" Bakker** proposed that dinosaurs were warm-blooded and much faster and smarter than previously assumed. Additional evidence for this reappraisal emerged in 1999, when oxygen isotope studies revealed that the temperature difference between core and extremities in *T. rex* was small enough to suggest a **constant body temperature**.

Traditionally shown standing upright like a kangaroo, *T. rex* is now thought to have held its body parallel to the ground, with tail extended behind to balance the head. The diminutive forelimbs, once dismissed as love handles, are now seen as muscle-bound arms that grasped struggling prey. Arguments rage as to how fast *T. rex* could run and over whether it was primarily a scavenger or a hunter. A 2006 paper suggested it possessed **binocular vision** sharp enough to match a hawk.

Since the 1980s over a dozen new specimens of *T. rex* have been described. The best preserved, from South Dakota, was found in 1990 by American fossil hunter Susan Hendrickson. Nicknamed Sue after its discoverer, the recovery of over 90 percent of the skeleton provided an unprecedented view of the whole animal. Sue is thought to have lived to 28, a record for the species, before perishing from a bite to the head, inflicted by a fellow tyrannosaur. Another well-preserved specimen, nicknamed Stan, also shows evidence of wounds and fractures probably delivered by another *T. rex*.

In 2004, a fossil of *Dilong*, an early relative of *T. rex*, was found to possess primitive feathers. Although skin impressions from adult specimens of *T. rex* show only pebbly scales, one cannot discount the possibility that juveniles of the species were feathered. In 2005, Mary Schweitzer and colleagues reported the remarkable recovery of **soft tissue** from the inside of a *T. rex* leg bone, complete with blood vessels, matrix tissue and even red blood cells. A 2007 study reported the recovery of a **collagen protein sequence** from a *T. rex* fossil, although this claim has been disputed. One thing is clear: new research on an old dinosaur continues to deliver fresh surprises.

Tyrannosaurus Rex, the Tyrant King Edited by Peter Larson and Kenneth Carpenter (Indiana University Press, 2008)

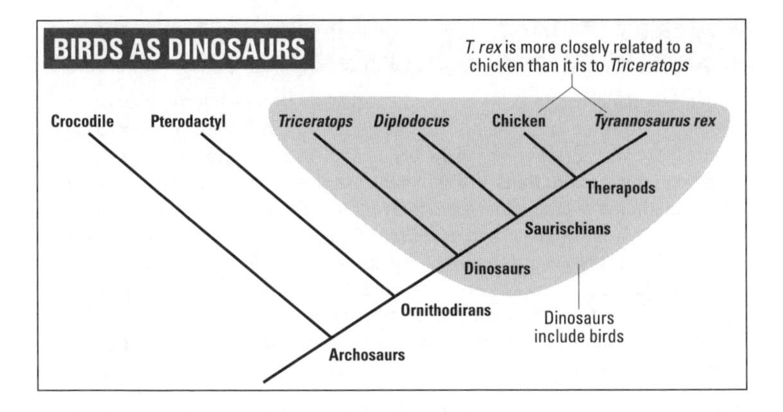

35 metres in length; the tallest sauropod, *Sauroposeidon* towered up to seventeen metres; *Argentinosaurus* weighed in at close to a hundred tons. **Ornithischian dinosaurs** were also herbivores, but generally less massive than sauropods. The best known include *Stegosaurus* (with its distinctive plates and tail spikes), *Ankylosaurus* (with its heavily-armoured body and substantial bony tail-club), *Triceratops* (with its three horns and large bony frill) and the duck-billed hadrosaurs.

During the Cretaceous, the oldest known termites, ants, aphids, grasshoppers and butterflies appear, along with flowering plants (angiosperms). Identifying the earliest true flowering plants in the fossil record has proven controversial. *Archaefructus,* from 125 million years ago, has been proposed as the most primitive flowering plant, although others have argued that it shows specialized features that link it to modern water lilies. Some researchers have even suggested that the middle-Jurassic *Schmeissneria* be classified as a flowering plant. However, whenever they first appeared, by 100 million years ago, the fossil record shows that flowering plants had undergone an explosive diversification.

The versatile first mammals

According to the popular view, the lineage leading to mammals quietly pottered on through the Jurassic and Cretaceous, confined to a shadowy existence as small nocturnal unspecialized tree- or ground-dwellers, similar to today's shrews, hedgehogs or tree shrews. A flurry of publications from 2005–06 highlight several notable exceptions. *Fruitafossor windscheffeli* is a chipmunk-sized mammal from 150 million years ago, first described by Zhe-Xi Luo and associates in 2005. A surprisingly complete skeleton

Catastrophism revisited 1: ET meets K-T

Uniformitarianism (see p.11) is now accepted as the default assumption in geology. However, business-as-usual evolution is sometimes interrupted by dramatic, even catastrophic, planetary changes. These appear in the fossil record as **mass extinctions**. In a seminal 1982 paper John Sepkoski and David Raup identified five big mass extinctions: the end-Ordovician, late Devonian, end-Permian, end-Triassic and end-Cretaceous. The end-Cretaceous **death-of-the-dinosaurs** event (often referred to as the **K-T event**, after a German abbreviation for the epochs on either side) features most prominently in the public imagination, although the **end-Permian** was the largest extinction event, eliminating 96 percent of marine species and around 70 percent of terrestrial species.

Attempts at a unified theory that explains all extinction events have proven unsuccessful. **Flood basalt events**, in which massive volcanic eruptions inundate large stretches of land or ocean floor with lava, are considered the most common cause. During these events, dust is likely to trigger immediate global cooling and choke off photosynthesis. Sulfur emissions are also likely to cause acid rain, while a release of carbon dioxide could set off long-term global warming. Several other environmental changes are associated with extinction events (sea level changes, changes in arrangements of continents, plummeting levels of oxygen in sea water, oceanic overturns in which deep and surface water switch places), although their relative contributions to mass extinctions are unclear. According to the **clathrate gun hypothesis**, massive release of methane encaged in water in oceanic sediments could have contributed to the end-Permian event.

Extra-terrestrial causes could have contributed to some extinction events. A burst of gamma rays from a nearby supernova has been suggested, perhaps implausibly, as a cause for the end-Ordovician event. In the 1980s, father-and-son team **Luis** and **Walter Alvarez** reported an increase in the rare metal **iridium** at the K-T boundary, prompting them to suggest an impact event – an asteroid fragment striking the earth – as the cause of the K-T extinctions. They went on to calculate the size of the object as six miles across, roughly the size of Manhattan. Subsequent investigations identified an impact crater of the right age and size, the **Chicxulub Crater**, buried under the Yucatan peninsula in Mexico and identified the culprit as a **carbonaceous chondrite**. The Chicxulub impact would have unleashed fearsome firestorms, followed by a decade of darkness and a longer period of global warming. Debate continues as to whether there might have been multiple impacts and whether other factors might have contributed to the K-T extinctions. In late 2007, a team of astronomers suggested that the break up of a large asteroid 150 million years ago, the **Baptistina disruption**, gave birth to the fragment that caused the K-T impact.

When Life Nearly Died Michael Benton (Thames & Hudson, 2005)

Extinction Douglas H. Erwin (Princeton University Press, 2006) of *Fruitafossor* from Colorado reveals that this mammal had massive forelimbs (earning it the nickname Popeye), which were specialized for digging, as well as open-rooted peg-like teeth resembling those of modern aardvarks and armadillos (suggesting it lived on termites or ants). Also in 2005, Yaoming Hu and his team described the largest known mammal from the Cretaceous: the 130-million-year-old, one-metre-long *Repenomamus giganticus,* which resembled a Tasmanian devil in size, shape and probably also behaviour. A closely related species, *R. robustus,* was described in 2000 with skeletal fragments from a small dinosaur, *Psittacosaurus*, preserved from its stomach – proof that some mammals ate dinosaurs. The **largest mammaliform** of the Jurassic, *Castorocauda lutrasimilis,* first described in 2006, lived as a semi-aquatic beaver- or otter-like animal. An astonishingly well-preserved *Castorocauda* fossil from China reveals skeletal adaptations to swimming and burrowing, teeth suited to aquatic feeding and, crucially, the **first hair** in any mammal-like animal. Also in 2006 came the first evidence of a **Mesozoic gliding mammal**, the 140–120 million-year-old *Volaticotherium antiquum*, which possessed a gliding membrane similar to that of modern-day flying squirrels.

Modern mammals are divided into **monotremes** (egg-laying mammals like the duck-billed platypus), **marsupials** (which incubate their young in pouches) and **placental mammals**. A fourth group, the **multituberculates**, went extinct in the Oligocene. The platypus-like mammal *Teinolophos* from the early Cretaceous ranks as the earliest known **fossil monotreme**. At around the same time lived *Sinodelphys*, the first known **marsupial**, and *Eomaia*, the earliest representative of the lineage leading to placental mammals – fossils of both species reveal imprints of their fur. *Didelphodon,* a predatory marsupial, lived in the Late Cretaceous. Molecular estimates for the origins of placental lineages (including primates) suggest dates in the Cretaceous, although analyses of the fossil record provide support for much later diversification, after the end of the Mesozoic.

Life in the Cenozoic

The Mesozoic ended with a bang, the **K-T extinction event** (see box, p.174), 65 million years ago. The era that followed, the Cenozoic, was a period of long-term cooling. With the demise of marine reptiles, sharks became the top predators in the Cenozoic oceans. On land, with the dinosaurs gone, the Cenozoic world was dominated by modern mammals and birds and by flowering plants and their insect partners.

Fossil evidence of mammals during the first epoch of the Cenozoic (the **Palaeocene**, 65.5–55.8Ma) is scarce. **Multituberculates** flourished, occupying rodent-like niches and making up more than half of the fossilized mammal species recovered from the northern hemisphere. In South America, marsupial relatives of today's opossums thrived, while another group of marsupials, the **sparassodonts**, took up a carnivorous lifestyle. Noteworthy Paleocene placental mammals include *Plesiadapis* (a kind of proto-primate), **condylarths** (primitive hoofed mammals), **mesonychids** (wolf-like meat-eaters, related to whales), and **creodonts** (close cousins to today's carnivores).

Reptiles occupied more of the Paleocene globe than they do at present. Snakes, turtles and carnivorous lizards did well, as did the **champsosaurs** (swamp-dwelling crocodile-like reptiles with long, thin snouts). The **Wannagan Creek** fossil beds in North Dakota have yielded evidence of two top Paleocene predators: a four-metre-long crocodile, *Leidysuchus formidabilis*, and *Champsosaurus gigas*, the largest champsosaur ever discovered, at

Catastrophism revisited 2: snowball Earth meets global warming

In addition to mass extinctions, our planet has suffered several dramatic climatic changes. The **ice ages** that have come and gone periodically during the last three million years provide one obvious example. Periods of extensive glaciation in the higher latitudes, with ice sheets covering most of Europe and North America have alternated with warmer inter-glacial periods. A simple explanation for the Ice Ages remains elusive. In the 1940s, Serbian engineer **Milutin Milankovitch** suggested a relationship between variations in the Earth's orbit (**Milankovitch cycles**) and patterns of global climate change. However, orbital variations are unlikely to provide the sole answer to what triggers an ice age.

In the 1960s, Cambridge geologist Brian Harland suggested an ice age occurred 850 to 630 million years ago. Two decades later, Joe Kirschvink suggested that this Precambrian glaciation was extensive enough to affect the whole planet, coining the catchy term **Snowball Earth**. Arguments continue as to whether this event was as dramatic as Kirschvink suggests (one alternative has been called **Slushball Earth**) and as to what effect it might have had on the evolution of life (did it trigger the Ediacaran explosion?).

At the opposite extreme, our planet's most dramatic example of sustained warming, the **Paleocene-Eocene Thermal Maximum** (or PETM), occurred 55.8 million years ago. In the space of 20,000 years, carbon dioxide levels soared and global temperatures rose by around 6°C, triggering a corresponding rise in sea level. Curiously, although some kinds of plankton were driven to extinction, the PETM was accompanied by a marked diversification among mammals.

During the Eocene, some mammals adapted to life in the sea, including the ancestor of today's manatees and sea-cows.

three metres long. Birds also began to diversify during the Palaeocene, with many modern forms (from pelicans to owls) appearing. Large carnivorous flightless birds, the phorusrhacids or **terror birds**, became top predators in South America, while another giant bird, *Gastornis*, flourished in Europe.

By early in the **Eocene epoch** (55.8–33.9Ma), most modern mammal and bird orders appear in the fossil record, including the first known **primate fossils**, scattered widely across the northern hemisphere. In the initially warmer climate, animals tended to be smaller than in the periods before or after. Ungulates of all sorts became prevalent, including carnivorous ungulates such as *Mesonyx*. Some mammals adapted to living in the sea, including *Basilosaurus*, an early whale (see box, p.84), and *Prorastomus sirenoides*, a Jamaican precursor of today's sea cows and manatees. **Grasses** also become evident in the fossil record. The Eocene ended with a major extinction event, known as the **Grande Coupure** (perhaps the result of an impact in Chesapeake Bay).

In the **Oligocene** (33.9–23Ma), oceans continued to cool while temperate deciduous woodlands, open plains and deserts became common features of terrestrial landscapes. Primitive whales flourished early in the Oligocene, with more modern baleen and toothed whales appearing towards the end of the epoch. A group of mammals, which included the progenitors of elephants, spread from Africa to the rest of the world. Monkeys and rodents spread to South America. Notable **large mammals** of the Oligocene include *Brontotherium* (a massive rhinosauros-like grazer), the fifteen-ton herbivore *Indricotherium* (the largest known land mammal), *Entelodont* (a pig-like animal with bony lumps its head), *Hyaenodon* (a horse-sized predator) and *Andrewsarchus* (a heavily-built, wolf-like mammal with hooves – possibly the largest mammalian predator ever seen).

In the **Miocene** (23–5.3Ma), grasslands expanded and grasses diversified as they co-evolved with herbivores. Mammals and birds grew more similar to those of today; sperm whales, apes and modern sharks appeared, including the huge forty-ton *Carcharodon megalodon*, the largest-ever carnivorous fish. Just before the end of the Miocene, the human lineage diverged from that of our closest living relative, the chimpanzee. The **Pliocene** (5.3–1.8Ma) saw a pattern of global biogeography similar to today's: rainforests limited to the tropics, with dry savannahs and deserts in parts of Asia and Africa. Collisions between wandering land masses provoked migrations and mixing between previous isolated ecologies. The most dramatic of these events was the **Great American Interchange**, as the volcanic Isthmus of Panama bridged the divide between North and South America. This brought an end to South America's isolated collection of animals; local marsupial predators soon went extinct. From North America into South America flowed mastodons, big cats (including saber-toothed tigers), wolves, llamas, tapirs, bears, horses and various rodents. In the reverse direction went armadillos, opossums and porcupines. Several species of giant sloth made it into what is now the United States, including *Megalonyx jeffersonii* (named after Jefferson, who described its bones at a meeting of the American Philosophical Society). One species of huge terror bird, *Titanis walleri,* established itself in North America, but went extinct before the end of the Pliocene.

During the **Pleistocene epoch** (the period from 1,808,000 to 11,550 years before present), the world was beset with repeated ice ages, which influenced the distribution of plants and animals. Spanning the Pleistocene and the subsequent **Holocene** (9600 BC–now) **megafaunal extinctions** occurred – selective losses of large animals from ecosystems. In North America this meant the end of the mammoths, mastodons, sabre-toothed cats, glyptodons, ground sloths, short-faced bears, horses and camels. Eurasia lost the woolly mammoth, the woolly rhinoceros, the Irish elk, and cave bears, lions and hyenas. In Australia, extinction struck *Thylacoleo carnifex* (a marsupial lion), the hippo-sized *Diprotodon* (the largest marsupial ever seen), *Macropus titan* and *Procoptodon goliah* (giant kangaroos), *Zygomaturus trilobus* (a large marsupial herbivore), *Palorchestes azael* (a marsupial tapir) and *Megalania prisca* (a giant monitor lizard).

Two main theories have been proposed for these megafaunal extinctions: climate change and the influence of humans, either indirect (for example, through use of fire or decimation of prey) or direct, through predation (the **overkill hypothesis**). Human-induced megafaunal extinctions in more recent times provide evidence for the overkill hypothesis. For example, the loss of moas in New Zealand after the arrival of the Maoris. But whether humans were always to blame for megafaunal extinctions remains a hotly debated topic. Dutch atmospheric chemist **Paul Crutzen** argues that the massive impact of humans on global ecosystems and environments since the Industrial Revolution has taken us into a new epoch, the **Anthropocene**, which he dates from the invention of the steam engine in 1784, a period marked by what some have called the **sixth mass extinction**.

7 Human evolution

Until well past the middle of the twentieth century, we humans, like some old aristocratic family, could depend on an ancient human pedigree stretching back fifteen million years or more. And although Linnaeus had lumped us in with lemurs, monkeys and apes as primates, in his taxonomic system we remained aloof from our closest relatives in a family of our own, the Hominidae, with gibbons and great apes banished to the Pongidae.

A change of status

But our special status wasn't set to last. Along with Beatlemania, sex and LSD, the 1960s saw the start of a **revolution** in our understanding of human evolution. In 1967, **Allan Wilson** and his student Vincent Sarich published the first molecular estimate of the time since humans diverged from chimps and gorillas. In a profound challenge to the accepted belief of an ancient human lineage, they placed the **last common ancestor** of humans and apes at just five million years ago. Although ridiculed at the time, many more recent molecular estimates have backed them up (but conflicts between molecules and fossils continue).

"In the distant future I see open fields for far more important researches … Light will be thrown on the origin of man and his history."

Darwin's only comment on human evolution in *The Origin of Species*

We humans now no longer stand alone on a lofty pinnacle of evolution: American evolutionary biologist Jared Diamond has provocatively described us as just a **third species of chimpanzee**, along with the common chimpanzee and the bonobo. Everyone now accepts that great apes and humans belong together in the **Hominidae**, and where anthropologists used to say hominid, they are now forced instead to use the term **hominin** (from the tribe Hominini) to describe fossils closer to us than to chimps.

The evolution of humans and our relatives is no longer seen as a tidy ladder but as a bushy, branching tree. Among palaeontologists, **lumpers** vie with **splitters** as to whether fossils should be grouped together in a few species or divided into many. But one thing is clear: while humans now stand alone on this planet as upright apes, the fossil record shows that, for millions of years, our lineage shared our world with many other species of ape – and often with other hominins – in what has been called the planet-of-the-apes or we-were-not-alone scenario. Furthermore, cladistics (see p.131) has challenged whether we should ever be viewing any fossil as the ancestor of any other species, instead, talk has turned to clades and orders of branching.

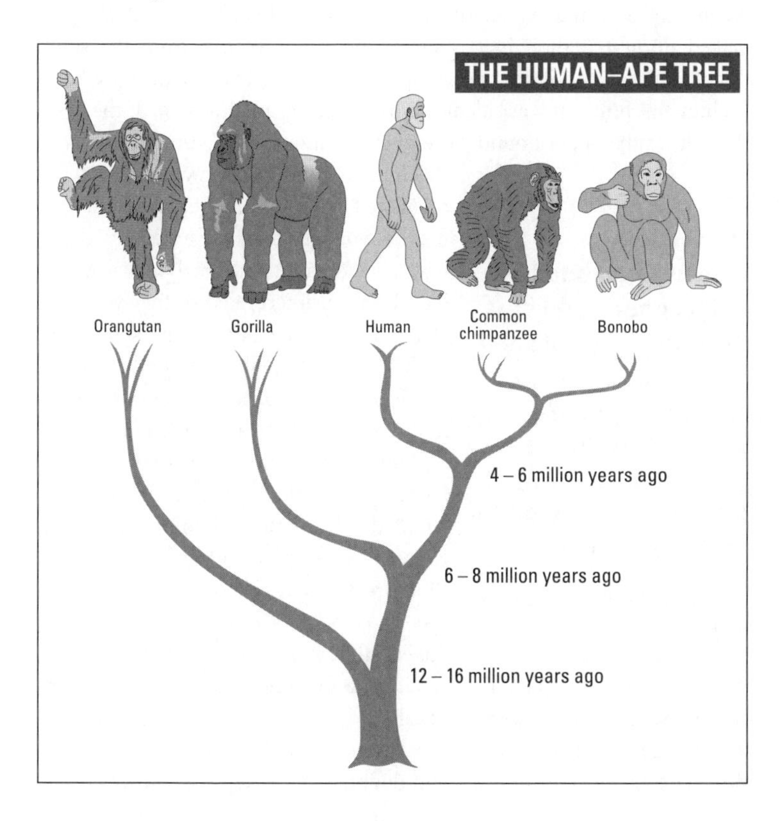

The first hominins: competing claimants

Attempting to identify the earliest hominins (the first human relatives after our divergence from chimps) is problematic. In 2002, **Michel Brunet** (b.1940) and a team of French and Chadian anthropologists described a new **six-million to seven-million-year-old** species, ***Sahelanthropus tchadensis***, on the basis of a cranium (skull without lower jaw) and jaw fragments. The cranium, discovered in the Djurab Desert in Chad by local field worker **Ahounta Djimdoumalbaye**, was soon nicknamed **Toumaï** ("hope of life" in the local language). Toumaï's brain is chimp-sized, but its brow-ridges, robust lower jaw and the wear on its teeth all link it with later hominins. To rectify distortions suffered by the cranium, a beautiful virtual reconstruction was created in 2005, which added weight to its assignment as a human relative and also suggested that *Sahelanthropus* walked upright (although it is impossible to prove this from the skull alone). Toumaï-as-hominin poses a problem in that this specimen predates the human-chimp split as calculated by molecular methods, or, at the very least, sits at the upper bound. Several explanations have been put forward: the human-chimp split was earlier than calculated by molecular approaches; the fossil is younger than suggested (its dates have been obtained indirectly); its features have been misinterpreted or the evolution at the time of the human-ape split is so bushy that, although apparently transitional, Toumaï belongs on an offshoot that leads neither to humans nor chimps.

"It is therefore probable that Africa was formerly inhabited by extinct apes closely allied to the gorilla and chimpanzee; as these two are now man's closest allies, it is … probable that our early progenitors lived on the African continent…"

Charles Darwin, *The Descent of Man* (1871)

Hailing from much further east in Africa, *Orrorin tugenensis* is another contender for earliest hominin. In 1974, Kenyan-born palaeontologist **Martin Pickford** found a solitary tooth from the species in 6 Ma sediments from Kenya's Tugen Hills. A quarter of a century later, after a bizarre legal squabble over access to fossil field sites, Pickford, with Brigitte Senut and co-workers, published a description of twelve fossils and named the new species, which was soon christened **Millennium Man**. The age of *Orrorin* is beyond doubt, and subsequent analyses of its

Michel Brunet holding the skull of Toumaï, chief contender for the earliest known hominin. Toumaï's discoverer, Ahounta Djimdoumalbaye, is on the left.

femur have suggested that it was a biped. Millennium Man thus presents the same challenges to the date of the human-chimp split as Toumaï. However, some experts (such as medic-turned-palaeontologist Bernard Wood) remain sceptical of *Orrorin*'s status as a hominin; its discoverers have courted controversy by claiming it as human ancestor in a simple two-branched view of human evolution.

From the Middle Awash valley in Ethiopia come fossils of *Ardipithecus*, another potential first hominin. In the early 1990s, a team led by American palaeoanthropologist **Tim White** (b.1950) found nearly half a skeleton's worth of material from what they called *Ardipithecus ramidus*, dated to around 4.4 Ma. Features of the skull and leg bones were suggestive of bipedalism. In 2001, White's colleague, keen-eyed Ethiopian fossil hunter **Yohannes Haile-Selassie** (b.1961) described some fossil teeth from 5.2–5.8 Ma, which, along with subsequent finds, were later assigned to a new species *Ardipithecus kadabba*. Curiously, *Ardipithecus* and *Orrorin* appear to have lived in forests rather than on the savannah, challenging the prevailing idea that human bipedalism evolved for life on the open plains. In a lively debate, fans of *Ardipithecus* have wanted to pull *Orrorin* towards the chimpanzee lineage, while *Orrorin* devotees wish to do the same to *Ardipithecus*. In a

Palaeoanthropology: Africans are doing it for themselves

In the early twentieth century, the search for human origins in Africa evoked images from Tarzan or Indiana Jones: white men (and occasionally white women) on expeditions to "the dark continent", magically pulling fossils from the dusty African soil. But now things have changed. One European family, the Leakeys, put down roots, became Africans, and started to employ local Kenyans as fossil hunters. In so doing, they trained one of the most successful fossil hunters in the world, **Kamoya Kimeu** (b.1940). Kimeu, a member of the Akamba tribe, started working for the Leakeys in the late 1950s and within a decade had become the family's right-hand man. His chief discoveries include a tibia from *Australopithecus anamensis* (the earliest leg-bone evidence for bipedalism), a skull of *H. habilis* and the Turkana Boy *H. erectus* skeleton. Two fossil apes have been named after Kimeu, *Kamoyapithecus hamiltoni* and *Cercopithecoides kimeui*, and the National Geographic Society awarded him the prestigious John Oliver La Gorce Medal. Kimeu retired recently after thirty years as curator of prehistoric sites at the National Museums of Kenya.

From the late 1970s, supporters of Louis Leakey have set up a foundation that funded fellowships for African palaeoanthropologists. One early beneficiary was **Berhane Asfaw** (b.1950), an Ethiopian who earned a PhD in physical anthropology at Berkeley in 1988. With his mentor Tim White and other collaborators, Berhane has described numerous hominin finds from Ethiopia, including two new species, *A. ramidus* and *A. garhi*, and the first skull of *P. boisei*. In 2003, Asfaw and colleagues described the oldest fossils of *H. sapiens* from Herto in Ethiopia, confirming Africa's status as the cradle of humanity. Two years earlier, fellow Ethiopian **Yohannes Haile-Selassie** (b.1961) wrote a single-author *Nature* paper, describing fossil teeth from a new species, *Ardipithecus kadabba*. Like Asfaw, Haile-Selassie earned his PhD under White's supervision at Berkeley. Haile-Selassie now works as curator at the Cleveland Museum of Natural History.

Several other Ethiopians have also blazed the trail for African palaeoanthropology. **Giday Wolde Gabriel** was part of the team that discovered Lucy (see p.186), has two first-author *Nature* papers to his name and now works as a geologist at the Los Alamos National Laboratory. **Zeresenay Alemseged** (b.1969) discovered "Lucy's child" and is now based at the Californian Academy of Sciences. **Sileshi Semaw** (b.1960), with a PhD from Rutgers and now based at the University of Indiana, has documented the oldest known stone tools and the largest ever find of *A. ramidus* fossils.

While Ethiopians lead the way, other Africans are not far behind. **Fredrick Kyalo Manthi**, a senior research scientist at the National Museums of Kenya, discovered a remarkably well-preserved *H. erectus* cranium. Even students can do their bit: in 2001, while still an undergraduate at the University of N'Djamena, Chadian **Ahounta Djimdoumalbaye** discovered Toumaï. Watch out for future contributions from Tanzanian and Kenyan palaeoanthropologists. African palaeoanthropology has come of age.

2004 paper, Haile-Selassie and colleagues suggest lumping *Orrorin* and *Sahelanthropus* into a single genus with *Ardipithecus.* Much hangs on the thickness of the enamel on fossilized teeth.

Chimp teeth and an Ethiopian ape

The use of fossils to investigate the human-ape split is frustratingly one-sided, with, until recently, no fossils on the chimpanzee or gorilla lines, a fact explained by the assumption that gorillas and chimps lived only in the forest. In fact, according to the **East Side Story** theory popularized by Yves Coppens (b.1934), it was a move from forest to the savannah east of the newly formed African Rift Valley that launched human evolution on a novel and unique trajectory, away from our chimpanzee brethren. However, a 2005 paper from two Connecticut anthropologists, Sally McBrearty and Nina Jablonksi filled the gap, but upset the applecart, by reporting three 545,000-year-old **fossil chimp teeth** from the Eastern Rift Valley. These findings show that the Rift Valley did not represent a barrier to chimpanzee occupation and so challenge the East Side Story (as does Toumaï way over in Chad). Another exciting recent fossil ape discovery was the report in late 2007 of a ten-million-year-old ape *Chororapithecus*

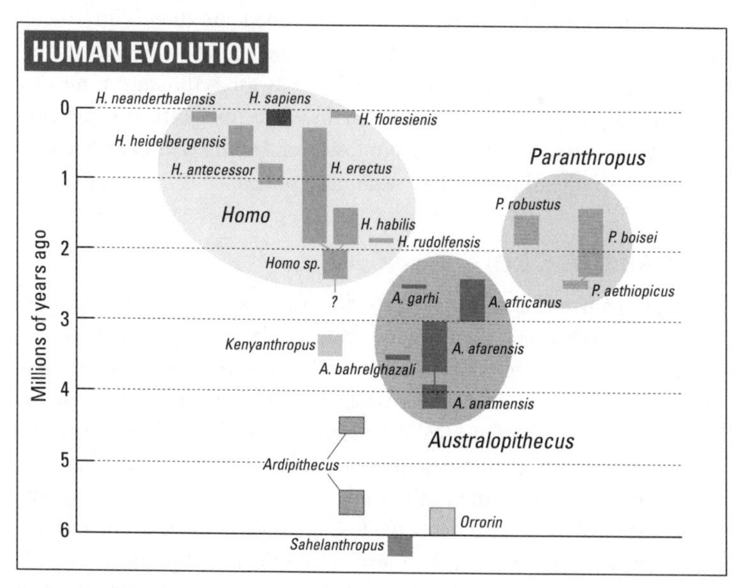

In the six million years since our last common ancestor with chimps, the evolution of the human lineage has been complex.

abyssinicus from the Afar Triangle of Ethiopia. Controversial suggestions that it represents an early member of the **gorilla lineage** challenge the molecular clock, which puts the split between the gorilla and the human-chimp lineages at around 8 Ma.

The australopithecines

The genus *Australopithecus* spans four small-brained African species (or maybe five) that lived between 4.2 Ma and 2.5 Ma. Their cranial capacities ranged from 350 to 550cm^3, almost overlapping with modern chimpanzees. In 1995, Kenyan palaeoanthropologist **Meave Leakey** (b.1942) and her associates described a collection of over twenty hominid fossils from Allia Bay and Kanapoi, near Lake Turkana in Kenya. The fossils included a lower jawbone, cranial fragments and parts of a tibia (a leg bone) and were sufficiently distinctive for Leakey to assign them to a new species, *Australopithecus anamensis* (anam means "lake" in the Turkana language). The tibia established that *A. anamensis* was bipedal. In retrospect, it turned out that a humerus (an arm bone) discovered in 1965 belonged to the same species and in 1998 new specimens from the same sites with more secure dates (4.12–4.17Ma) were described. In 2006, Tim White and his colleagues described thirty fossils, belonging to at least eight individuals of *A. anamensis* from the Middle Awash sediments of north-eastern Ethiopia. Dated to about 4.1 Ma, these specimens are separated by only 300,000 years and six miles from fossils of *Ardipithecus ramidus*, and are set in a woodland rather than savannah context.

Lucy and her kin

Although many palaeanthropologists are wary of suggesting ancestor-descendent relationships, there is good evidence placing *A. anamensis* on an evolutionary continuum with the later species ***Australopithecus afarensis***, which also occurred in the Middle Awash, but also in many other localities in Ethiopia, Kenya and Tanzania. American anthropologist **Don Johanson** (b.1943) and colleagues first described the species in 1979 on the basis of fossils from Laetoli in Tanzania and Hadar in Ethiopia. These included not only a 3 Ma fossilized knee joint (the first unexpected evidence of ancient bipedalism) but also probably the most famous fossil find in the history of palaeoanthropology: a remarkable near-half skeleton from a single female, nicknamed **Lucy**, because the Beatles song "Lucy in the Sky with Diamonds" was playing from a tape-recorder the night the skeleton was extricated. Lucy stood just over a metre tall (small for her species), weighed around half as much as the average adult human female, was small-brained, but possessed a pelvis and leg bones that were clearly adapted to a **bipedal upright stance**. Lucy quickly rose to stardom, and, commensurate with her celebrity status, is currently on tour in US museums, raising funds to modernize Ethiopia's museums (somewhat controversially, given that she is even more fragile than a flakey pop star).

The discovery of Lucy was followed by more good luck for Johanson. In 1975, he discovered the **First Family**, two hundred fragments from thirteen individuals that died together, perhaps from a flash flood. Despite considerable size differences, Johanson viewed them all as members of *A. afarensis*. Uncertainties remain as to how far size differences between the sexes (sexual dimorphism) influence the biology of this species. The **Laetoli footprints**, a 23-metre-long trail of 3.7 Ma imprints in Tanzania, were discovered in 1978. The fossilized footmarks from a remote part of the Ngorogoro National Park appear to have been made by two or three bipedal individuals and have been tentatively assigned to *A. afarensis* (the only known hominin species from that time). For many, these iconic footprints represent an evocative symbolic link to our upright, bipedal predecessors and epitomize humanity's long walk to self-discovery. After fears that the footprints might be harmed, by erosion, animals or humans, they were buried in the mid-1990s. However, now that their covering is starting to erode, fresh concerns have been raised about the footprints' long-term conservation, along with controversial suggestions that they belong in a museum.

Abel is the nickname given to a 3 Ma australopithecine jaw fragment discovered in Chad by Michel Brunet and his colleagues in 1993. Although Brunet's claim that it belongs in a separate species (*Australopithecus bahrelghazali*) remains tendentious, Abel does provide startling evidence of early australopithecines living over a thousand miles west of the Rift Valley. Equally controversial are some Kenyan fossils, comprising over thirty skull and tooth fragments, dated 3.2–3.5 Ma and described as a new species, *Kenyanthropus platyops*, by Meave Leakey and her team in 2001. The skull provides evidence of a small brain with a broad flat face, and Leakey suggests that this cousin and rival to Lucy provides additional evidence of the bushy nature of the hominin family tree. However, Tim White has argued that it simply represents a mangled specimen of *A. afarensis*.

Lucy's child or **Selam** (Amharic for "peace") is the name given to the partial skeleton of a three-year-old *A. afarensis* girl buried 3.3 Ma in Dikika,

Walking the walk

For several decades, it has been clear that our ancestors walked on two legs long before they evolved large brains. But the puzzling questions remains why did our ancestors evolve a two-legged, or **bipedal**, upright stance. Numerous explanations have been put forward, many depending on the assumption that walking originated after a move to the savannah. Under this scenario, standing upright might help lessen exposure to the sun or walking might prove more energy-efficient for travelling over long distances. However, now that we know that the earliest bipedal hominins, *Orrorin*, *Ardipithecus*, and the early australopithecines, lived in forested environments, savannah-centred explanations seem much less plausible. Other hypotheses have focused on bipedalism as an adaptation for feeding on the forest floor or facilitating displays of strength or sexual suitability (humans do have larger breasts and longer penises than our chimpanzee cousins).

Another assumption that our common ancestor with the African apes was a **four-legged knuckle-walker**, like today's chimps and gorillas. In fact, this assumption is so ingrained that suggesting that someone drags his knuckles is used as a term of abuse! But over the years there have been quiet murmurs of dissent. Maybe we learnt to walk in the trees? Maybe the human-ape ancestor did not knuckle-walk, but stood upright? Maybe knuckle-walking is a specialized innovation unique to the chimp and gorilla lineages?

This dissenting view received a boost in 2007 with a controversial publication in *Science* magazine. **Susannah Thorpe** (a colleague of the author at the University of Birmingham) and her co-workers reported extensive field studies of Sumatran orangutans, apes that spend far more time in the trees than do chimps or gorillas. Thorpe's team observed that orangutans often adopt an upright hand-assisted bipedal posture when traversing thin flexible branches, reverting to a four-legged approach only on the thickest, least challenging branches. Crucially, they saw orangutans adopt a human-like straightened hip-and-knee posture, distinct from the bent-knee hunched-hip pose adopted by chimps when upright.

These examinations of ape behaviour support a hypothesis of **ancestral arboreal bipedalism**, where "walking in the air" evolved to help our ancestors move around the forest canopy. The air-walking orangutans have also prompted fresh examinations of contemporary and fossil bones. Arguments rage as to whether subtle features of early hominin wrists and fingertips are indicative of knuckle-walking ancestry, or whether the presence of similar features in orangutans support the air-walker hypothesis. We can hope that more fossils will resolve the argument. But then, if we allow for early arboreal bipedalism, it becomes harder to identify fossils as early hominins, given the loss of a feature previously thought unique to the human ancestral line. What we need are more fossils of chimp and gorilla ancestors. Watch this space!

Ethiopia. First described in 2006 by her discover, Ethiopian evolutionary anthropologist **Zeresenay Alemseged**, Salem was found a few miles south, across the Awash river, from Lucy, but pre-dates her by more than 100,000 years. As expected, Selam's foot and leg bones provide clear evidence for bipedalism, but her shoulder blades resemble those of a gorilla and her fingers are curved like those of a chimpanzee, suggesting that she retained the ability to climb trees.

The Taung Child and Mrs Ples

Three quarters of a century before Lucy's child, in the mid-1920s, remains of another fossilized infant, the **Taung Child**, provided the first evidence that, as Darwin predicted, humanity has a fossil history of its own and that history is predominantly African. Taung is a small town in the North West Province of South Africa. In 1924, the skull of a three-year-old child was discovered here by a quarryman and passed on to **Raymond Dart**, an anatomy expert at the University of Witwatersrand. Dart recognized the specimen's importance, describing it as the type specimen of a new species of bipedal "ape man", which he named *Australopithecus africanus* (southern ape of Africa). With its small brain (uniquely preserved as a brain cast) but human-like teeth, the Taung skull was at odds with the 1920s establishment view, seduced by what is now known to be a forgery, the large-brained skull, with ape-like teeth, of **Piltdown man** (gathered from an English gravel pit in 1912, this was in fact a mixture of human skull, an orangutan jaw and some chimp

This orangutan's upright stance suggests that humans did not evolve from a knuckle-walking ancestor.

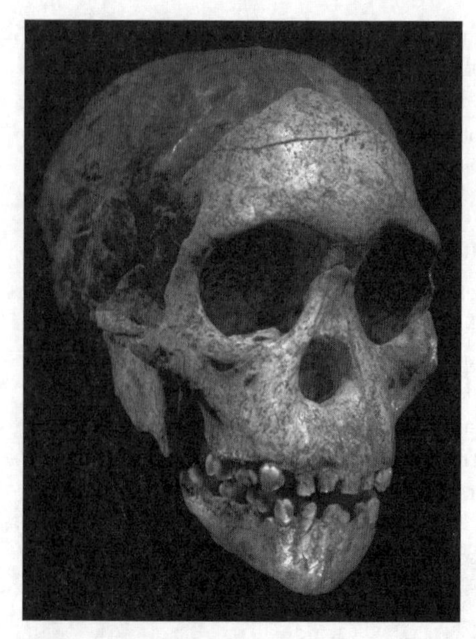

The Taung child skull. Marks in the eye sockets suggest an attack by an eagle.

teeth). However, the true status of the Taung skull was recognized after similar specimens were recovered later in the twentieth century. Recent re-examinations suggest that the Taung child was the victim of an eagle, which collected the child's head and other animal remains in the cave: gruesome evidence for this scenario include marks in the base of the child's eye sockets, made when the eagle ripped out its eyes.

Australopithecus africanus, thought to have lived 2.5–3.0 Ma, was clearly more human-like than *A. afarensis*, in terms of brain size and facial features. In addition to the Taung child, *A. africanus* fossils have been recovered from three other sites in southern Africa, most notably in caves in the **Sterkfontein** area (now designated the **Cradle of Humanity World Heritage Site**). Here, in the 1940s, Robert Bloom and John Robinson discovered a well preserved *A. africanus* skull, nicknamed **Mrs Ples**. A fossilized pelvis, spine, femur and ribs (specimen STS14) found nearby may represent Mrs Ples's body. In the mid-1990s, the Sterkfontein cave system also yielded the first glimpses of an extraordinary near-complete skeleton, nicknamed **Little Foot**. Although clearly australopithecine, the species of the fossil remains unclear and the date controversial, with estimates of its age ranging from 4 to 2.2 Ma. Painstaking excavations on the Little Foot skeleton continue. Quite how *A. africanus* fits into the human lineage is unclear, given that it lived so far away from other australopithecines and early *Homo*.

Robust relatives

South African locations have also yielded fossils of broad-faced, heavy-chewing **robust australopithecines**, that have now been placed in a separate genus, *Paranthropus*. The first species in this genus, *P. robustus*, described by

Scots-born South African doctor **Robert Broom** (1866–1951) in the 1930s, helped overturn the view that the evolutionary trail leading to humanity was a straight line. In addition to its more robust jaw and back teeth, male specimens of *R. robustus* are instantly recognizable by a crest running along

What made us human: genome sequences and their consequences

The arrival of genomics has facilitated a new approach to the study of human evolution, independent of the vagaries of the fossil record – as the controversial anthropologist Vincent Sarich once stated: "I know my sequences have ancestors, you can only hope your fossils have descendents". Driven by the completion of a human genome sequence in 2002 and the recent availability of draft sequences of a **chimp genome** (in 2005) and a **macaque genome** (in 2006), primate evolutionary genomics is now underway in earnest. Genomic comparisons have confirmed that humans have one less pair of chromosomes than chimps (our chromosome 2 is a fusion of two ancestral chromosomes), and have revealed how astonishingly similar we are to chimps at the molecular level. On average, just two changes distinguish each of our proteins from their chimp equivalent, in fact, 29 percent of our proteins are identical to their chimpanzee counterparts. Important differences in disease susceptibility (chimps don't get Alzheimer's or sick with AIDS when infected with HIV) will, it is hoped, soon be linked to genomic differences. But bizarrely, some of the mutations that cause genetic disease in humans represent the normal form of the gene in a healthy chimp or macaque.

Genomic comparisons provide no easy answer to the question of what makes us human, significant changes are buried in a sea of irrelevant differences. Attention has focused on finding genes influenced by natural selection since the human-chimp divergence. Curiously, there appear to be more such genes changing on the chimp lineage than on the human, provoking the claim that chimps are more evolved than humans! Genes that have evolved quickly in the human lineage include **FOXP2** (possibly involved in the evolution of speech) and genes involved in the production of natural opiates. We humans have also lost genes, particularly those involved in the sense of smell.

However, it is not just loss or gain of genes that have driven our evolution but also changes in gene expression. Some studies have attempted to compare the expression of genes in humans, chimps and other primates. In an affront to human pride, gene expression in the chimp brain looks more like that of humans than that of other "animals". One gene highly expressed in the human brain, **HAR1**, also shows a higher-than-expected number of changes in sequence, suggesting that it might be important in the evolution of the human brain. Although the search for the genes that make us human has just begun, advances in genomics mean that it is no longer a vain quest – one day we *will* identify and understand the changes that made us human.

the top of the skull, a bony anchor for their exaggerated chewing muscles. Numerous specimens of *P. robustus*, which lived 1.2–2.6 Ma, have been found in South Africa. One notable specimen, a ridgeless female skull nicknamed **Eurydice**, was described in the early 1990s.

Mary Leakey (1913–96) described a second kind of robust australopithecine from Olduvai Gorge, Tanzania in 1959. Initially designated *Zinjanthropus boisei* and nicknamed **Zinj** or **Nutcracker Man**, the sample is now classified in the species *P. boisei*. Raymond Dart is said to have wept for joy in seeing the beautiful skull of Zinj. Specimens of *P. boisei* have been recovered from several sites in Tanzania, Kenya and Ethiopia, their ages ranging from 1.2 to 2.3 Ma. The **Black Skull** from West Turkana in Kenya (so called because of dark colouration of the fossilized bone) provides the best evidence of a third species of robust australopithecine, *P. aethiopicus*. Discovered in 1985, dated to 2.5 Ma and equipped with a small brain and a primitive face, the skull provides the earliest evidence of the robust lineage.

Australopithecus garhi (after the Afar word for "surprise") is a species that lived 2.5 Ma in Ethiopia. The type specimen, described in 1999 by Ethiopian Asfaw Berhane, is a partial cranium; from nearby sites come additional partial skeletons. In terms of its skull and the proportions of its limbs, *A. garhi* looks like a possible intermediate between *Australopithecus afarensis* and *Homo*, but experts still squabble as to how the early hominin, australopithecine and *Homo* lineages relate to one another.

Our own genus: Homo

In the early 1960s, **Louis Leakey** described the new species *Homo habilis*, or **Handy Man**, based on fossils from Olduvai Gorge, in Tanzania. The type specimen, a 1.75-million-year-old jaw and other bones from a late juvenile, was nicknamed **Jonny's child**, after the Leakey's oldest son, who had helped discover it. According to most experts, handy man still qualifies as the earliest member of our own genus (some prefer to re-assign it to *Australopithecus*). One early *Homo* skull, "1470", discovered in 1972 has been a source of continuing controversy. Initially dated at 3 Ma, it appeared too modern for its time; more recent estimates place its age at 1.9 Ma, with some now seeing it as a variant of *H. habilis*, others as a separate species, *H. rudolfensis*.

Compared to modern humans, *Homo habilis* was short and had disproportionately long arms; however, its larger brain (nearly half the size of modern humans) and small molars set it apart from the australopithecines.

In the 1960s–70s, additional *H. habilis* fossils were described in Tanzania and Kenya, often accompanied by primitive stone tools assigned to the **Oldowan** technology (perhaps also mastered by some australopithecines). It is likely that *H. habilis* used these tools for butchering scavenged meat rather than for defence or hunting. From its inception as a species, it was known that *H. habilis* co-existed with the robust australopithecines. However, a paper published in 2007 by British palaeoanthropologist **Fred Spoor** and others suggested that *H. habilis* survived long enough – until less than 1.5 Ma – to overlap with a supposedly descendent species, *H. erectus* (another example of the bushiness of human evolution).

Lumpers and splitters disagree as to what to do with African *Homo* fossils that immediately follow handy man (1.9 to 1.4Ma). Some split them off into a separate species *Homo ergaster;* others lump them into a single species with much later specimens, calling them early African *Homo erectus*. Such fossils have been found as far a field as Tanzania, Ethiopia, Kenya and South Africa and are often associated with a culture of stone tool use known as the **Acheulean**.

Kenyan veteran fossil hunter **Kamoya Kimeu** (b.1940) discovered the most complete *Homo ergaster* skeleton in 1984: **Turkana Boy** (also called Nariokotome Boy), a pre-pubescent boy from 1.6 Ma. Projections from Turkana Boy's 1.6 metre height suggest he would have reached an impressive 1.85 metres (over six foot) had he survived to adulthood. Despite his long legs and impressive brain size (880cm^3), Turkana Boy's sloping forehead, marked brow ridges, and lack of a chin mark him out as not-yet-human.

Long before Turkana Boy's premature death, some humans had already left Africa. For over a decade after his country's independence in 1991, Georgian anthropologist **David Lordkipanidze** (b.1963) has worked with an international team of scientists to recover and study a series of spectacular fossils from **Dmanisi, Georgia**, which provides the earliest evidence of *Homo* in the temperate zones of Eurasia. Seen as an intermediate between African *H. habilis* and later Asian *H. erectus*, the name *Homo georgicus* has been suggested for these 1.8 Ma humans. Georgian fossils of an adolescent and three adults, described in late 2007, revealed a curious mix of ancestral and modern features, with diminutive bodies and 600cm^3 brains.

By one million years ago, *Homo erectus* was a far-flung traveller. The classical descriptions of the *H. erectus* fossils hail from Java, where Dutch anatomist Eugene Dubois described **Java Man** in the 1890s, and from China, where Canadian geologist Davidson Black described **Peking Man** in the 1920s (these Chinese specimens were entrusted to the US Navy during World War II but disappeared, fueling various conspiracy theories as to where they are now). *Homo erectus* appears to have persisted for a long time outside Africa, with some fossils from **Ngandong** in Java dating from possibly as recent as 50,000 years ago. Several large jaw and skull

The Leakeys: a legendary dynasty and its legacy

Like Charles Darwin, **Louis Leakey** (1903–72) made an intellectual and dynastic contribution to science. Born to missionary parents, Louis grew up as a white member of Kenya's **Kikuyu** tribe. Despite an indifferent school career, Louis won a place at Cambridge University in 1922. At first intending to become a missionary, he switched his attention to anthropology and archaeology after a fossil hunting expedition to East Africa in 1924. He received his PhD from Cambridge in 1930. The following year he made his first visit to **Olduvai Gorge,** now in Tanzania, the site of his greatest discoveries. Scandal overtook him in 1933, when he became romantically involved with illustrator Mary Nicol, while his wife Frida was expecting their first child. He and Mary were married following his divorce in 1936.

During the low point of his life in the mid-1930s, Louis lived with Mary in an English cottage without power or plumbing. In 1937, he returned to Africa. World War II saw Louis working as an intelligence agent for the Kenyan government while Mary continued to excavate sites and work at the Coryndon Memorial Museum, where her husband later joined her. In the 1950s the Mau Mau Rebellion against British colonial rule placed him in a difficult and dangerous position. While siding with the white settlers – acting as their spokesman and doing intelligence work – he also advocated greater rights for the Kikuyu. The same period saw the Leakey husband and wife team at its best, with the legendary discoveries in Olduvai of nutcracker man, handy man and numerous others; their nineteen-year-old son **Jonathan Leakey** was director of one of the excavation camps. Around this time Louis encouraged three women (later known as **Leakey's angels**) to study primates in their natural environments: **Jane Goodall** (chimpanzees), **Dian Fossey** (gorillas) and **Biruté Galdikas** (orangutans). Louis's final years were marked by misguided opinions on American archaeology and growing estrangement from his wife. He died in London, but was buried in his beloved Kikuyuland. Like Darwin, he is memorialized by a crater on the Moon.

After Louis's death, **Mary Leakey** (1913–96) continued working, discovering many new fossils and helping uncover the Laetoli footprints. **Colin Leakey** (b.1933), Louis's son by his first wife, became a leading English botanist specializing in beans. His second son **Richard Leakey** (b.1944) became a world-famous palaeoanthropologist, with numerous finds from Kenya and Ethiopia to his credit, and second and third careers as conservationist and Kenyan politician. Richard's wife, **Meave Leakey** (b.1942) is also a renowned palaeoanthropologist, with *Kenyanthropus platyops* to her name. Richard and Maeve's daughter **Louise Leakey** (b.1972) continues the family tradition – at the age of six, she was the youngest person ever to discover primate fossils!

The husband and wife team of Louis and Mary Leakey digging for hominin bones in Tanzania's Olduvai Gorge, scene of many of their best-known discoveries.

fragments from Sangiran, Central Java, have sometimes been classified as a robust Asian hominin, *Meganthropus*, although it is more likely that they belong to *H. erectus*.

The first humans

From several sites in Africa and Europe comes evidence of a successor to *H. ergaster* and *H. erectus*: *Homo heidelbergensis*. Sporting a bigger brain (up to 1200cm^3) than *H. erectus*, *H. heidelbergensis* remained substantially more robust than modern humans. In Africa, *H. heidelbergensis* is represented by fossils from Ethiopia and Zambia (these latter sometimes called **Rhodesian Man** or *H. rhodesiensis*). Fossils from a complex of caves in **Atapuerca**, northern Spain, document the earliest hominins in Western Europe. One Atapuerca tooth has been dated as 1 Ma, while a juvenile face is thought to be at least 0.75 Ma. As always, what counts as one species and what another is controversial. Some bracket the Spanish fossils with *H. heidelbergensis*, while their discoverers, **Eudald Carbonell** (b.1953) and **Juan Luis Arsuaga Ferreras** (b.1954), view them as a separate species, *H. antecessor*. Bone cuts on some of the Spanish fossils provide chilling evidence of early cannibalism. Other evidence of early human occupation in Europe includes an incomplete adult skull cap discovered near Ceprano, in Italy, dated at 800,000–900,000 years old; some 700,000-year-old stone tools from Pakeham in Suffolk,

England; and a half-million-year-old *H. heidelbergensis* shin bone and some teeth from **Boxgrove Quarry**, in southern England.

Despite these Eurasian excursions, Africa remains the cradle of humanity. The earliest undisputed fossils from our own species ***Homo sapiens*** come from Ethiopia. In 2003, Tim White and colleagues described three cranial fossils, dated at around 160,000 years old, from **Herto** in Ethiopia that were intermediate in time and anatomy between earlier African fossils and later anatomically modern humans, providing strong evidence for the idea that modern humans emerged in Africa. In 2005 two modern-looking *H. sapiens* crania, found by Richard Leakey's team in 1967 at Omo Kibish (also in Ethiopia), were re-dated to 195,000 years. Early remains of anatomically modern humans, dated at 125,000 years old, have also been recovered

From tools to war: nothing uniquely human?

Plato tried to define humans as featherless bipeds; fellow philosopher Diogenes responded by conjuring up a plucked chicken. In *The Descent of Man*, Darwin wrote, "It has often been said that no animal uses any tool; but the chimpanzee in a state of nature cracks a native fruit, somewhat like a walnut, with a stone." One might look to tool creation or the transmission of culture to salvage human uniqueness. But in yet another revolution, even here we humans have been knocked off our pedestal.

The revolution began in 1960 in the **Gombe National Park** in Tanzania, where British primatologist **Jane Goodall** (b.1934) observed a chimp, who she had named **David Greybeard**, using a piece of grass to fish termites out of a hole. Subsequent observations showed chimps stripping sticks of leaves to provide improved fishing rods: the first evidence of chimp tool creation. At the time, Louis Leakey put it, "Now we must redefine tool, redefine man, or accept chimpanzees as humans." Since then a flood of data has confirmed that chimps and other great apes routinely modify and use rocks and items of vegetation as tools in feeding, drinking, digging, grooming, fetching and fighting. And variations in tool use, as well as transmission of information from one generation to the next, provide convincing evidence of a **primitive culture** among chimpanzees. For example, in the Taï rainforest in West Africa (but not in Gombe) young chimps learn how to use stone hammers to crack open nuts.

In 2007, Canadian researchers reported excavations in West Africa that provide the first evidence of ancient chimpanzee **tool use**, from over four thousand years (predating the arrival of agricultural villages in the region). They made the bold suggestion that chimps and humans have inherited stone tool use from our common ancestor. Also in 2007 came a report from American anthropologist **Jill Pruetz** and her colleagues of chimps from Senegal using their teeth to sharpen spears, which were subsequently used to impale bushbabies.

from the **Klasies River Caves** in South Africa. Fossils of anatomically modern humans dated to 100,000 years ago have been recovered from a cave in Israel at at **Qafzeh**, but this incursion into Asia appears to have been a dead-end, leaving no modern descendents outside of Africa.

Our Neanderthal cousins

The **Neander valley** (or Neanderthal) near Düsseldorf in Germany is named after a local hymn-writing pastor Joachim Neumann (i.e. "new man", which translates as "neander" in Greek). There are numerous limestone caves and rock shelters along the valley. In the 1850s, limestone quarrying began in the

Tool use and manufacture are not even restricted to humans and great apes. The list of tool-using animals gets ever longer – from badgers to bottle-nosed dolphins, from owls to otters. Today's Diogenes would hold up a crow rather than a chicken. These birds have proven adept tool users and makers in the lab and in the wild. A recent provocative study on New Caledonian crows even identified cumulative progress in their creation of tools.

Jane Goodall and friend.

Is there anything unique in the human propensities to make love or war? Extreme and gratuitous violence is certainly not confined to humans. In the 1970s, Goodall observed the "four-year war" (or **Gombe Holocaust**) in which one group of chimps relentlessly attacked and killed members of another group until the whole group – seven males, three females and their young – was wiped out. While some dismissed this chimpanzee war as the result of human encroachment, numerous reports of similar "murderous" chimpanzee incidents have followed. Since the 1950s, the free-love sexuality of the lesser-known chimpanzee species, the bonobo, has amused and embarrassed human observers. Bonobos, like humans, have sex for fun and throughout most of the females' oestrus cycle, but they draw on a repertoire of sexual acts that would put the Karma Sutra to shame. And one last insult to a Victorian view of human uniqueness – bonobos even have sex in the face-to-face "missionary" position!

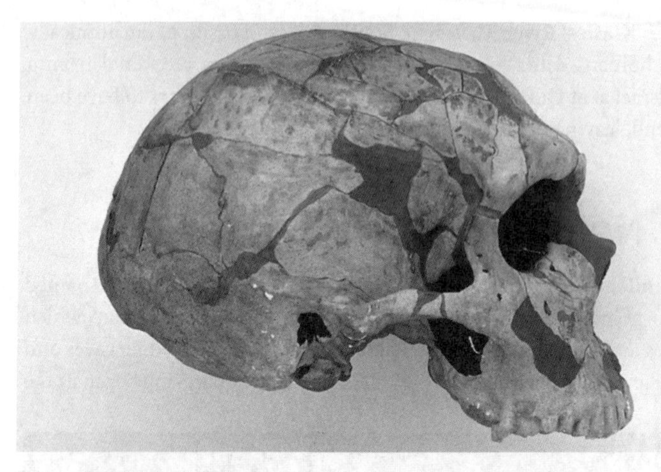

A Neanderthal skull with characteristic brow ridges.

valley, so that cave and valley walls were excavated, and in August 1856, the top of a skull and fifteen other postcranial bones were recovered from one particular cave (the *Kleine Feldhof* grotto). At first, the bones were thought to have come from a cave bear, but local teacher and natural historian **Johann Carl Fuhlrott** (1803–77) soon identified them as human. Fuhlrott, along with local anatomy professor **Hermann Schaaffhausen**, described the bones in a paper in 1857, suggesting that they represented the remains of an ancient extinct race of humans. Now known as **Neanderthal 1**, Fuhlrott's fossil bones became the type specimen for a new species, *Homo neanderthalensis* (although it was found in retrospect that earlier finds from Belgium and Gibraltar belonged to the same species).

Since Fuhlrott's time, over four hundred Neanderthal specimens have been discovered, from sites scattered across Europe, the Middle East and Asia: from Spain to Siberia, from Israel to England. The Neanderthals were **anatomically distinct** from modern humans: they were more robust, with a low flat elongated skull, a pronounced brow ridge, a projecting mid-face and an absent chin. But unexpectedly, they had slightly larger brains. They were like us, but they were unlike us: readers brought up on a diet of *Star Trek* might like to think of them as ancient Klingons or Vulcans!

Neanderthal characters began to show up in *H. heidelbergensis* fossils in Europe 400,000 years ago (most clearly at Sima de los Huesos); full-blown Neanderthals flourished from 130,000 years ago to less than thirty thousand years ago. By fifty thousand years ago, they appear to have died out in Asia, but they probably survived in Europe until as late as twenty-four thousand years ago. The Neanderthals overlapped in time and range with anatomically modern humans for at least fifteen thousand years, but there is no evidence that the two types of human co-existed at any one site. They had their own culture, which included stone tools (hand axes, spears from the **Mousterian culture**), skinning and fire. A bone fragment from Slovenia has even been identified, rather controversially, as a **Neanderthal flute**. Nine Neanderthal skeletons from **Shanidar** in northern Iraq document withered arms and healed fractures – strong evidence that Neanderthals cared for their injured. A flower-filled grave from Shanidar has been taken as evidence that Neanderthals buried their dead (collection of flowers by a gerbil-like rodent provides an alternative explanation). Neanderthals ate meat and may even have engaged in cannibalism (alternatively, butchered bones of Neanderthals may indicate ritual defleshing). They possessed a tongue bone, the **hyoid**, similar in shape to ours, which some interpret as evidence that they could produce a similar repertoire of sounds, although it is uncertain whether they possessed complex language.

Neanderthal endgames

The fate of the Neanderthals, like that of the dinosaurs, carries an enduring fascination for the expert and layperson alike. Three options have been put forward: genocide (hunted to death by modern humans), gradual extinction (unable to adapt after the end of the Ice Ages; unable to compete with humans in, say, intelligence or long distance running) or assimilation (of which more later). Although there was no happy ending for the Neanderthals, a recent piece of archaeological detective work has brought a satisfying conclusion to their initial discovery. The cave where Neanderthal 1 was discovered was destroyed as the area was excavated and flattened to provide the local steel industry with limestone. It was generally assumed that the first, and most important, Neanderthal site was lost to science. But in the late 1990s, German archaeologists **Ralf Schmitz** and **Jurgen Thissenset** were determined to find it. Poring over nineteenth-century paintings and old maps, they recognized a rock that still stood in the valley. They dug nearby and found cave debris. After careful examination, a team of archaeologists found over sixty human skeletal fragments, from at least three individuals. Their efforts culminated in a triumphant solution to a skeletal jigsaw puzzle: three of the newfound bony fragments fitted perfectly on to the original specimen, excavated nearly a century and a half before.

Waking the dead: Neanderthal genes and genomes

Some palaeoanthropologists believe that Neanderthals interbred with modern humans and look to the fossil record for support. For example, in 1999, Neanderthal expert **Eric Trinkaus** (b.1948) suggested that a 24,500-year-old skeleton from **Lagar Velho**, Portugal exhibits a mixture of features from modern humans and Neanderthals that are best explained by interbreeding. But the fossil record is not the only source of information on this issue. In 1997, pioneering Swedish molecular archaeologist **Svante Pääbo** (b.1955) obtained the first mitochondrial DNA sequence from the original Neanderthal specimen. As the Neanderthal sequence fell well outside the range of variation seen in modern human sequences, he concluded that the Neanderthals were indeed a separate species with no input into modern human populations. Several studies since then have confirmed that Neanderthal mitochondrial sequences are more similar to each other than to any human sequence. Although this is generally taken as evidence against any significant Neanderthal input into the human gene pool, it is impossible to prove a negative, i.e. show that there was never a single instance of interbreeding.

Neanderthal genetics soon turned into **stone-age genomics**,, when powerful new sequencing approaches were applied to a 38,000-year-old leg bone from the Vindija cave in Croatia. Two papers appeared in November 2006 revealing analyses of up to a million base pairs of Neanderthal genome sequence. They provided estimates from non-mitochondrial genes of the date of the most recent common ancestor of modern humans and Neanderthals that ranged from 500,000–700,000 years ago. One of the two papers interpreted the genome data as ruling out Neanderthal admixture in the human gene pool, while the other was more equivocal. Perhaps, not surprisingly, a subsequent critique suggested conclusions from the two papers were mutually incompatible. More recent analyses have established that the **Neanderthal FOXP2** gene (potentially involved in speech) was identical to that in modern humans, that at least some Neanderthals had pale skin and red hair, but that Neanderthal Y chromosomes differed from those of modern humans. With Pääbo promising a full draft of Neanderthal genome sequencing project within the next year or two, we can expect a lot more stone-age genomics.

Scatterlings of Africa

And now we come to one of the greatest evolutionary discoveries of modern times, the fact that we are all **Africans** under the skin. There are at least three alternative evolutionary views of human origins and diversity. According to proponents of the now abandoned idea of **polygenic evolution**, different human populations or races have independent origins and represent ancient lineages that have evolved in isolation from each other. A more sophisticated variation on this theme was proposed by anthropologists

Franz Weidenreich (1873–1948) and **Carleton Coon** (1904–81), in which five geographically distinct sub-species of *H. erectus* each "passed a critical threshold from a more brutal to a more *sapient* state" to become the modern "races" of *H. sapiens*. According to Weidenreich, gene flow could still occur between different populations, so, far example, genes for intelligence could spread around the globe. However, Coon was an out-and-out racist, who, even as late the 1960s saw "superior races" benefiting from having crossed the threshold to humanity before their less fortunate "inferiors".

Continuity or replacement?

In the early 1980s, the idea of gene flow in evolving human populations resurfaced, but shorn of all racist overtones. University of Michigan anthropologist **Milford Wolpoff** (b.1942) outlined the **Multiregional Continuity Hypothesis** to express his belief that the transition from *H. erectus* to *H. sapiens* occurred across multiple regions, allowing for genetic continuities between Asian *H. erectus* and modern Asians and between *H. neanderthalensis* and modern Europeans. In fact, for Wolpoff, classifying Neanderthals, and even *H. erectus* fossils, as a separate species is tantamount to racism: central to his hypothesis is the belief that modern humans and Neanderthals inter-bred and that Neanderthals contributed to the modern European gene pool. The 1980s also saw the first clear articulations by English palaeoanthropologist **Chris Stringer** (b.1947) and others of a competing **Out-of-Africa model** (also called the recent single-origin or replacement hypothesis). According to this model, modern humans evolved in Africa around 200,000 years ago, and then spread around the world replacing any pre-existing populations of *H. erectus* or *H. neanderthalensis*.

Several lines of evidence can be brought to bear when the two competing theories are evaluated according to the testable predictions that flow from them. If multiregionalism were true, the oldest human fossils could be found anywhere and one would not necessarily expect to see any human cultural universals. Instead, as one would predict from the Out-of-Africa model, the oldest human fossils are, in fact, found in Africa; and anthropologists have documented that all human populations, even the most isolated indigenous groups, share a common repertoire of behaviours (language, art, music, cooking, games, religious rituals, mythology, marriage, gender roles, incest taboos and humour). However, the most compelling evidence for the Out-of-Africa model comes from molecular genetic studies. According to multiregionalism, the greatest genetic

Domestication: evolution made by man

We humans have brought numerous plants and animals under our control in the last 15,000 years, whether for food, companionship, protection, work, or purely aesthetic reasons. In *The Origin*, Darwin discussed the astonishing variation seen in our older cultivated plants and animals. Since Darwin's time, archaeology and molecular biology have clarified the when, how and where of most instances of domestication.

Cereal crops were first domesticated 11,000 years ago in the Middle East (in the Fertile Crescent that includes Egypt, Palestine and Mesopotamia). Since then different crops have been domesticated in different parts of the world: for example, squash, maize, beans, potatoes in the Americas; millets, rice and soy

Carrying wheat and winnowing grain. Modern humans domesticated cereal crops during the transition to agriculture.

diversity could be anywhere; instead, there is more genetic diversity in Africa than anywhere outside.

Mitochondrial studies

Modern molecular phylogenetic studies have focused on two lines of descent. Analysis of DNA sequences from sub-cellular organelles called mitochondria, inherited only from the mother, provides insight into the line of exclusively female ancestors. Studies on **mitochondrial DNA** sequences in eastern Asia. Today's domesticated species have become quite distinct from their natural ancestors, showing exaggerated growth in the parts that interest humans, for example, cultivated maize cobs are now many times larger than their wild relatives.

Evolutionary biologist **Jared Diamond** has suggested several criteria required for **domestication** of animals: flexible diet, fast growth rate, an ability to breed in captivity, a pleasant disposition, without an uncontrollable urge to panic, and a modifiable social hierarchy. The **dog** was the first animal to be domesticated. Molecular approaches have shown that dogs are descended exclusively from wolves, but it remains unclear whether domestication happened just once or on more than one occasion. One recent study suggests a common origin for all dogs in East Asia around 15,000 years ago.

Sheep were domesticated in Mesopotamia between 9000 and 11,000 BC, probably from wild progenitors called mouflons. The first evidence of domesticated **goats** comes from 10,000 years ago in the Zagros Mountains of western Iran. **Pigs** were apparently domesticated independently in the Middle East and China. **Cattle** were domesticated from **aurochs** around 10,000 years ago independently in India and the Middle East (and perhaps also in Sub-Saharan Africa and East Asia). A 2007 study showed that **cats** were first tamed around 9000 years ago in the first agricultural villages of the Fertile Crescent, with most domestic cats descended from five **matrilineal ancestors** from this region. Archaeological discoveries in the Indus Valley and in Hebei Province, China, suggest that **chickens** were probably domesticated from the red jungle fowl (*Gallus gallus*), as early as 5400 BC. However, the authorities differ as to whether chickens have a single ancestor or multiple origins. Curiously, the **horse** wasn't domesticated until as late as 4500 BC, in Central Asia.

Humans have also **domesticated microorganisms**: for example, the **yeast** *Saccahromyces cerevisiae*, favoured by brewers and bakers alike. A French study from 2007 showed that yeast genetic diversity reflects human history, but it remains unclear how often yeast was domesticated and quite what its wild ancestor looked like. But let's all raise a glass to man's best microbial friend!

from ethnically and geographically diverse populations have shown that there is more variation in Africa than anywhere else: all humans from outside Africa fit into a single branch of the African tree. Such studies place the most recent common ancestor of all current mitochondrial sequences in Africa less than 150,000 years ago (evocatively nicknamed **African** or **Mitochondrial Eve**). Similar studies on the Y chromosome provide insight into lines of exclusively male ancestors. The deepest branches from the Y chromosomal tree are also all in Africa. **Y Adam** was also an African, although he lived more recently than Mitochondrial Eve.

Out of Africa

For non-Africans, the most poignant conclusion from these studies is that we are all derived from a single small band of humans (perhaps as few as 150) who crossed the Red Sea around 60,000–70,000 years ago in an **African Exodus**. Why those first modern human pioneers left Africa when they did is uncertain; one explanation (the **Sahara Pump theory**) suggests that periodic climate changes led to an extension of the African

The human hobbits: *Homo floresiensis*

In 2003, a joint team of Australian and Indonesian archaeologists and anthropologists, headed by **Mike Morwood**, made a remarkable discovery in the Liang Bua limestone cave on the Indonesian island of Flores. They unearthed an almost complete skeleton, dated at around 18,000 years old. In a paper published in *Nature* in 2004, they argued that a mixture of features, primitive and derived, identified the skeleton as a new species of human, *Homo floresiensis*, that had lived at the same time as modern humans. Fragments from seven similar individuals were recovered from the same site. The diminutive stature of these alternative humans (around one metre tall) has led to their popular designation as **hobbits**. It has been argued that their small size (smaller than any other hominin) is the result of insular dwarfism, a well-documented evolutionary process that leads to reduced size of large animals isolated on islands (other examples include dwarf elephants and their extinct relatives, the stegodons). Remarkably, at 417cm^2, the *H. floresiensis* brain appears diminutive even compared to the earliest representatives of the genus *Homo*. Nonetheless, their discoverers found evidence of advanced behaviours, such as tool use and use of fire.

The new species designation has proven controversial, with several anthropologists arguing that the remains are derived from modern humans suffering from a developmental abnormality known as microcephaly (characterized by a small head) or from a form of dwarfism known as Laron syndrome. The original discoverers have mounted a fierce defence against these claims, with additional studies shoring up the new species claim. Palaeoneurologist Dean Falk has reported two studies in which he found marked differences between the *H. floresiensis* brain and those of microcephalic humans. A team from the Smithsonian found that the *H. floresiensis* wrist bones look decidedly primitive and non-human, closely resembling their equivalents in chimpanzees or australopithecines. Another report suggested that the *H. floresiensis* shoulder was closer to that of *H. erectus* than to modern humans. While evidence is mounting against the simple explanation that the Flores finds represent the remains of medically abnormal humans, a definitive answer to the question of whether this really is a new species of human will have to wait until the discovery of new specimens, especially more skulls, or the recovery of DNA from the existing specimens (to date, attempts at DNA extraction have proven unsuccessful).

savannah into the Sahara and Arabia, facilitating this migration and the earlier hominin movements out of Africa. But we now recognize that some men alive today in Eritrea, Ethiopia or Sudan are the closest African relatives of the pioneers who left Africa – those familiar with the Bob Marley song "Exodus" should savour the irony that Marley was probably more closely related to Haile Selassie through his European paternal ancestry than via his mother's West African ancestry. If the mitochondrial ancestors of any non-African were made to stand in line, it would take us just an hour to walk the two miles back to our African mother. Whether Chinese or Indian, European or Native American, Inuit or Fuegian, Pacific Islander or New Guinea Highlander, from the palest Scandinavian to the darkest-skinned Aboriginal Australian, we are all scatterlings of Africa.

Another legacy from this time means that the whole of humanity shows less genetic variation than a single subspecies of chimpanzee. This relative lack of genetic diversity is taken as evidence of an ancient human population bottleneck, with just a thousand to ten thousand people alive. In a provocative theory, **Stanley Ambrose** lays the blame for the bottleneck on the eruption of the Sumatran volcano **Toba** 70,000–75,000 years ago, an event that represents the largest volcanic eruption in the last two million years.

Detailed analysis of non-African mitochondrial and Y chromosome sequences has allowed geneticists to reconstruct the subsequent **peopling of the world**, showing a migration from eastern Africa into the Middle East, then into Southern and Southeast Asia and then into New Guinea and Australia. The oldest post-exodus fossils of modern humans outside of Africa are found in Australia and have been dated to about 42,000 years ago. One explanation for the arrival of modern humans in New Guinea and Australia, relatively soon after leaving Africa, is the **beachcomber hypothesis**, whereby seafood-fed humans spread east all the way along the Asian shoreline. Europe is thought to have been colonized by migrants moving northwest from India and the Middle East, while Native Americans share a common ancestry with Eurasians and East Asians. The last places on Earth to be colonized by humans were the Pacific islands, peopled by the intrepid Polynesians, who spread across the Pacific from the west, even reaching South America. Italian population geneticist **Luigi Luca Cavalli-Sforza** (b.1922) has drawn on a wide range of evidence for human migrations, including pioneering but controversial analyses of large-scale language families.

Archaeologists and anthropologists working in Europe have detected a series of apparently revolutionary changes in human behaviour that occurred around 40,000-50,000 years ago. These changes, which saw technological advances combined with the first symbolic paintings and carvings,

Of lice and men

Scientists look for evidence of human evolution in the strangest of places. Molecular bacteriologist **Mark Achtman** has used sequences from a stomach bacterium, *Helicobacter pylori*, to affirm an African origin for humanity and reconstruct the peopling of the world. Others have used subtypes of a virus, **human polyoma JCV**, for similar purposes. Molecular analyses of rats, pigs and chickens have been used to track human migrations. And the bacteria that cause tuberculosis appear to be most diverse in East Africa, home to the earliest humans.

But the most bizarre source of evidence on human evolution must surely hail from comparative studies of lice! Human head and body lice are, as expected, most closely related to lice from chimpanzees. However, human pubic lice are closer to lice from the gorilla (evidence of human-ape hanky panky?). **Mark Stoneking** and his colleagues report that lice from Ethiopia show more diversity than all non-African lice, and has suggested the recent divergence of human head and body lice (around 72,000 years ago) dates to the first use of clothes. The existence of two ancient lineages of lice among humans has been taken as evidence of contact between *H. sapiens* and archaic humans. And following the flea-bites-flea paradigm, a team in Florida has used a bacterial companion of lice, *Riesia*, to investigate the evolutionary relationships between bacteria, parasites and owners. Lice might also have played an important role not just in reconstructing human evolution but in fashioning it. **Mark Pagel** and **Walter Bodmer** have suggested that avoidance of the disease burden associated with lice, reinforced by sexual selection, may have been a major force for humans to lose their body hair. Think of that when next combing head lice out of your children's hair.

have been described, somewhat colourfully, as the Human Revolution, the Great Leap Forward, the **Upper Palaeolithic Revolution** or **Behavioural Modernity**. However, a more sceptical, less Eurocentric view is that such changes were simply better documented in the dry climates and caves of Europe and that the underlying advances in human intellectual ability, including language, must have been well underway before our last common ancestor in Africa.

Less than two hundred generations ago, humans learnt to write. And the rest, as they say, is history!

Human evolution: the future?

As the old saying goes, predictions are always difficult, especially about the future. British evolutionist **Steve Jones** (b.1944) has suggested human evolution has come to an end, at least in the West, citing improved child survival rates and an increase in marriages between people of different geographical or ethnic origins. But natural selection is clearly still at work in developing countries, where infectious diseases such as **AIDS, tuberculosis** and **malaria** continue to kill before reproductive age. And, in an era of widespread **contraception**, surviving to reproductive age is no guarantee of reproduction, and it's a matter of speculation what affect fertility control will have on the human gene pool. When procreation is a matter of choice, rather than an inevitable consequence of passion, perhaps there will be a selective pressure for children to become steadily more manageable: if your first child is a terror, you might choose not to have any more.

The long-term future for humanity can be seen as widely optimistic or depressingly gloomy, depending on the commentator's outlook. Futurologist **Ray Kurzweil** (b.1948) predicts an imminent **technological singularity**, when machine intelligence overtakes human and the rate of progress soars off the scale in an **intelligence explosion**. Advocates of **transhumanism** articulate a proactive view of future human evolution, supporting the use of the latest advances in biology and other science to enhance human mental and physical abilities and ameliorate undesirable aspects of the human condition (such as disease, aging and death). Human life spans in industrial societies are already markedly longer than those in the past. Delaying procreation is known to select for longevity in model organisms, so an additional increase in longevity might be a side effect of the deferred parenthood seen in industrialized societies. However, proponents of **life extension** or **technological immortality** predict more impressive progress in the near future, drawing on advances in **gerontology** (the study of the molecular and cellular causes of aging). As a precedent of what might be possible, a gene that protects against mitochondrial DNA damage, when introduced into laboratory mice, extends the animals' lifespans by around 20 per cent. Even more dramatic effects are seen with **calorific restriction** (limiting dietary energy intake), which has been shown to extend the lifespan of organisms as diverse as worms and mammals (doubling lifespan in rats). If research discoveries can be translated into practical outcomes, then readers of this book may be among the last humans to face unwanted death from the effects of aging.

Chief among the merchants of doom is British Astronomer Royal **Martin Rees** (b.1942) who, in his 2003 book *Our Final Century*, estimates the chances of **human extinction** before 2100AD as around 50 per cent. His chief concern is with a **bioerror** or **bioterror** event causing the release of a highly transmissible lethal agent. Two recent developments add fuel to his arguments: the unanticipated discovery that a genetically engineered **monkeypox virus** was uniformly lethal to mice and the

Talking the talk: evolution and the origins of language

Use of language is the most obvious trait that distinguishes humans from all other species. When, why and how humans evolved the ability to speak remains unclear. Unfortunately, neither speech nor the soft tissues that create it, fossilize. Palaeoanthropologists have tried instead to draw inferences from tenuous bony evidence. The **hyoid bone** at the base of the tongue may shed light on how the voicebox was built. Fossil evidence shows that Neanderthals possessed a hyoid bone indistinguishable from modern humans, but the implications for language abilities are not clear. Attempts have also been made to use fossils to assess the size of components of the nervous system involved in speech, but have so far proven inconclusive as well.

Koko, a Californian-born lowland gorilla, who is said to understand more than 1,000 signs and 2,000 English words.

synthesis of an **entire bacterial genome** from chemical precursors by American molecular biologist **Craig Venter** (b.1946) and his colleagues (proof of principle for anyone wanting to make smallpox from scratch).

Linguistic investigations have established that all human languages possess grammars of equivalent sophistication, all capable of what Wilhelm von Humboldt (brother of the explorer-naturalist) called **infinite use of finite means**. American linguist **Noam Chomsky** (b.1928) argues that language is a universal human trait, mediated by what he calls a **language-acquisition device**, preloaded with an innate **universal grammar**. Psychological investigations now focus on teasing out the linguistic, behavioural and cognitive precursors and prerequisites for human speech, such as the ability, shared with parrots, to imitate the vocal sounds of others. Ever-more sophisticated imaging techniques are likely to reveal the regions and circuits of the brain involved.

An alternative approach is to search for genes involved in language. One noteworthy candidate is **FOXP2**. In one human family, the **KE family**, mutations in this gene are associated with deficient coordination of the movements required for speech, despite normal intelligence. Brain scans have shown that these language deficits are not simply the result of poor muscle control; the impairments also include difficulties in comprehension. Crucially, FOXP2 has experienced an unexpectedly high number of gene changes since humans diverged from chimps.

Even with a human brain in charge, the chimp vocal tract could not produce sounds crucial to human speech. But language is not confined to speech. Several **apes** have been taught to use **American Sign Language**, starting in the 1960s with a female chimpanzee **Washoe** (1965–2007), and followed by a pair of gorillas **Koko** (b.1971) and **Michael** (1973–2000), male chimp **Nim Chimpsky** (1973–2000) and a bonobo, **Kanzi** (b.1980). Poignantly, Michael's keepers reported what they interpreted as his description of his own capture and the death of his mother. However, vociferous arguments have raged as to whether ape use of signs and symbols really constitutes language (key features like grammar and recursion seem to be missing).

Sign language also gained a prominent place in the origins of language debate from another angle. In the 1970s and 1980s deaf children in a number of schools in western Nicaragua spontaneously developed their own sign language. American sign-language expert Judy Shepard-Kegl has documented a remarkable transition in **Nicaraguan sign language** from crude pidgin to a higher level of sophistication, with verb agreement and other complex grammatical features. Evidence from apes and Nicaragua have led some to suggest that sign language might have preceded speech in the evolution of human language.

Wandering genes and the Jewish genome

Molecular geneticists have exploited mitochondrial (matrilineal) and Y chromosome (patrilineal) lineages to unravel the patterns of human migration inside Africa and across the world. Examples include the **Bantu expansion** into Africa south of the equator, the easterly spread of the **Polynesians** across the Pacific (chicken genetics suggest they made it from as far as South America), the **peopling of the Americas** from Siberia and the **Indian origins** of the **Romani people**. However, probably the most attention-grabbing collision of molecules with myths, flows from the application of these approaches to Jewish communities. Although scattered around the world after the destruction of the Jewish temple in AD 70, in a dispersal known as the **diaspora**, Jewish traditions place the historical and spiritual homeland of all Jews in the Middle East.

Y chromosomal studies support a predominantly **Middle Eastern ancestry** for modern Jewish populations, casting doubt on the theory championed by British novelist Arthur Koestler that most Eastern European (**Ashkenazi**) Jews were descendants of the **Khazars** (a Turkic people from the Caucasus who converted to Judaism in the eighth or ninth century); instead **Jews** and **Palestinians** are chromosomal brothers. However, although Jewish identity in modern times is defined according to maternal inheritance, mitochondrial DNA studies show multiple origins for Jewish **matrilineal lines**, suggesting that most Jewish communities were founded by Jewish men and local women.

In the late 1990s, two research groups, including a team led by British genetic anthropologist **Mark Thomas** (b.1964), showed that a large proportion of men from the Jewish priestly class (the Cohens or **Kohanim**) share a common set of Y chromosomal markers, known as the **Cohen Modal Haplotype**. Similar

Rees also evaluates the so-called **Doomsday Argument**. This is a probabilistic argument, first articulated by Australian astrophysicist **Brandon Carter** (b.1942) and later championed by Canadian philosopher **John Leslie** (b.1940). In simple terms, it states that if we assume that humans alive today occupy a "nothing-special" place in the timeline of human history, then the chances are that we are about halfway through humanity's lifespan. Whether this argument should be a cause for sleepless nights or whether it simply represents armchair sophistry remains a matter of lively debate. But even if true, the argument does not necessarily predict humanity's mass destruction – **pseudo-extinction** (our evolution into another species) remains a possibility.

The most precarious aspect of future human survival is the limitation of our species to a single planet. There are sound arguments for results were found in both Ashkenazim and Sephardic Jews (who fled persecution in the Iberian peninsula) confirming ancestral links between communities separated for over half a millennium. Crucially, Thomas's analysis places the most recent common ancestor of all Kohanim at around 3000 years ago, a date consistent with biblical accounts identifying that ancestor as Moses' brother **Aaron**. However, a similar study on the broader priestly caste, the Levites, suggested multiple male ancestries for these Jews.

Several far-flung populations claim Jewish ancestry. The most remarkable folk history comes from the **Lemba**, a black African Bantu-speaking people from Southern Africa, who follow apparently Jewish religious traditions and claim their ancestors left Judea 2500 years ago. Thomas has provided genetic support for Lemba beliefs by showing that nearly 10 per cent of Lemba males and over 50 per cent of Lemba priests carry the Cohen Modal Haplotype. Similar studies on the **Bene Israel** people from India, who claim to have left Galilee in the second century BC, support a Jewish ancestry. However, another group of black Jews, the **Beta Israel**, who were spectacularly airlifted to Israel during a series of covert operations, appear from genetics studies to be descendants of Africans who converted to Judaism. Clearly, mixing population genetics with the politics of personal identity remains a hazardous pursuit!

Jacob's Legacy: A Genetic View of Jewish History David Goldstein (Yale University Press, 2008)

Genes, Peoples, and Languages Luigi Luca Cavalli-Sforza (University of California Press, 2000)

establishing a programme of detection of, and even defence against, future asteroid impacts. However, although a few humans might survive a dinosaur-killer-sized impact, an impact of the size that created the Moon, capable of melting the crust, would signal humanity's end. As Martin Rees and innumerable science fiction writers have pointed out, humanity's best hopes for survival lie in the **colonization** of other **bodies** (Mars, the Moon, Venus). American astronomer **Carl Sagan** (1934–99) once outlined the possibility of **terraforming Venus** (rendering its atmosphere and environment like those of Earth's) by seeding its atmosphere with photosynthetic bacteria. Although Sagan's suggestions for Venus are now considered unworkable, the practicalities of **terraforming Mars** remain a subject of ongoing discussion and even research. The last word on future human evolution should go to **Konstantin Tsiolkovsky** (1857–1935), Russian pioneer of the space rocket, who, in 1911 (with Alfred Russel Wallace still alive), prophetically proclaimed: "The Earth is the cradle of humanity, but mankind cannot stay in the cradle forever."

A view of Mars's Victoria crater with the tracks of NASA's Exploration Rover. Could this bleak terrain be a possible new home for mankind?

Part 3
Impact

8 Other sciences

Although obviously rooted in biology, Darwin's theory of evolution had strong informative links with other scientific disciplines – such as geology and linguistics – from the very outset. Today, the influence of evolutionary thinking can be discerned in such diverse fields as cosmology, computer science, comparative psychology and economics.

Evolutionary physics and cosmology

In the closing words of *The Origin*, Darwin contrasted the evolutionary malleability of life with the apparently fixed laws of Newtonian physics. On a couple of occasions in his letters, Darwin brushes aside speculation on the origin of life with what he considered the ultimate put-down: "one might as well think of origin of matter". Today, in the wake of Einstein's relativity and the mysteries of quantum mechanics, the universe revealed by physics is no longer an eternal, unchanging, monotonous cosmic grandfather clock. Instead, matter, energy and the laws of physics have origins and histories; like life, they evolve over time. Some even champion a universal Darwinism that encompasses the physics of the very small (quantum mechanics) and the physics of the very large and very old (cosmology).

Quantum mechanics

The term "**quantum mechanics**" was coined in 1924 by Max Born to describe a puzzling new view of physics at the atomic and subatomic scale that emerged in the early twentieth century. The predictions made by this theory have been verified in astonishing detail. However, according to quantum mechanics, the classical laws of physics do not apply on the atomic scale. Instead, the world of the very small runs according to principles that seem utterly bizarre from the everyday perspective. For example, in the quantum world, energy comes in discrete packets called quanta, and particles and waves are seen as equivalent. Attempts to make sense of quantum mechanics and the paradoxes it entails have resulted in a number of alternative interpretations, the most widely accepted of which is the **Copenhagen interpretation**, articulated by Niels Bohr and Werner Heisenberg. According to this view, the act of measuring has a profound effect on what is being measured and observers get entangled in their own experiments. As one can never determine a particle's position and energy at the same time, the future can never been predicted with certainty, only probabilistically; until things are measured, they exist in some twilight zone, where alternative options are superimposed on top of one another. Einstein recoiled in horror from the Copenhagen interpretation, stating that "God does not play dice with the universe".

In the late 1950s, American physicist Hugh Everett (1930–82) proposed an alternative view, the **many worlds interpretation**, in which quantum paradoxes are resolved by a continuous splitting of the universe into parallel alternative realities. Proponents of Everett's interpretation include Israeli-born Oxford physicist **David Deutsch** (b.1953), a pioneer in the application of quantum mechanics to computing. In his 1997 book *The Fabric of Reality*, Deutsch proposes a "theory of everything" with four strands: Everett's many worlds interpretation, Karl Popper's epistemology (see p.239), Alan Turing's theory of computation and Darwin's theory of evolution as interpreted by Richard Dawkins.

Another leading expert on quantum mechanics, Polish-born physicist **Wojciech Zurek** (b.1951), has proposed a theory of **quantum Darwinism** (complete with quantum natural selection), to explain how the classical world emerges from the quantum world. According to Zurek, the environment selects a restricted set of "pointer states" of classical reality from the vast potentialities of the quantum world; the information in these pointer states undergoes reproduction and further evolution within the macroscopic realm. Thus, according to Zurek, his theory interweaves the crucial Darwinian processes of selection and reproduction.

If physical laws can change over time, then, one might also question whether the fundamental physical constants might have been different. In fact, these values appear to represent a series of remarkable coincidences. For example, if Newton's gravitational constant, or the values of the strong and electromagnetic forces in atomic nuclei, were only very slightly different, the universe would have been unable to support life. Cosmologist and science writer Paul Davies has coined the term "**Goldilocks enigma**" to encapsulate the unsettling conclusion that the laws of physics appear to be "just right" for life to exist and evolve in the universe. One response is to invoke the **anthropic principle** (a term coined by Australian physicist **Brandon Carter** in 1973), which explains away these apparent coincidences by pointing out that had the universe been different in any important way, we (i.e. intelligent life) wouldn't be here to observe and ask the question. This argument is often made against the backdrop of the **multiverse hypothesis**, which posits the existence of a very large, or even infinite, number of universes (an idea which is taken seriously by experts in cosmology but which, some argue, is unfalsifiable and therefore unscientific).

American theoretical physicist **Lee Smolin** (b.1955) has proposed **cosmological natural selection** (or the theory of fecund universes) as one rather speculative answer to the Goldilocks enigma. According to Smolin, new universes are born when stars collapse down into black holes, regions of space where gravity is so powerful that nothing can

Evolutionary chemistry: from Darwinism to drugs

Most therapeutic drugs work by binding to proteins and interfering with their function. A key challenge for chemists working in the pharmaceutical industry is to discover new medicinal chemicals that fit important protein targets, rather like a key fits a lock. The traditional way to do this is to take a long hard look at the protein "lock" and then rationally design a chemical "key" that fits it. However, analogies with biological evolution have recently inspired an alternative approach: **evolutionary chemistry**. Instead of attempting rational drug design, the evolutionary chemist simply generates a massive pool of variable DNA-like starting molecules (representing the variation that underlies biological natural selection). When these are then introduced to the target protein, only a small fraction of the molecules bind (the selection step). However, various chemical tricks then allow the chemist to amplify this population of molecules (the reproduction step). The amplified molecules are then used as the starting point for a subsequent round of selection and amplification. After several rounds of selection, the molecular mixture is greatly enriched for **aptamers**, molecules that bind tightly and specifically to the chosen target. This evolutionary approach has already led to the development of one useful drug, **Pegaptanib** (with the trade name Macugen), which has been licensed as a medicine to treat a common cause of blindness (age-related macular degeneration). But this is just the start: evolutionary chemistry is all set to deliver additional medically useful aptamers in the next few years that will target heart disease or cancer.

escape. The collapse of a black hole thus represents a kind of mirror image of the origin of our universe in the Big Bang, when space expanded from a point of infinite temperature and density. In addition to this form of reproduction, Smolin proposes a source of variation whereby new universes inherit some, but not all, of the properties (laws, constants, etc) of their parent universes. Those universes that are most fecund in terms of producing black holes which spawn new universes will, in an echo of Darwinian natural selection, out-compete universes that cannot produce black holes. As both life and the ability to form black holes are thought to emerge solely or predominantly in universes in which atoms and stars can form, cosmological natural selection might explain why the universe looks primed for life.

Evolution and the earth sciences

As science historian **Sandra Herbert** documents in *Charles Darwin, Geologist* (2005), Darwin made major contributions to geology. Today evolutionary thinking obviously permeates palaeontology and the distribution of primeval life forms helps geologists map the location and movements of ancient continents. However, when we consider the Earth's surface environment over geological time, we encounter the Goldilocks enigma on a planetary scale. For example, life depends on liquid water

The Earth remains hospitable to life despite changes in energy from the Sun.

on the Earth's surface. But as astronomer **Carl Sagan** pointed out in the 1970s, our planet was first formed under a **cool faint sun**, which provided only 70 percent of the energy we receive today. Had current atmospheric conditions prevailed at the outset, all the planet's water would have been trapped as ice. Instead, it seems that within an oxygen-free atmosphere, carbon dioxide and/or methane sustained a powerful greenhouse effect, warming the early Earth. However, without the changes in atmospheric composition wrought by life (especially the production of oxygen), as the Sun warmed up, temperatures on Earth might have soared way beyond those compatible with life. Also, how is it that our current atmosphere, with so much chemically reactive oxygen, has wandered so far from the equilibrium conditions sustained by chemistry alone, yet remains relatively constant in composition?

Gaia

In the 1970s **James Lovelock** (b.1919) proposed the **Gaia hypothesis** to explain how conditions on Earth have remained suitable for life for so long. His suggestion is that the whole planet, including the biosphere, acts as a single organism (named Gaia after the Greek goddess of the Earth), maintaining constant conditions just as humans and other animals do through feedback mechanisms. In animals this phenomenon is known as **homeostasis**; to take an example from human physiology, when cold, we shiver to warm ourselves up; when hot, we sweat to cool us down. Lovelock backed up his claim with a simple model, **Daisyworld**, in which the proportion of white and black daisies on a world maintains a constant temperature despite an increasingly hotter sun. The Gaia hypothesis has been criticized many counts, and at best, it implies a teleology at odds with Darwinian evolution (how does an impersonal planet *know* what temperature is required for life?). At worst, it substitutes unscientific New Age religious metaphor and anthropomorphism for testable rigorous mechanistic explanations (a problem epitomized in the title of Lovelock's latest book, *The Revenge of Gaia*). One could argue that the anthropic principle provides a simpler explanation for the prolonged existence of a life-sustaining planetary environment – without such conditions, we would not be here to ask the question! However, there is no doubt that Lovelock's hypothesis-metaphor has stimulated scientific discourse and provoked a re-examination of humanity's effect on the planet and prospects for a sustainable future; it has also permeated the creative arts, from science fiction to computer games.

Earth's non-equilibrium atmosphere represents a signature of life that could be detected by an alien civilization light years away. But what about life's influence on the geography of our home world? In a 2006 *Nature* paper, two Californian Earth scientists, William Dietrich and Taylor Perron, looked for **signatures of life** among the landscapes of our planet. While they argue that life clearly did influence some of the processes that shape our landscape (weathering, erosion, sediment transport), they conclude from comparisons between surface features on our planet and those on Mars that there is no unique topographical signature shaped by life on Earth; instead, life's effects are manifested only in changes in the distribution and scale of features drawn from a common topographical repertoire.

Evolutionary computation

Several concepts from evolutionary biology have been carried over into computer science, inspiring the birth of evolutionary computation. The roots of this movement reach back into the 1960s and 70s, to pioneering efforts in the USA by **Lawrence Fogel** (1928–2007) and **John Henry Holland** (b.1929) and in Germany by **Ingo Rechenberg** (b.1934) and **Hans-Paul Schwefel** (b.1940).

Evolutionary algorithms

For several decades, computer scientists have used evolutionary approaches (**evolutionary** and **genetic algorithms**) to "breed" solutions to problems or to improve designs. These methods draw heavily on concepts from biological evolution, such as reproduction, mutation, recombination, chromosomes, natural selection and fitness. A key feature of such evolutionary approaches is that they can deliver solutions to complex problems way beyond the intuition of their creators. Evolutionary algorithms have proven particularly useful in engineering, robotics and economics. Early examples of their use include improvements to the wing, fin and flap profiles of aircraft; the streamlining of cars; the design of jet turbines and control systems for gas pipelines. Stock traders have used them to predict the stock market. More recent applications include improvements to the design of aerials, lenses and optical fibres, speedier fine-tuning of cochlear implants and production of a USB memory stick that lasted thirty times longer than its progenitor.

How do these genetic algorithms work? In effect, the computer program creates a virtual world in which solutions to a problem are engaged in a computational struggle for existence. The first step is to encode possible solutions and their features in a mathematical representation that can be exploited by a computer program. A set of progenitor solutions plays the role of individuals in a founder population, which are then left to breed and reproduce over a series of generations. Novelty is injected into the system through **mutation** (random change in a feature) and **recombination** (where features from existing solutions are combined to produce new candidate solutions). The part of natural selection is played by a mathematical **fitness function**, which is used to evaluate which of the solutions best address the problem and so should be allowed to breed. Over the course of thousands, or even billions, of generations, useful features are combined in ways that would never have occurred to a human designer. At the end of the process, the fittest solutions represented within the computer program provide strikingly effective answers to real-world problems. Darwinism reaches designs that intelligent, human designers cannot reach!

Getting Alife

The term "**artificial life**" was coined in the late 1980s by American biologist Chris Langton. Artificial life (or **Alife**) exploits evolutionary algorithms and computer models to capture the logic of living systems, to simulate ecological and biological evolution and to evaluate theories about how

Life in black and white

In 1970, British mathematician **John Conway** (b.1937) described what has become the iconic cellular automaton, **the game of life**. Conway's game is played on an infinite two-dimensional grid of square cells, each cell a live or dead. The game is seeded with an initial pattern. Like squares on a chessboard, each cell interacts with eight neighbours. At each step forward in time, the state of the grid is determined by simple rules: any live cell with two or three live neighbours survives unchanged to the next generation; any live cell with fewer than two or more than three live neighbours dies; any dead cell with exactly three live neighbours comes to life. Conway's game of life has attracted much interest because of the surprising ways in which complex and unpredictable patterns evolve, without design, in a universe built from rules scarcely more complex than those of tic-tac-toe – even amateurs dabbling in the game recognize patterns that appear to move through the game of life universe, while experts have shown that the universe can even be used to build a theoretical general purpose computer.

evolution works, at the micro-evolutionary and macro-evolutionary levels. The simplest models rely on **cellular automata**, theoretical worlds consisting of cells arranged in a grid, where each cell can adopt one of two or more states (say white or black) and with each step forward in time changes its state according to rules that take as input the states of neighbouring cells.

In the early 1990s, Alife emerged as a fully fledged discipline, with the development of sophisticated computer simulations such as **Tierra**, developed by ecologist **Tom Ray** (b.1954) and **Avida**, developed by Charles Ofria, Chris Adami and Titus Brown at the California Institute of Technology (Caltech). In these digital ecosystems, computer programs act as **digital organisms**, evolving through mutation and recombination while competing for computer time and memory. Numerous other digital organism simulators have been developed more recently, with colourful names like **Darwinbots**, **Evolve 4.0**, **Darwin Pond**, **TechnoSphere** and **Noble Ape**. Life simulation games have also become a key part of the commercial computer games landscape. Alife enthusiasts such as **Richard Lenski** (b.1956) and Chris Adami have replicated many key evolutionary phenomena using digital organisms, including host-parasite co-evolution, adaptive radiations, punctuated equilibrium and the evolution of complex features.

Proponents of Alife divide into those that believe that it merely mimics biological systems (the **soft Alife** position) and those that propose that digital organisms really are alive (the **hard Alife** position, enunciated by Tom Ray). The term **computer virus** was coined in 1983 to describe self-replicating computer programs that can infect computers. Since then, a

Adventures in synthetic biology

Synthetic biology is a new fusion of biology, chemistry, computer science and engineering in which novel biological systems are designed and built from the bottom up, ignoring the constraints of naturally evolved systems. Unlike conventional genetic engineering, synthetic biology starts with the equivalent of a blank sheet of paper, before going on to assemble an artificial **molecular toolkit** of "biological parts". For example, Boston University's Tim Gardner and Jim Collins have built a genetic toggle switch, while researchers at the Massachusetts Institute of Technology maintain a **Registry of Standard Biological Parts** (parts.mit.edu), complete with measurement systems and signalling devices. A team from Texas has re-engineered bacteria to produce living photographic film. Another recent headline achievement was the synthesis in the test tube of an entire bacterial chromosome from chemical precursors by American molecular biologist Craig Venter and his team. It is likely that future efforts in this field will extend and even transcend the capabilities of natural organisms.

relentless arms race has occurred between the creators of computer viruses and those devising virus protection software. Physicist Stephen Hawking has even suggested that computer viruses count as living organisms: in other words, Alife in the wild. However, computer viruses are clearly designed entities, stemming from the human intellect. The probability of their evolving through natural selection is low because the underlying computer programs are "brittle" (i.e. almost any random change to a functional program will render it non-functional) and copying is usually perfect. However, we cannot rule out the possibility that someone will one day devise a free-living computer virus that can evolve (hackers, don't take that as a challenge!).

Evolutionary thinking in economics

Just as Darwin is seen as the father of modern biology, the Scottish economist **Adam Smith** (1723–90) is credited with laying the foundations of economics with his magnum opus *The Wealth of Nations* (1776). In this highly influential book, Smith analyzed and defended free market economics, arguing that, although a free market appears chaotic and free of restraints, with each man acting for himself, the market is in fact steered by an **invisible hand** to a rational outcome, producing just the right amount and variety of goods at the right price. Although Smith used the phrase only once in his work on economics, the invisible hand proved a powerful metaphor that influenced many subsequent thinkers, including Darwin (who read Smith and repeatedly used

Adam Smith, whose economics and ethics influenced Darwin.

the phrase "the economy of nature"). There is a clear analogy between the undirected effects of Smith's invisible hand in the creation of wealth and the designer-free biological adaptations forged by Darwin's natural selection. Similarly, competition can be seen as a driving force in both contexts – a fact seized on in social Darwinism (see p.264) – and diversification in an economic setting is analogous to adaptive radiations seen during biological evolution.

Evolutionary economics

In recent years, evolutionary biology has re-paid its debt to economics by priming the birth of **evolutionary economics**. In contrast to conventional economics, which draws on metaphors from Newtonian physics (labour force, elasticity, velocity of money, economic equilibrium, etc), evolutionary economics draws on a more fluid and open-ended interpretation of human nature, rational decision-making and the objects of choice. Following **Joseph Schumpeter** (1883–1950), who argued that any economic equilibrium is subject to repeated disruption by innovation, Liverpool-born maverick economist **Kenneth Boulding** (1910–93) articulated a more explicitly evolutionary model, particularly through his books *Ecodynamics* (1978) and *Evolutionary Economics* (1981). According to Boulding, "economics and evolution, are both examples of a larger process ... the development of structures of increasing complexity and improbability. The evolutionary process always operates through mutation and selection and has involved some distinction between the genotype which mutates and the phenotype which is selected ... Economic development manifests itself largely in the production of commodities, that is, goods and services. It originates, however, in ideas, plans, and attitudes in the human mind. These are the genotypes in economic development." In recent years, several other experts have argued for an explicitly Darwinian paradigm in economics, including American duo **Richard Nelson** and **Sidney Winter** and **Geoffrey Hodgson** from the UK's University of Hertfordshire. At the interface between sociology and economics, **Howard Aldrich** has explored the implications of evolutionary thinking in understanding organizations. In his influential *Organizations and Environments* (1979) and *Organizations Evolving* (1999), Aldrich weighs up the contributions of variation, reproduction and selection to produce non-deterministic outcomes to the question of which organizations disappear and which survive and thrive.

For more than half a century, evolution and economics have shared common ground in their exploitation of **game theory** to explain behavioural strategies. More recently, evolutionary biologists and students of economics, together with those in the political and social sciences have begun to exploit **agent-based models** to understand the origins of complex behaviours and systems. These approaches employ simulations of multiple autonomous decision-making agents engaged in dynamic interactions to show how complexity can emerge from relatively simple rules. One key advantage of such approaches is that they can model emergent phenomena: counterintuitive situations where the system is more than the sum of its parts (for example, a traffic jam that is propagated in the opposite direction to the movement of the cars that form it). In a recent *Nature* paper, Russian evolutionary biologists **Mikhail Burtsev** and **Peter Turchin** have exploited agent-based models in an attempt to explain the evolution of cooperative strategies from first principles. Elsewhere in biology, agent-based models have been used in settings as diverse as bacterial genetics, cancer biology and the evolution of human birth weights and mating behaviour.

Psychology and social behaviour

In the concluding chapter of *The Origin*, Darwin made a pregnant prediction: "Psychology will be based on a new foundation, that of the necessary acquirement of each mental power and capacity by gradation." In his later works *The Descent of Man* and *Expression of Emotions in Man and Animals*, Darwin argued for psychological continuities among all races of humans and between humans and animals. Darwin's young friend and colleague **George Romanes** (1848–94) laid the foundations of **comparative psychology** with works such as *Animal Intelligence* (1881), *Mental Evolution in Animals* (1883) and *Mental Evolution in Man* (1888). **Behavourism**, a school of psychology dominant early in the twentieth century, relied on the premise that studies of animal behaviour had relevance to human psychology – encapsulated in the old joke, "what is the difference between a behaviourist and a conjuror?", "One pulls rabbits out of hats, the other habits out of rats!". However, the behaviourists dismissed the subjective inferences about mental states reported by Darwin' and Romanes' efforts as mere anecdote and instead emphasized the need to observe reproducible behaviour in controlled settings.

Ethology

The science of animal behaviour (**ethology**) blossomed in the 1920s through the work of Dutch zoologist **Nikolaas Tinbergen** (1907–88) and Austrian animal psychologist **Konrad Lorenz** (1903–89). Tinbergen placed evolution centre-stage in ethology with what are now known as **Tinbergen's four questions** (adapted from an analysis of causality by Aristotle). Two of these look to evolutionary (ultimate) explanations. Question one relates to **adaptation**: what is the selective advantage of the behaviour? As in the fact that birds migrate in winter to find food. The second question to **phylogeny**: what is the evolutionary history of the behaviour? How does it compare with what is seen in related species? For example, why do Siberian stonechats migrate, while European stonechats do not? The next two questions look for proximate explanations, closer to the facts. Number three deals with **causation**: what are the stimuli that elicit the response and the mechanisms that underlie it? Migratory birds show a restlessness (named Zugenruhe) even when enclosed, suggesting an in-built clock. Question number four is about **development**: does behaviour change with age, or depend on early experience? The reintroduction to the USA of endangered whooping cranes suggests the answer is the latter, since they had to be shown how to migrate by following a microlite plane.

Konrad Lorenz made several important contributions to ethology. He is best remembered for his studies on **imprinting**, a kind of rapid learning restricted to a short phase in an organism's life, for example, when a young animal learns the characteristics of its parent. Lorenz showed that greylag geese would imprint on the first moving object they saw – as a result, he is often depicted being followed by a gaggle of geese which had imprinted on him! Lorenz also described **fixed action patterns** – instinctive stereotyped sequences of behaviour, evoked by what he called a signal stimulus (for example, yawning in mammals, evoked by seeing others yawn).

In 1966 Lorenz published *On Aggression,* in which he described not only animal violence but also human aggression. Other science writers also evaluated human behaviour in the light of animal studies. In *The Naked Ape* (1967), English zoologist **Desmond Morris** (b.1928) described humans as an unknown species of ape, complete with a detached account of human sexual behaviour. In a series of books spanning the 1960s and 70s, American writer **Robert Ardrey** (1908–80) proposed the **hunting hypothesis** – the need to hunt on the savannah drove human evolution – and the **killer ape hypothesis** – the idea that the aggression needed for hunting was a key feature that distinguished humans from their ancestors. Although neither

Austrian animal psychologist Konrad Lorenz showed that geese would imprint on the first moving object they saw, which was sometimes even Lorenz himself.

hypotheses holds much weight for today's palaeanthropologists, Ardrey, as well as Desmond, helped to focus attention on evolutionary explanations for contemporary human behaviour.

Sociobiology

In 1975, **Edward O. Wilson** (b.1929), an American entomologist specializing in ants, tossed an intellectual grenade into the marketplace of ideas with his book ***Sociobiology***. Wilson defined the subject of his book as "the systematic study of the biological basis of all social behavior" and in it he attempted to explain animal behaviour in the light of evolution and genetics. Where Darwin speculated in his notebooks that even an oyster might have free will, Wilson proposed instead that the behaviour of all animals is restrained by "a genetic leash", whereby their behaviour is in large part predetermined by their genes (a position known as **biological** or **genetic determinism**). Had he stuck to ants, Wilson's views would not have upset anyone. Instead, in the final chapter of *Sociobiology* and in a subsequent book, *On Human Nature* (1979), Wilson applied his ideas to humans, challenging the prevailing dogma that the human mind is a blank slate, unfettered by genetic constraints. In so doing, Wilson unleashed a barrage of criticism, chiefly because his conclusions appeared to chime with the conservative political ideology.

In 1984, a geneticist-neurobiologist-psychologist trio, **Richard Lewontin**, **Stephen Rose** and **Leon Kamin**, responded with a left-wing critique of sociobiology and biological reductionism, ***Not in Our Genes***, in which they pointed out that scientists are at risk of political and cultural bias and that many of the recent trends in genetics, psychology and evolutionary biology could be seen as handy rallying points for an emerging right-wing political agenda. They also launched a methodological attack on the twin studies and adoption studies beloved of psychologists and took sociobiology to task for its reliance on just-so stories and over-emphasis on genes and adaptations. Drawing on their Marxist roots, they proposed a "dialectical" approach to human nature, capable of integrating many different disciplines.

In response, **Richard Dawkins** lambasted Lewontin, Rose and Kamin's efforts. In a scathing review in *New Scientist*, he dismissed *Not in Our Genes* as a "silly, pretentious, obscurantist and mendacious book", attacking its leaden left-wing language and dismissing its straw-man views of genetic reductionism and determinism as unrepresentative of anything actually said by the professional scientists who might call themselves sociobiologists.

Evolutionary medicine

In an ideal world, no one would ever get sick. Proponents of **evolutionary** or **Darwinian medicine** integrate medicine with evolutionary biology in asking why a disease occurs, when natural selection "should" have weeded it out. In fact, some argue that training in evolutionary thinking is so good at getting researchers and clinicians to think "outside the box" that evolution should be a key part of any medical school's curriculum. We have seen how evolutionary thinking has primed advances in our thinking about the immune system and cancer (see p.113). But the scope of evolutionary medicine is vast, impinging on any disease that apparently decreases reproductive fitness but nonetheless occurs commonly in human populations – everything from pre-eclampsia and polycystic ovaries to schizophrenia and depression.

One obvious example of evolution in the medical arena is the evolutionary arms race underway between pathogens on the one hand and our immune systems, antibiotics and infection control policies on the other. But evolutionary medicine challenges an oft-quoted old myth that long-standing pathogens always evolve towards lower virulence. Instead, in cases where insects carry disease, keeping hosts so ill that they can't avoid insect bites might be advantageous to the pathogen. Many diseases represent the results of evolutionary trade-offs of costs versus benefits. Thus, **sickle-cell anaemia** is common among people of West African descent because a mild form of the condition once provided a selective advantage in areas where severe malaria was common.

Evolutionary medics, like evolutionary psychologists, look to understand modern human problems in the context of the **environment of evolutionary adaptedness (EEA)**. Thus, the current high rate of heart disease, obesity, hypertension and diabetes in Western societies has to be explained against the backdrop of an adult and even foetal physiology crafted by the demands of a Pleistocene lifestyle and diet. Controversially, on these grounds, some have even advocated a return not just to a pre-industrial diet, but even to a **Palaeolithic diet**. The **hygiene hypothesis** suggests an evolutionary explanation for an increase in allergies and asthma: in the EEA, our immune systems encountered a range of bacteria and parasites that have been eliminated from our modern hygienic lifestyles. According to this hypothesis, this recent departure from our ancestral state has provoked a deregulation of the immune system triggering a dysfunctional response to harmless allergens.

However, despite its broad scope and fashionable appeal, evolutionary medicine is not yet a mature discipline. One central challenge is to replace just-so stories with fertile testable hypotheses. Watch this space!

Evolutionary psychology

The discipline of **evolutionary psychology** is a contemporary successor to Wilson's sociobiology, and elicits similar intellectual and emotional responses from its critics (for example Hilary and Steven Rose's *Alas, Poor Darwin: Arguments against evolutionary psychology*). The overarching aim of the movement (to explain mental and psychological features as adaptations fashioned by natural selection) is less controversial than some of its working assumptions. One such assumption is the Swiss army knife view of the human mind, i.e. that the mind consists of a number of distinct and specialized modules, each equipped to perform a discrete task. For example, in his 1994 book *The Language Instinct*, American experimental psychologist **Steven Pinker** (b.1954), argues eloquently that the human capacity for language is a specialized adaptive instinct, crafted by natural selection. However, critics, including Canadian psychologist Merlin Donald, suggest that natural selection, twinned with the emergence of culture, has instead favoured the evolution of the mind as a general problem-solving device (perhaps like a computer which can run an almost infinite number of programs).

A second working assumption of evolutionary psychology is that human nature is best explained as a set of evolved psychological adaptations to recurrent problems in the environment of our pre-agricultural hunter-gatherer ancestors (the **environment of evolutionary adaptedness** or EEA). Husband-and-wife pioneers of the field, **Leda Cosmides** and **John Tooby** put it bluntly: our modern skulls house a stone-age mind. Critics complain that we know little about the environment in which our ancestors lived. However, in many cases we know enough for the EEA concept to work. For example, our liking for sugary, salty and fatty foods was clearly adaptive in a past when such foods were scarce, even though this predisposition is now responsible for epidemics of obesity, hypertension and tooth decay. Similarly, the fact that most of us have a greater fear of spiders or snakes than of motor vehicles (which today kill far more people) is best explained as an adaptation that worked well in the EEA.

Evolutionary psychology has encompassed a wide range of fascinating and provocative topics, including evolutionary explanations for mental illness or for the abuse of step-children; numerous issues in mate choice and preference (for example the **cad-dad dichotomy**, where women prefer masculine men, cads, when ovulating, but go for less butch partners, dads, at other times in their menstrual cycles); the detection of cheating and the origins of altruism; and even the evolutionary origins of religion, ethics and the law. However, the most controversial publication in this field must be *A Natural History of Rape* (2000) by Randy Thornhill and Craig Palmer, where the authors speculate as to whether male-on-female **rape** is an adaptation in its own right (i.e. increases reproductive fitness) or whether it represents a byproduct of another adaptive behaviour (such as aggression or dominance). While the question itself may be admissible and much of the book is concerned with how to prevent rape, many have understandably objected to Thornhill and Palmer's strident claim that rape is "a natural, biological phenomenon that is a product of the human evolutionary heritage…"

Nonetheless, recent research studies have produced unsettling evidence of links between human biology and the risks and costs of rape: a 1998 study concluded that female students were less likely to engage in risk-taking behaviour while ovulating, while a 2001 survey of rape victims suggested that women who were raped were twice as likely to get pregnant as those who engaged in a single act of consensual unprotected sex.

Spicy selection: Darwinian gastronomy

One of the goals of evolutionary psychology is to explain food preferences and aversions. Why we like sweet foods seems obvious – they provided scarce calories in the EEA. But why do we like spicy foods? Cornell ecology professor Paul W. Sherman and his former student Jennifer Billing suggest a Darwinian anti-microbial hypothesis: spices help rid foods of pathogens and so in reducing the risk of food poisoning contribute to the health and reproductive success of people who find their flavours enjoyable. In defence of their hypothesis, they point out that many spices do indeed have anti-microbial properties. They have also shown the use of spices is commonest, regionally or globally speaking, where the ambient temperature (and thus risk of food spoilage) is highest. Thus, one could argue that the use of spices represents domestication of a plant's defensive antimicrobial compounds. More speculatively, Sherman and Billing suggest that spicy recipes might evolve over time in evolutionary arms races between recipe and pathogen. The fact that spices in today's societies often act as vehicles for food-poisoning bacteria is perhaps evidence against the anti-microbial hypothesis. But Sherman and Billing's suggestion is the best explanation yet for Britain's new national dish – the vindaloo curry!

Spice: a domesticated weapon against bacteria.

Linguistics

In the 1780s, **William Jones** (1746–94), an English lawyer stationed in India, noticed the remarkable similarities between the ancient Indian language, Sanskrit, and the European classical languages Latin and Greek. In an insight that marks the birth of **comparative linguistics**, Jones proposed that all three languages "have sprung from some common source, which, perhaps, no longer exists". Systematic comparisons in the early nineteenth century by Bavarian linguist **Franz Bopp** (1797–1861) and German orientalist **Max Müller** (1823–1900) fleshed out the theory that most European and Indian languages belonged in a single family of **Indo-European languages**. Curiously, Darwin's first appreciation that the Earth was more than 6,000 years old came, not from geology but from comparative linguistics: just before the *Beagle* voyage, one of his letters records speculation that Chinese and European languages share an common ancestry so ancient as to invalidate biblical chronologies.

August Schleicher and family tree theory

In the mid-nineteenth century, German language expert **August Schleicher** (1821–68) saw languages as organisms and started to borrow ideas from Linnean taxonomy in his linguistic classifications. In the decade before Darwin published *The Origin*, Schleicher embarked on a genealogical classification of languages (**Stammbaumtheorie**, or family tree theory) that incorporated a branching-tree model, uncannily similar to Darwin's and Wallace's view of biological evolution.

In *The Origin* Darwin makes analogies between biological and linguistic evolution: "It may be worthwhile to illustrate this view of classification, by taking the case of languages. If we possessed a perfect pedigree of mankind, a genealogical arrangement of the races of man would afford the best classification of the various languages now spoken throughout the world … The various degrees of difference in the languages from the same stock, would have to be expressed by groups subordinate to groups; but the proper or even only possible arrangement would still be genealogical…"

When Schleicher read *The Origin*, he immediately saw its relevance to his own work on languages. In 1863, Schleicher published a paper, *Darwinism and the Science of Language*, in which he drew a tree of Indo-European languages and commented on Darwin's work: "First, as regards Darwin's assertion that species change in course of time, a process repeated time and again which results in one form arising from another, this same

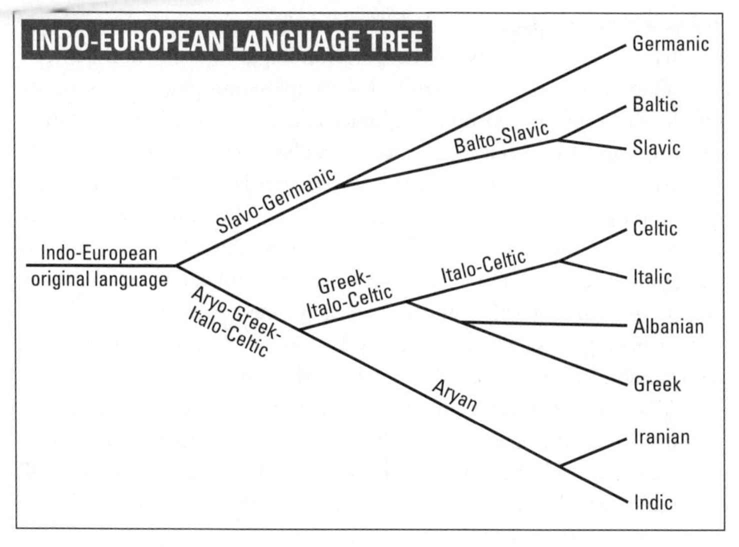

Schleicher's original language tree (modern versions differ).

process has long been generally assumed for linguistic organisms... We set up family trees of languages known to us in precisely the same way as Darwin has attempted to do for plant and animal species."

In return, Darwin referenced Schleicher and Müller in *The Descent of Man*, and elaborated at length the parallel between languages and species: "We find in distinct languages striking homologies due to community of descent, and analogies due to a similar process of formation. The manner in which certain letters or sounds change when others change is very like correlated growth. We have in both cases the reduplication of parts, the effects of long-continued use, and so forth. The frequent presence of rudiments, both in languages and in species, is still more remarkable." Bizarrely, Darwin's intellectual opponent, Swiss-born Harvard geologist **Louis Agassiz** (1807–73) accepted the analogy between biological and linguistic evolution, but was then driven to deny that Latin, Greek and Sanskrit shared a common ancestor, because he refused to accept biological evolution!

Now that molecular biology has revealed the language of life in the form of protein and DNA sequences, the parallels between historical linguistics and biological evolution are more cogent than ever. Where biologists talk of **homologous** protein sequences, linguists talk of **cognate** words – in both cases, similarity reflects common ancestry. In both settings, a branching-tree model of lineage splitting applies, consistent with a hierarchical groups-within-groups classification. The what-is-a-species question parallels the what-is-a-language debate. The historical linguist's distinction between innovations and retentions mirrors the symplesiomorphies and synapomorphies of cladistics. In both life and language, transfer of information (words or genes) between lineages can complicate the simple genealogical picture (for example, English, although a Germanic language, contains many **word borrowings** from French, Latin and Greek). In both fields, attempts have been made to devise clocks, based on assumptions of constant rates of change, and use these to time key historical events, although applications of this approach in linguistics (**glottochronology**) remain more controversial than in biology. Just like evolutionary biologists, historical linguists can frame testable hypotheses about the origin and evolution of the features they study. The most celebrated example comes from Swiss linguist **Ferdinand de Saussure** (1857–1913), who in 1879 hypothesized that early Indo-European languages possessed a particular kind of consonant (the laryngeal), a prediction confirmed when texts of the ancient Hittite language were deciphered in the early twentieth century.

As with taxonomists, those who classify languages can be categorized as either lumpers or splitters. The most controversial lumper was American linguist **Joseph Greenberg** (1915–2001), who classified all the native languages of the Americas into just three groups, one of which (Eskimo-Aleut) he later lumped in with Indo-European and several other large groups into one massive **Eurasiatic** language family. Although criticized by linguists, Greenberg's views have found favour with Italian geneticist **Luigi Cavilli-Sforza** (b.1922), who has for twenty years integrated linguistic and genetic approaches in an attempt to unravel the history of human migrations.

A new synthesis of biology and linguistics?

In recent years, algorithmic and computational approaches from molecular phylogenetics have been applied to history-of-language studies in what Auckland psychologist Russell Gray has called the new synthesis of biology and linguistics. However, papers on this topic have drawn sceptical responses from more traditional linguists. One thing is clear, by contrast with biologists, experts in historical linguistics have been thwarted in their attempts to show a common ancestry for all human languages by the shallow look-back time of current methods (<10,000 years). Thus, the ancestral language **Proto-World** will forever remain more elusive than life's Last Universal Common Ancestor.

In addition to sharing descent with modification with living organisms, do languages also experience natural selection? Darwin thought so, writing in *The Descent of Man*: "The survival or preservation of certain favoured words in the struggle for existence is natural selection." More recently, this idea has been explored by **Steven Pinker**, who in his *Words and Rules* (1999) examines the struggle between rules/symbolic computation and irregularities/memory in language, and by Mark Pagel and his colleagues, who in a 2007 paper showed how the frequency with which a word is used predicts the rate of its evolution.

One final resemblance haunts linguistics and biology: **extinction**. As Darwin noted in *The Descent of Man*: "Dominant languages and dialects spread widely, and lead to the gradual extinction of other tongues. A language, like a species, when once extinct, never, as Sir C. Lyell remarks, reappears" (the revival of Hebrew just after Darwin's death is the only credible counter-example to this assertion). Sadly, thanks to globalization, the rate of extinction has in recent years increased dramatically for both languages and biological species; half of the world's six thousand or so languages will be extinct within a hundred years.

Manuscript evolution

Before the invention of printing, texts were not accessible through the original manuscripts written and approved by the authors themselves. Instead, texts were preserved through copies of the originals or more commonly by copies of copies of copies, many steps away from the original. During repeated copying, texts accumulate errors, with the result that manuscripts exhibit "descent with modification", similar to that seen with living organisms. A tree that captures the branching patterns of manuscript genealogy is called a **stemma** (plural stemmata) and the construction and study of such trees is termed **stemmatics**.

From stemmatics to phylogenetics

Although as long ago as 1508, Dutch humanist theologian **Erasmus** (c.1469–1536) argued that manuscripts were genetically related, the first stemmata appeared in the early nineteenth century. In 1827, Swedish jurist **Carl Johan Schlyter** (1795–1888) presented an analysis of manuscripts of a Swedish legal document in which he attempted to "present their affinities, as far as we could determine them from mutual agreements and differences, in a kind of family-tree". A few years later, German philologist **Karl Zumpt** (1792–1849) published a genealogy of the known copies of Cicero's *Verrine Orations*. However, the credit for the birth of stemmatics usually goes to another German philologist, **Karl Lachmann** (1793–1851), who pioneered a scientific systematic approach to textual criticism, culminating in his 1850 edition of the works of the Roman philosopher Lucretius. In his book *Textual Criticism* (1927), German classicist **Paul Maas** (1880–1964) described the Lachmannian approach as a two-step process: an initial phase, in which all "errors" are collated and a stemma generated according to the principle "community of error implies community of origin", followed by a second phase in which the original text is reconstructed.

Stemmatics did not meet with universal acclaim among textual editors. One potential problem is that material can be copied from one manuscript tradition to another, which, analogous to horizontal gene transfer, can invalidate the assumption of branching evolution. However, most opposition centred on contradictory views as to the role of the expert in stemmatics: some recoiled emotionally from an emphasis on mechanical procedures that threatened to take human expertise out of the loop, while others complained that the method was not mechanical enough, in relying on expert opinion to distinguish original from erroneous readings.

Two scholars at the University of Chicago, **John Manly** and **Edith Rickert**, improved on Lachmann's approach in their eight-volume edition of *The Text of the Canterbury Tales* (1940). Instead of making subjective judgements as to which readings were original and which the result of copying errors, Manly and Rickert suggested that manuscripts should be grouped "according to their readings without reference to whether the readings are correct or incorrect". In a 1977 paper, comparative biologist **Norman Platnick** (an expert on spiders) and classicist **Don Cameron** (an expert on ancient Greek historian Thucydides) highlighted deep analogies between cladistics in biology and textual stemmatics, pointing out that both rely on recognition of kinship by detection of shared innovation (synapomorphy and community of error, respectively). Towards the end of the twentieth century, computer analyses of manuscripts became possible, fuelling a creative transfer of methodologies used to draw trees from molecular sequences to a **neo-Lachmannian approach** to manuscript evolution (also called the **new stemmatics**).

In 1991, Australian-born textual scholar **Peter Robinson** set a *Textual Criticism Challenge*, in which computer-based approaches were pitted against his own laborious traditional efforts in creation of a stemma of the Old Norse narrative *Svipdagsmal*. American evolutionary biologist **Robert**

The evolution of biblical manuscripts

In an ironic twist of fate that might infuriate creationist fundamentalist Christians, evolutionary thinking dominates scholarly studies of biblical manuscripts, particularly attempts to reconstruct original texts of the New Testament in the face of copying errors. The New Testament of the King James Bible is a seventeenth-century English translation of the *Textus Receptus*, a Greek text prepared by Dutch theologian Erasmus in the sixteenth century from a few late-medieval manuscripts. In the late nineteenth century, Birmingham-born theologian **Brook Westcott** and his Dublin-born collaborator **Fenton Hort** tried to improve on the *Textus Receptus*, publishing *The New Testament in The Original Greek* (1881), which incorporated information from a wide range of manuscripts, including the oldest fragments known at the time. Crucially, they adopted a genealogical view of **manuscript affiliation** that directly parallels the tree-like branching descent with modification seen in Darwin's theory of evolution. In their own words: "All trustworthy restoration of corrupted texts is founded on the study of their history, that is, of the relations of descent or affinity which connect the several documents." However, Westcott and Hort also recognized the potential for horizontal transfer between lineages, viewing the Byzantine textual lineage as a fusion of the two earlier traditions (the western and Alexandrian).

In the early twentieth century, British theologian **Burnett Streeter** proposed a **theory of local texts**, in which textual traditions diverged as a result of geographical separation – a parallel with allopatric speciation in evolutionary biology. From the 1950s onwards, American biblical scholar Ernest Colwell attempted to bring quantitative methods into the analysis of New Testament textual traditions. Cladistic approaches borrowed from evolutionary biology now sit at the cutting edge of studies of New Testament manuscripts: exponents include **David Parker,** a theologian at the University of Birmingham, **Gerd Mink** at the Institute for New Testament Textual Research, Münster, Germany and, among American scholars, **Stephen Carlson**.

O'Hara won the challenge hands down using software developed for phylogenetic analysis of biological sequence data (PAUP). Subsequently, many textual scholars abandoned attempts at guessing what the ancestral manuscript looked like and stopped trying to add a root to their phylogenetic trees. Robinson (now a colleague and collaborator of the author at the University of Birmingham) and Cambridge-biologist **Chris Howe** have exploited phylogenetic approaches in their analyses of *The Canterbury Tales* (www.canterburytalesproject.org). In a provocative fusion of biology and textual scholarship, plans are now afoot at Birmingham to analyse Darwin's six editions of *The Origin of Species* as if they were variant "textual genomes".

Philosophy and the arts

From the very beginning, Darwin's ideas fired the imaginations of not just scientists and political thinkers but also of philosophers, novelists, poets and even musicians. Their responses have ranged from the serious and profound to the bizarre and frequently humorous. The following chapter is a selection of some of the more interesting ways in which evolutionary ideas have been developed, mangled and generally played around with.

Philosophy

The pragmatist philosophers were the first to draw parallels between advances in knowledge and Darwinism. American philosopher **William James** (1842–1910) started his 1880 lecture *Great Men and their Environment* with the statement "A remarkable parallel … obtains between the facts of social evolution on the one hand, and of zoölogical evolution as expounded by Mr. Darwin on the other." Fellow pragmatist **John Dewey** (1859–1952) closed his 1909 essay *The Influence of Darwinism on Philosophy*, "Old questions are solved by disappearing, evaporating, while new questions … take their place. Doubtless the greatest dissolvent in contemporary thought of old questions, the greatest precipitant of new methods, new intentions, new problems, is the one effected by the scientific revolution that found its climax in *The Origin of Species*."

Evolutionary epistemology

Evolution has impacted on one of the central disciplines of philosophy, epistemology – the study of what it means to know something and how we know what we know and why. In the 1930s, Austrian-born philosopher of science **Karl Popper** (1902–94) suggested that scientific theories were tested by repeated attempts at falsification, with those that survive longest having the best explanatory power. In later life, Popper made explicit comparisons between the advance of scientific knowledge and evolution by natural selection, most explicitly in his 1972 book *Objective Knowledge: An Evolutionary Approach*. In Popper's view, an initial generation of diversity (imaginative formulation of conjectures) is followed by winnowing (elimination of erroneous theories by falsification) leading to increased fitness (better explanations and more interesting problems). However, as in biology, current fitness is no guarantee of long-term future survival – an apparently adequate theory can still go "extinct" in the light of new findings. In fact, rather fancifully, Popper suggests that "our theories ... suffer in our stead in the struggle for survival of the fittest".

Karl Popper, who made falsifiability a hallmark of scientific theories.

Several others have championed a selectionist approach to human intellectual progress. American social scientist **Donald Campbell** (1916–96), coined the term "evolutionary epistemology" to describe approaches to epistemology that draw on natural selection. However, it is important to distinguish between two alternative approaches: those that follow Popper in seeing natural selection as an analogy for how science works (sometimes termed the **evolutionary epistemology of theories** or **EET**) and those that use natural selection as an explanation for how we interpret the world (the **evolution of epistemological mechanisms** or **EEM**).

Within the realm of EET, British philosopher **Stephen Toulmin** (b.1922) sees concepts struggling in a "forum of competitions" while for American philosopher **David Hull** (b.1935), Darwinian principles apply not just to scientific ideas but also to the careers of the scientists that formulate and test them: "Science works as well as it does because the selfish goals of individual scientists happen to coincide with the manifest goal of the institution, the increase of empirical knowledge." Hull, alongside Richard Dawkins, generalizes a selfish-gene view of natural selection to conceptual selection. For Hull, all selection processes, natural selection and conceptual selection, involve **replicators** and **interactors**. Replicators are the things that copy themselves (genes or memes). An interactor is an "entity that interacts as a cohesive whole with its environment in such a way that this interaction causes replication to be differential", in other words something that causes the replicators to appear in different proportions from one generation to the next, for example animal bodies or scientific papers and the judgements of scientists.

British science writer **Richard Dawkins** (b.1941) coined the term **meme** to depict an idea or action that spreads from one person to another through imitation, outlining what he saw as the conceptual analogue of the gene. Although the meme has become highly fashionable, the concept has not met with universal approval. One criticism of the meme/gene analogy, as admitted by Dawkins himself, is that memes do not resemble genes all that closely: for example, memes are more likely to undergo blending inheritance rather then particulate inheritance. Also, ideas do not come in discrete digital packets, so it is unclear where to apply the concept in a hierarchy of ideas (is Islam a meme in its own right, or merely a collection of memes, such as monotheism or submission to God?). And where is the genotype-phenotype distinction for memes? Crucially, how does calling things memes rather than simply ideas or concepts take us further forward in framing testable hypotheses? Why are Abba songs, comedians' catchphrases or religious rituals so catchy?

The second strand in evolutionary epistemology, the evolution of epistemological mechanisms, deals with the problem of whether we are born knowing some things (**innate knowledge**) or whether all knowledge is acquired through learning. This latter view is known as **empiricism** and was championed by the English philosopher **John Locke** (1632–1704). When discussing how we come to know what we know, it is important to demarcate the **content** of a belief (for example, big hairy spiders are dangerous) from our empirical **grounds** for believing it (perhaps because we know of someone bitten by a spider). We must then distinguish both of these from the **origins** of the sensory and cerebral mechanisms that predispose us to that belief (how do we recognize spiders and categorize them as dangerous) and the **value** of that belief (which in evolutionary terms equates to its benefits to survival).

Crucially for the philosopher, natural selection provides an alternative to individual learning in shaping beliefs and predispositions, providing an evolutionary rather than empirical justification for the belief. The evolution of epistemological mechanisms is now a lively research field, including topics such as how and when human and chimpanzee infants gain concepts such as number or self or begin to recognize goal-directed behaviour and the individuality and continuous existence of objects. However, the survival value of an innate cognitive or emotional predisposition sculpted by natural selection may not always map neatly on to the "truth". For example, an innate fear of spiders may lead an individual to categorize many harmless species as dangerous. Some philosophers object to evolutionary epistemology because they see it as a descriptive approach, best characterized as a branch of psychology or cognitive neuroscience, that sidesteps the central philosophical challenge of providing a normative analysis of knowledge (i.e. explaining and evaluating the grounds for our beliefs).

The logic of evolution

Some have argued that Darwin's natural selection is a tautology (a circular definition). The simplest argument proceeds along the lines of "natural selection is defined as survival of the fittest, but fitness is defined in terms of how well the fittest survive, so the whole argument is circular". In fact, this argument fails because fitness is not defined by survival or reproductive success, but by the comparative goodness-of-fit to the environment of one individual versus all the others in the species. In addition, natural selection requires that adaptive features are heritable. Imagine a population of bears living on the polar ice in which half happened to have black fur and half white. The white ones, by being camouflaged against the ice, would survive best and have more offspring, but if coat colour were set randomly in each generation, instead of being inherited, there would still be no natural selection.

Curiously, the natural-selection-as-tautology argument has bewitched many authorities, even including the philosopher Karl Popper who initially claimed that natural selection did not count as a scientific theory, but was still useful as a "metaphysical research programme". However, in later years, Popper recanted: "I have changed my mind about the testability and the logical status of the theory of natural selection; and I am glad to have an opportunity to make a recantation. My recantation may, I hope, contribute a little to the understanding of the status of natural selection."

Evolutionary ethics

Evolutionary biologists, starting with Darwin, have attempted to provide naturalistic explanations for our ethical beliefs and preferences. In *The Descent of Man*, Darwin gives an account of how what he called our **moral sense** might have evolved. Darwin starts with the social instincts of animals and then posits the evolution of **sympathy**, the inclination to suffer when another individual feels pain. Next, Darwin saw the evolution of the **conscience** as a means of reflecting upon and anticipating the clashes between social instincts and those that purely favour personal survival. Then, with the arrival of intelligence, language and human society came rules indicating how an individual should behave for the public good. Darwin points out that such rules are not always accurate in promoting what he sees as the public good (for example, he disapproved of the caste system in India). He also claimed that advances in civilization led to an **expanding circle** of sympathy: "As man advances in civilisation, and small tribes are united into larger communities, the simplest reason would tell each individual that he ought to extend his social instincts and sympathies to all the members of the same nation, though personally unknown to him. This point being once reached, there is only an artificial barrier to prevent his sympathies extending to the men of all nations and races."

In recent years, Australian philosopher **Peter Singer** (b.1946) has combined Darwin's expanding-circle argument with the assertion of evolutionary continuity between humans and other animals to argue that denying other animals moral status merely because they are "animals" equates to **speciesism**. Instead, we have to take into account animal suffering in any calculation of the moral consequences of our actions. Singer's logic has led him (and many others, including the author) to the conclusion that **vegetarianism** is preferable on ethical grounds to the consumption of meat. Similarly, the proponents of the **Great Ape Project** aim to have non-human hominoids granted the same basic moral and legal protection that humans enjoy. More generally, insights from palaeoanthropology that all humans share a recent common ancestor and that we are all Africans under the skin help fuel an appreciation of our common humanity and a respect for our mother continent.

A key challenge in evolutionary ethics has been how to explain the evolution of **altruism**, when natural selection is driven by the competition between individuals and their genes. As noted, plausible explanations have been advanced based on kin selection. In addition, evolutionary game theory has provided evaluations of the costs and benefits of **reciprocal**

altruism, a concept first proposed by American evolutionary biologist **Robert Trivers** (b.1943) in his classic 1971 paper. American political scientist **Robert Axelrod** (b.1943), working with Bill Hamilton, has shown how natural selection could favour the evolution of a simple **tit-for-tat strategy**, in which an organism helps a partner at the first opportunity and then helps on subsequent opportunities only if the partner helped on the previous opportunity. According to these explanations, psychological altruism does not really equate to evolutionary altruism, in that apparently altruistic behaviours may still benefit an individual's genes. More controversially, some philosophers, including **Elliot Sober**, look to group selection as an explanation for altruism.

Equally controversial is the impact of evolutionary thinking on **meta-ethics**, the study of ethical facts, such as what it means to say something is right or wrong. Do evolutionary explanations for the origins of morality undermine the validity or meaning of our ethical beliefs and intuitions? A comparison with mathematics suggests that this need not be true: investigations into the evolutionary origins and neurobiological basis of our innate understanding of arithmetic in no way challenge the truth of mathematical statements (one plus one remains two, irrespective of whether the ability to understand this statement emerged last week or two million years ago). As Thomas Huxley argued in his 1893 book *Evolution and Ethics,* "The thief and the murderer follow nature just as much as the philanthropist. Cosmic evolution may teach us how the good and the evil tendencies of man may have come about; but, in itself, it is incompetent to furnish any better reason why what we call good is preferable to what we call evil than we had before." Similarly, most philosophers would agree that attempts to shift moral responsibility to evolutionary imperatives are flawed: evolutionary studies investigating why men cheat on their wives should not be used to excuse or justify such behaviour, nor should evolution by natural selection be used a guide to how humans should behave (see p.266).

There are some evolutionists who remain sceptical of meta-ethics, claiming that natural selection has driven us to believe that there are ethical facts, when in fact there are none. According to **E.O. Wilson** and British-born philosopher of science **Michael Ruse** (b.1940) "ethics is a collective illusion of the human race, fashioned and maintained by natural selection in order to promote individual reproduction … ethics is illusory inasmuch as it persuades us that it has an objective reference". Whether such evolutionary determinism signals any advance in moral philosophy is unclear. Genome biologist Craig Venter takes such determinism to absurd lengths in his autobiography, claiming that his Y chromosome led him to have sex with a teenage girl against his father's wishes; but by the same logic, Venter Senior could claim that *his* evolutionary heritage had left him no choice but to hold a gun to his son's head in punishment!

Literature

In his later years Darwin's youthful delight in the arts largely abandoned him, although he always retained his enthusiasm for novels. English literature, in turn, felt the impact of Darwinism as soon as *The Origin of Species* was published and many writers were quick to grapple with its implications. In the ensuing 150 years, writers of all types have continued to take inspiration from his work in a variety of ways, from Charles Kingsley's children's classic *The Water-Babies* to the contemporary "palaeofiction" of Jean Auel.

Evolution and literary fiction

The English clergyman-novelist **Charles Kingsley** (1819–75) was among the first establishment figures to praise *The Origin* and was quoted by Darwin in the book's second edition. Three years after *The Origin* first appeared, Kingsley began work on his children's novel, *The Water-Babies, A Fairy Tale for a Land Baby*, completing it in 1863. Its hero Tom is a chimney sweep who after drowning is transformed into an amphibious

An illustration by Linley Sambourne, from the 1881 edition of *The Water-Babies*, showing Huxley and Richard Owen scrutinizing a water-baby in a specimen jar.

Literary Darwinism

Until recently, those interested in the links between evolution and literature were satisfied with fathoming the *direct* influences of evolutionary thinking on the *conscious* efforts of writers. Not any more! Proponents of the new discipline of literary Darwinism (or Darwinian literary studies) analyze and interpret the great themes of literature, and even the purpose of literature itself, in the light of evolution. In looking for biological universals behind variable plots, literary Darwinism places itself at odds with the prevailing poststructuralist and postmodernist views in literary studies that claim that culture alone constructs human goals, values and behaviours.

> **"...novels which are works of imagination ... have been for years a wonderful relief and pleasure to me ... I like all if moderately good, and if they do not end unhappily – against which a law ought to be passed".**
>
> **Charles Darwin**

Joseph Carroll, an English professor at the University of Missouri and author of *Literary Darwinism* (2004) argues passionately for an adaptationist view of literature: "It helps us to regulate our complex psychological organization, and it helps us cultivate our socially adaptive capacity for entering mentally into the experience of other people." According to Carroll, the theme of Jane Austen's *Pride and Prejudice* boils down to the fundamental biological problem of mate choice, where Darcy and Elizabeth's preference for partner based on honesty, kindness and intelligence sits at odds with the established social order. From a Darwinian viewpoint, the book helps us make sense of the conflicting impulses that characterize romantic relationships.

Jonathan Gottschall lays out an evolutionary perspective of Homeric warfare in *The Rape of Troy: Evolution, Violence, and the World of Homer* (2007), claiming "Darwin's powerful lens brought sudden coherence to my experience of Homer's *Iliad*". Gottschall argues that romantic love, far from being the recent invention that some claim, rates as human universal permeating literary traditions from all around the world. In *Comeuppance: Costly Signaling, Altruistic Punishment, And Other Biological Components Of Fiction* (2008), Brandeis professor William Flesch exploits evolutionary theory to show how fiction satisfies our desire to see the good vindicated and the wicked get their comeuppance.

At the more populist end of the spectrum sits *Madame Bovary's Ovaries*, written by evolutionary psychologist **David Barash** and his daughter **Nanelle**. The two authors take the reader on a breathless Darwinian romp through literature, exploiting elephant seals to explain Achilles and Agamemnon, vampire bats to illustrate *The Grapes of Wrath*, peacock's tails to illuminate Austen, and Bovary's ovaries to excuse the eponymous heroine's adultery. We look forward to the time when literary Darwinism eats itself by explaining Darwin's own preference for romantic novels with pretty women and happy endings!

water baby (complete with gills) and experiences a series of character-building adventures. The novel combines ideas of Christian redemption; the opposing forces of degeneration and regeneration (human evolution is contrasted with the degeneration of "Doasyoulikes" into gorillas); a critique of Victorian child labour and treatment of the poor with a satire on the reception of evolutionary thinking and praise for scientific progress that even mentions Darwin, Huxley and Owen by name. Among the numerous allusions to evolution are a conversation between Tom and a stand-in for Mother Nature, in which the boy states, "I heard, ma'am, that you were always busy making new beasts out of old." To which the woman replies: "I am not going to trouble myself to make things ... I sit here and make them make themselves."

According to English scholar Gillian Beer, the novelist **George Eliot** (1819–80) fell under Darwin's influence, particularly after publication of *The Descent of Man*. In *Darwin's Plots* (1983), Beer argues that there is a Darwinian flavour to many of Eliot's themes. In the prelude to *Middlemarch* (1871–72), Eliot outlines her theme as "the history of man, and how the mysterious mixture behaves under the varying experiments of Time" and, echoing Darwin's emphasis on variation in biological evolution, welcomes unpredictable variability in feminine behaviour: "...the limits of variation are really much wider than any one would imagine from the sameness of women's coiffure and the favorite love stories in prose and verse."

"Mr. Crichton ... left a sum of money in the hands of trustees ... to send out a man with a thousand fine qualifications, to make a scientific voyage, with a view to bringing back specimens of the fauna of distant lands ..."

from Elizabeth Gaskell's *Wives and Daughters*

George Eliot's contemporary, **Elizabeth Gaskell** (1810–65) was a distant relative of Darwin. Her final unfinished novel, *Wives and Daughters* (1866), is a love story in which scientific endeavour is presented in a positive light, epitomized by the character of Roger Hamley, an evolutionary explorer who travels to Africa, and who was based – at least in part – on Darwin.

Anthony Trollope (1815–82) seems an unlikely Darwinian, he even wrote: "I am afraid of the subject of Darwin. I am myself so ignorant on it..." Nonetheless, in *Darwin and the Novelists* (1988), **George Levine** devotes a whole chapter to Trollope and finds parallels between the two men's lives. These include an emphasis on uniformitarianism (in biology or society) where large effects result only from the cumulative action of numerous small causes; a self-deprecating approach to autobiography and a *laissez-faire* view of biological or social interactions, where freedom from design, constraint or ideology guarantees order but eschews predictability.

Of the next generation of writers, **Thomas Hardy** (1840–1928) claimed to have "been among the earliest acclaimers of *The Origin of Species*". Towards the end of his life Hardy listed Darwin as one of the thinkers "whom I used to read" and he attended Darwin's funeral. Several critics suggest a link between Darwin's ideas and the gloomy, fatalistic plots of Hardy's novels, in which deep time ("long beyond chronology") and the impersonal laws of nature are indifferent to the lot of ephemeral humans. Darwin also underpins Hardy's fascination with genealogical destiny and the balance between the inevitabilities and contingencies of history. In *A Pair of Blue Eyes* (1873), the geologist-hero Henry Knight comes face to face with a fossil trilobite and confronts a vision of the evolutionary past: "He saw himself at one extremity of the years, face to face with the beginning and all the intermediate centuries simultaneously. Fierce men, clothed in the hides of beasts, and carrying, for defence and attack, huge clubs and pointed spears, rose from the rock, like the phantoms before the doomed Macbeth..."

Evolution also informs the background to the fiction of **Rudyard Kipling** (1865–1936). In *The Jungle Book* (1894), wild boy Mowgli is raised by wolves and grows up with divided loyalties – part-human, part-animal. In the same book, a monkey troupe, the Bandar-Log, sing: "Now we are talking just like men!" Kipling's *Just So Stories* (1902) provide gently parodying alternatives to Darwinian evolution as explanations for the origins of animal behaviour or anatomy. The book's title was later adopted by evolutionary biologists to describe unverifiable narrative explanations for biological or psychological traits.

Darwin's immediate influence extended to novelists outside of England. In his novel *Anna Karenina* (1877), Russian writer **Leo Tolstoy** (1828–1910) portrays a crisis of faith for property owner Levin: "And in all of us, as well as in the aspens and the clouds and the misty patches, there was a process of evolution. Evolution from what? into what? – Eternal evolution and struggle ... As though there could be any sort of tendency and struggle in the eternal!"

At least two critics have argued for a Darwinian influence on the work of **Joseph Conrad** (1857–1924). Allan Hunter's *Joseph Conrad and the Ethics of Darwinism* (1983) suggests that Conrad was well versed in contemporary debates about evolution and was preoccupied by its impact on ethics and what motivated human behaviour. In his *Joseph Conrad* *and Charles Darwin* (1984), Redmond O'Hanlon sees the later writings of Darwin as having a profound effect on Conrad's tales of empire and psychological disintegration. In particular, he focuses on *Lord Jim* (1900), exploring how Jim's action in abandoning the passengers on a floundering ship and his subsequent attempts to redeem himself lead to a regressive withdrawal into his own unconscious.

D.H. Lawrence (1885–1930) was another early twentieth-century writer fascinated by the question of what it means to be human, but he had an ambivalent attitude to evolution, stating in *Mornings in Mexico* (1927): "Myself, I don't believe in evolution, like a string hooked on to a First Cause ... I prefer to believe in what the Aztecs called Suns: that is, Worlds successively created and destroyed." However, according to critic Ronald Granofsky, an interest in evolution drove the transition between his well-received early work, such as *Sons and Lovers* (1913) and *The Rainbow* (1915), and the less well-viewed work of his middle and later

Darwin plots

In recent years several contemporary novelists have woven Darwin and evolution into their work, either directly, as in Harry Thompson's 2005 *This Thing of Darkness* (see p.25), or more obliquely, for instance Margaret Drabble's *The Peppered Moth*. The following are ten of the best recent "Darwinian" works of fiction.

- **The French Lieutenant's Woman** (1969), John Fowles. This pastiche Victorian novel is set in the fossil-rich English town of Lyme Regis. The relationship of the mysterious Sarah Woodruff and gentleman-palaeontologist Charles reveal the implications of evolution for the human condition.
- **Galápagos** (1985), Kurt Vonnegut. Written after the author visited the islands that enchanted Darwin. In the novel, a small band of mismatched celebrities and tourists, shipwrecked on the Galápagos, end up repopulating the planet while evolving flippers, beaks and smaller brains.
- **Morpho Eugenia** (1992), A.S. Byatt. The first of two richly detailed novellas in the collection *Angels and Insects*, *Morpho Eugenia* tells of William Adamson, a Victorian entomologist and explorer who, back in England, finds himself caught up by the suffocating Alabaster family.
- **Ship Fever** (1996), Andrea Barrett. A collection by biology-graduate-turned-historical-fiction-writer of eight short stories on the theme of natural history and travel in the nineteenth century. Like Byatt, Barrett effortlessly weaves together fact and fiction.
- **The Evolution of Jane** (1998), Cathleen Schine. This witty and perceptive novel takes place on a nature tour of the Galápagos islands, where Darwin's

years. In *D.H. Lawrence and Survival: Darwinism in the Fiction of the Transitional Period* (2003), Granofsky also argues that Lawrence not only applied Darwinism *in* his work, but also *to* it, in an attempt to rid his writing of weak and undesirable elements.

Evolution and poetry

The two nineteenth-century poems most often linked with evolution actually pre-date the publication of *The Origin,* proving that a widespread interest in scientific developments and the crisis of faith were well underway before 1859. **Alfred Tennyson** (1809–92) completed "In Memoriam AHH" in 1849. This long, mournful work took seventeen years to write and was prompted by the death of Tennyson's friend Arthur Hallam. Featuring the famous line about "Nature, red in tooth and claw", it drew its evolutionary themes in part from readings of Lyell and the

observations and ideas guide 25-year-old New Yorker Jane Barlow Schwartz in her search for identity and the key to her relationships.

- **Mr Darwin's Shooter** (1998), Roger McDonald. Focusing on the clash between religion and science that *The Origin* provoked, the novel fleshes out what little is known about Syms Covington, Darwin's assistant on the *Beagle* and later his servant.
- **The Peppered Moth** (2000), Margaret Drabble. Evolution and mitochondrial DNA genealogies play prominent roles in this tale of a young woman attempting to escape her working class roots. The title comes from Kettlewell's famous experiments on natural selection (see p.79).
- **Confessing a Murder** (2002), Nicholas Drayson. The author is a naturalist and brings a scientific authority to this oblique fictional take on Darwin's life, which purports to be a memoir written in old age by a friend of Darwin marooned on an island.
- **Saturday** (2005), Ian McEwan. Steeped in allusions to Darwin, evolution, the Victorian crisis of faith and the limits to human and personal progress, McEwan's novel charts a day in the life of London brain surgeon Henry Perowne.
- **Charles Darwin in Cyberspace** (2005), Claire Burch. A fragmented narrative in which a deranged, ergot-intoxicated Emma Darwin, grieving for her dead daughter, is imagined slipping between nineteenth-century England and twentieth-century America.

Tennyson reading his poem "Maud", a drawing by D.G. Rossetti.

anonymous *Vestiges of the Natural History of Creation* (1844). Two stanzas are particularly revealing:

Are God and Nature then at strife,
That Nature lends such evil dreams?
So careful of the type she seems,
So careless of the single life;

That I, considering everywhere
Her secret meaning in her deeds,
And finding that of fifty seeds
She often brings but one to bear.

The poet and cultural critic **Matthew Arnold** (1822–88) probably wrote his poem "Dover Beach" in 1851, although it was not published until 1867. Like Tennyson, Arnold emphasizes the loss of old certainties in the face of new scientific ideas, especially in the third of its four stanzas:

The Sea of Faith
Was once, too, at the full, and round earth's shore
Lay like the folds of a bright girdle furled.
But now I only hear its melancholy, long, withdrawing roar,
Retreating, to the breath
Of the night-wind, down the vast edges drear
And naked shingles of the world.

The religious poet **Gerard Manley Hopkins** (1844–89) was familiar with evolutionary theory but did not see it as conflicting with his transcendent vision. "I do not think … that Darwinism implies necessarily that man is descended from any ape or ascidian or maggot or what not but only from the common ancestor of ascidians, common ancestor of maggots, and so on: these common ancestors, if lower animals, need not have been repulsive animals." In his richly religious verse, Hopkins celebrates the wonder of nature, and in "The Sea and the Skylark" contrasts the joyousness of birdsong with the way "We … [mankind]

Have lost that cheer and charm of earth's past prime:
Our make and making break, are breaking, down
To man's last dust, drain fast towards man's first slime.

Robert Browning (1812–89) discussed natural theology in letters with Charles Darwin's niece, Julia Wedgwood. Browning touches on Darwinian themes in his 1863 poem "Caliban Upon Setebos", portraying Caliban as half-man half-beast, while alluding to Darwin's observations during the *Beagle* voyage. His wife Elizabeth Barrett Browning, after reading *Vestiges of the Natural History of Creation*, wrote of "our condition of fully developed monkeyhood". The German-born English poet **Mathilde Blind** (1841–96) weaves evolution into verse in a style reminiscent of Erasmus Darwin in her long poem "The Ascent of Man" (1888), complete with graphic depictions of the struggle for existence:

War rages on the teeming earth;
The hot and sanguinary fight
Begins with each new creature's birth:
A dreadful war where might is right;
Where still the strongest slay and win,
Where weakness is the only sin.

A witty take on the idea that Darwin's view of evolution was synonymous with progress can be found in **May Kendall**'s "Lay of the Trilobite" (1887), an exchange between a man and a fossil which ends with the man declaring:

"I wish our brains were not so good,
I wish our skulls were thicker,
I wish that Evolution could
Have stopped a little quicker;
For oh, it was a happy plight,
Of liberty and ease,
To be a simple Trilobite
In the Silurian seas!"

As with his prose, **Thomas Hardy**'s poetry was influenced by Darwin, his late poem "Drinking Song" (1930) explicitly so:

Next thing this strange message Darwin brings,
Though saying his say in a quiet way;
We all one with creeping things;
And apes and men blood-brethren,
Likewise reptile forms with stings.

Robert Frost (1874–1963) was another twentieth-century poet who admitted a debt to Darwin, an idea explored in Robert Faggen's *Robert Frost and the Challenge of Darwin* (2001). Interestingly, in his poem "Accidently on Purpose", Frost suggests a certain ambivalence about the implications of evolution, acknowledging Darwin's significance but hankering after a sense of purpose:

They mean to tell us all was rolling blind
Till accidentally it hit on mind
In an albino monkey in the jungle,
And even then it had to grope and bungle,

Till Darwin came to earth upon a year
To show the evolution how to steer.
They mean to tell us, though, the Omnibus
Had no real purpose until it got to us.

The (un)natural selection of poetry

Most poets and lyricists attempt to combine linguistic inventiveness with a degree of compositional refinement. There are exceptions, such as Dadaist Tristan Tzara who created poetry by drawing random newspaper cuttings from a hat; or David Bowie who – inspired by novelist William Burroughs – has used similar "cut-up" techniques to create some of his lyrics. Steering a course between the totally random and the consciously designed, American genetic algorithm enthusiast **David Rea** recently created a website (www.codeasart.com/poetry/darwin.html) that allowed poetry to evolve through a process akin to natural selection. Rea stacked the odds in favour of something interesting emerging by using a word set derived from a thousand poems created the conventional way.

The website started off with a thousand proto-poems, each consisting of four lines of five randomly chosen words from the set. Visitors to the site provided the fitness function, choosing between two randomly selected verses from the population. Only the fittest poems – those with the most votes – were selected for survival and further evolution. Recombination between poems provided a source of variation. Just as in biological evolution, most changes did nothing to improve fitness and almost all the poems were incomprehensible drivel. But by the end of a year, after tens of thousands of generations, the standard of grammar, style and even poetry, had noticeably improved. Contrast this not-quite grammatical early example...

cannot the this muse
little fluster my which
the china married such we scents patterns
buying
roll
time awake

...with this more evolved successor:

first snowfall
beating beyond the head
that
with cold knowledge
revealing
one dream is and
was again

Most people would accept this as poetry, if not of the prize-winning variety. Sadly, Rea scarcely maintains the site and has made no attempt to formalize the study (for example, by measuring the improvement and investigating the dynamics of change). Nonetheless, his efforts provide an interesting case study illustrating how even something as intangible and subtle as poetry can emerge relatively quickly from the cumulative effect of selection acting on numerous small changes (Richard Dawkins did something similar in *The Blind Watchmaker* showing how cumulative selection could create the phrase "Methinks it is like a weasel").

Evolution in science fiction

Evolution is a dominant theme throughout science fiction, beginning in Darwin's own lifetime with the pioneer of the genre, **Jules Verne**. In *Journey to the Centre of the Earth* (1864), Verne put evolution centre stage with petrified forests, a battle between an ichthyosaur and a plesiosaur and an ape-man tending a herd of mastodons. In *20,000 Leagues Under the Sea* (1869), Verne even revealed an awareness of the *Beagle* voyage, referring to "Keeling Island … which had been visited by Mr. Darwin and Captain Fitzroy."

Evolution erupts in the work of English writer **H. G. Wells**, who was also an advocate of eugenics. In *The Time Machine* (1895), he projects human evolution over 800,000 years into the future, to a time when humanity has evolved into two species, the docile Eloi and the bestial Morlocks. In *The Island of Dr Moreau* (1896), Wells explores the boundary between human and animal, while an inter-planetary "struggle for existence" between man and Martian animates *The War of the Worlds* (1898). His contemporary **Arthur Conan Doyle** follows the Jules Verne path with *The Lost World* (1912), an adventure novel about an expedition to a remote plateau in South America where prehistoric animals and ape-men have somehow survived. On a rather more epic scale is *Last and First Men* (1930) by **Olaf Stapledon**, a panorama of future evolution, spanning two billion years and eighteen human species. Stapledon's masterpiece is cited by several science fiction writers, and by evolutionary biologist John Maynard Smith, as a key influence on their choice of career.

In 1953, **Arthur C. Clarke** (1917–2008) published *Childhood's End*, a novel in which humanity's evolutionary progression to an intergalactic "hive mind" is aided by an alien species, the Overlords. Clarke returns to the theme of aliens playing God with human evolution, in his most famous work, *2001: A Space Odyssey* (1968). At the start of the novel, Clarke portrays an alien monolith as instrumental in awakening human intelligence in an ape-like ancestor, triggering tool use for hunting and inter-human violence.

Of the many SF works to explore human evolution, *Planet of the Apes* (1963) is arguably the most famous. Written by Frenchman **Pierre Boule**, it holds a mirror up to humanity when astronauts travel from Earth to a distant planet where apes rule and humans are mere animals. In his collection of short stories *A Different Flesh* (1988), alternative-history writer **Harry Turtledove** imagines a world in which *Homo erectus* survived in the Americas, only to be enslaved by modern humans. British writer and science graduate **Stephen Baxter** tackles something far more ambitious in his 2003 novel *Evolution*, which attempts to provide the definitive

George Taylor (Charlton Heston) and Zira (Kim Hunter) share a tender moment in the 1968 film version of *Planet of the Apes*.

fictionalized account of human evolution, from our shrew-like ancestors 65 million years ago to our biological and non-biological descendants 500 million years in the future.

Another hugely popular subject with science fiction writers is Neanderthal man. In his 1955 novel *The Inheritors*, **William Golding** sets up a bleak and poignant confrontation between a group of Neanderthals and the first humans, the inheritors of the title. One of **Isaac Asimov**'s best short stories, *The Ugly Little Boy* (1958) tells of a Neanderthal child transported to the present through time travel. **Michael Crichton**'s *Eaters of the Dead* (1976) centres on a group of Vikings who are terrorized by a relict band of Neanderthals who have survived into the tenth century. Finnish palaeontologist **Bjorn Kurtén** brought scientific authenticity to the Neanderthal world in his 1980 "palaeofiction" novel *Dance of the Tiger*, while **Jean Auel** explores similar ground in her *Earth's Children* series of novels. *Ember from the Sun* (1996) by **Mark Canter** tells the story of a Neanderthal child born from a frozen embryo implanted into a surrogate mother. In the same year **John Darnton**'s *Neanderthal* had a pair of archaeologists investigating an apparently recent Neanderthal skull found in Tadjikistan, while a year later **Petru Popescu** covered similar terrain, this time about *Australopithecus*, in his novel *Almost Adam*. The most recent addition to this genre is *The Neanderthal Parallax* (2003–04) a trilogy of novels (*Hominids, Humans, Hybrids*) by **Robert Sawyer**, which explores the contrasts between our own world and an alternative universe where Neanderthals are the dominant human species.

Frank Herbert weaves evolution and ecology into his richly imagined *Dune* universe, set in the far future and the backdrop for a six-book series. In the Darwinian struggle for existence in harsh environments across the galaxy, races of humans (the Fremen and the Sardaukar) achieve almost superhuman resilience. The evolutionary future is the theme of several novels by **Greg Bear**: in *Blood Music* (1985), re-engineered lymphocytes evolve rapidly not just into a new organism, but a new civilization. In *Darwin's Radio* (1999) and its sequel *Darwin's Children* (2003), a new form of endogenous retrovirus manipulates human evolution through effects on intra-uterine development that trigger speciation.

Robert Charles Wilson explores the possibility of alternative paths in evolution in his 1998 novel *Darwinia*, in which part of the Earth is replaced with an alien landmass of roughly similar geography, but a strange new flora and fauna. Parallel-world evolution is also the subject of *West of Eden* (1984) by **Harry Harrison** (b.1925), in which the K-T mass extinction never occurred, leaving the way clear for the evolution of an intelligent species of mosasaur. Evolution features in several works by techno-thriller writer **Michael Crichton**. In *Congo* (1980), expeditionaries, aided by a gorilla adept in sign language, seek out a lost tribe of hybrid gorillas. In *Jurassic Park* (1990) and *The Lost World* (1995), Crichton imagines the recreation of dinosaurs by contemporary molecular biologists, and in his 2006 novel *Next* returns to the theme of transgenic apes.

Terry Pratchett (b.1948) claims that *The Origin of Species* changed his life when he read it at the age of thirteen. In his 1981 novel *Strata*, he explores the "what if?" of someone creating worlds and seeding them with fossils to create a fake history. His subsequent highly successful Discworld series of novels gently satirize creationism. Discworld itself is balanced on the backs of four elephants; the Discworld universe runs not by physical laws but by magic and according to what people believe; our own universe, Roundworld, is created by the wizards of the Unseen University. In *Darwin's Watch, the Science of Discworld III* (2005), co-authored with science writers Ian Stewart and Jack Cohen, Pratchett mixes serious science with fanciful fiction, with a sidesplitting romp through an alternative universe in which Darwin is prevented from travelling on HMS *Beagle* by Discworld deities and ends up writing a religious book entitled *Theology of Species*!

Darwin at the theatre

Darwin and evolution have proved suitable material for dramatic retelling, with playwrights tending either to concentrate on Darwin's own life or on one of the controversies engendered by his theories. Here are three of the best:

- **Inherit the Wind** 1955, Jerome Lawrence and Robert Edwin Lee. The twentieth century's classic text of Darwinian drama provides an engaging fictionalized account of the 1925 **Scopes monkey trial** (see p.281), while conveying a covert criticism of McCarthyism. The play quotes extensively from the trial transcript, but engages in considerable poetic license. The 1960 Hollywood film version starred Spencer Tracy, Frederic March and Gene Kelly.
- **After Darwin** 1988, Timberlake Wertenbaker. Darwin's conflict with Robert Fitzroy is presented as a play-within-a-play being rehearsed by a group of actors with the director and the play's author. All of the characters, in their own ways, are struggling to survive: Tom, the gay actor playing Darwin, is an intuitive performer; Ian as FitzRoy has to work much harder at his role. Eventually bogged down by additional, extraneous characters, this is, nevertheless, a compellingly theatrical experience.
- **Darwin in Malibu** 2003, Crispin Whittell. Features an elderly Charles Darwin, together with Huxley, Wilberforce and blonde bimbo companion, in a present-day Californian beach resort. The play is stuffed full of creative anachronisms and clever juxtapositions of ideas: Darwin as beach bum; Wilberforce on a mission to convert him; Huxley cross-examining Wilberforce over the logistics of Noah's ark; Huxley and Darwin do Crick and Watson; culminating in a "what-does-it-all-mean" climax.

Music

Although Darwin himself was tone deaf, he may have been amused had he known just how much music his work was later to inspire. Songs about Darwin and his theory of evolution permeate just about every musical genre, from classical to heavy metal. The following is a far from exhaustive trawl of evolutionary music, from the sublime to the ridiculous.

New wave, hard rock, prog rock and heavy metal

As their name suggests, American new wave group **Devo** were preoccupied with "de-evolution", the idea that instead of evolving, humankind has regressed. Their zany theme song "Jocko Homo", first released in 1977, interweaves evolutionary themes with a satirical view of social progress. The song's refrain drills itself into the listener's consciousness: "Are we not men?;

no, we are Devo!" The song's title was taken from a 1920s anti-evolution tract called *Jocko-Homo Heavenbound*. **XTC**, Swindon's contribution to new wave, included "The Smartest Monkeys" on their 1992 album *Nonsuch* and in the same year Talking Heads front man **David Byrne**'s devolutionary "Monkey Man" track appeared on the album *Uh-oh*. English band **Blur** included "Beagle 2" as a B-side to their 1999 hit "No Distance Left to Run". As a result they were invited to contribute to the ill-fated Beagle 2 Mars mission, writing the call sign that the probe was supposed to send back after landing on Mars.

Heavy metal is riddled with disturbing evolutionary themes: noble apemen and bestial humans, beset with degeneration and a dystopic future. The best of "evolutionary metal" has to be **Motörhead**'s "Line in the Sand", theme song for a professional wrestling group named Evolution. Other notable tracks include "Natural Selection" by English 1980s band Wildfire, "Neanderthal" by American death-metallists Demolition Hammer, "Neanderthal Sands" by Italian metal-heads Novembre, "The Missing Link" by German band "Rage" and "Darwin Was Right" by the Norwegian band, Odium.

Darwin! (1972) by Italian prog rock band **Banco del Mutuo Soccorso**, is evolution's premier concept album and arguably one of the greatest prog rock albums ever released. In a reference to Paley's watchmaker, the original album cover sports a psychedelic fob watch. The fourteen-minute opening track, "L'Evoluzione", a glorious celebration of the evolution of life, interweaves oboes, clarinet and harpsichord with synthesizers, piano and electric guitar.

The greatest ever Darwinian concept album?

Of similar vintage is **Rick Wakeman**'s *Journey to the Centre of the Earth*, recorded live in 1974 and loosely based on the Jules Verne book of the same name. The album's protagonists encounter "impressions of rock weeds and mosses from the Silurian epoch", a terrifying battle between an ichthyosaur and a plesiosaur, and a herd of mastodon marshalled by a primitive human. A progressive view of evolution mixed with religious metaphor pervades *God's Monkey* (1993), a collaboration between **David Sylvian** (ex-Japan) and **Robert Fripp** (from King Crimson).

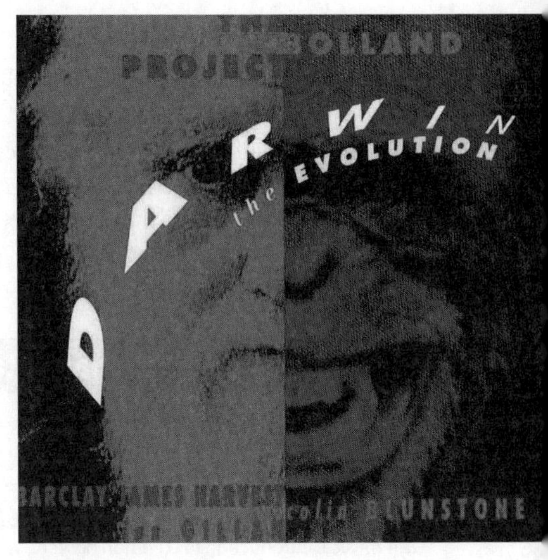

Or is it this one?

In 1991, brothers **Rob** and **Ferdi Bolland** created another contender for the title of ultimate evolutionary concept album with *Darwin (The Evolution)*, an eclectic mix of prog rock, rock ballardry, electronica and euro-pop. The album features such diverse talents as glam-rocker Suzi Quatro, the British band Barclay James Harvest (who had a hit single with one of the songs, "Stand Up"), Austrian rapper Falco, and Colin Blunstone (who belts out a rock ballad love song from Charles to wife Emma) and Ian Gillan (ex-Deep Purple), who sings and provides the linking narration. Sadly, talk of a video and a live show never came to anything and at the time of writing the album is hard to find.

Soft rock and pop

Created for an American TV series of the same name, **The Monkees** were hugely popular 1960s pop group. One of their bizarrest moments came in a chaotic TV special entitled "33 Revolutions Per Monkee", broadcast in 1969. During the show Charles Darwin (played by jazz organist Brian Auger) forces The Monkees through various stages of evolution until they become the greatest rock 'n' roll singers in the world. At one stage Darwin sings in a four-part piano harmony with Jerry Lee Lewis, Little Richard and Fats Domino! A year later, English rock band **The Kinks** had a hit single with "Apeman", a tragic-comic song bemoaning the fact that, despite evolution, man was basically destructive and wasteful. That same year, the short-lived band **Hotlegs** (a precursor of 10cc) achieved chart success with "Neanderthal Man", a strange mix of chorus chants, acoustic guitar, flute and a distinctive drum beat.

Hip hop and nerdcore

In 1970, **Gil Scott-Heron**, poet and godfather of hip-hop, included an ironic poem, "Evolution (and Flashback)", on his debut album, linking evolution with the legacy of slavery. More recently, the struggle against creationism has enlivened **nerdcore** (a variety of hip-hop created by and for nerds). **MC Hawking** is a nerdcore artist who parodies gangster rap, while adopting the persona of physicist Stephen Hawking. In his award-winning "What We Need More Of Is Science", the "Hawkman" pulls no

The ultimate evolutionary playlist

1 **PART MAN, PART MONKEY BRUCE SPRINGSTEEN** from ***Tracks***
A song about the Scopes trial that went on to become the song of the Dover trial.

2 **WE COME FROM MONKEYS EMERALD ROSE** from ***Archives of Ages***
A musical introduction to common descent, with some sideswipes at creationists.

3 **DANCE LIKE A MONKEY NEW YORK DOLLS** from ***One Day it Will Please Us to Remember Even This***
A deliciously ironic song and the video is a must-have: featuring Darwin, Dick Cheney, Pat Robertson and the flying spaghetti monster.

4 **HEY, CHARLY SUZIE QUATRO** from ***Darwin (The Evolution)***
The queen of glam rock's tribute to the great man of evolution, complete with poignant voice-over.

5 **DO THE EVOLUTION PEARL JAM** from ***Rearviewmirror***
"… I'm a man, I'm the first mammal to wear pants…" A great song but an even better video; a vividly disturbing dystopic view of where evolution has taken us.

6 **DARWIN THIRD EYE BLIND** from ***Blue***
An alienated and cock-eyed view of human evolution that is more Spencer than Darwin, but superb anyway.

7 **CHARLES ROBERT DARWIN ARTICHOKE** from ***26 Scientists Volume One Anning – Malthus***
No need to read any biographies, this track summarizes the great man's life in song!

8 **SCATTERLINGS OF AFRICA JOHNNY CLEGG & SAVUKA** from ***Premium Gold Collection***
South African anthropologist-turned-musician creates uplifting world music from the out-of-Africa hypothesis.

9 **I WANNA BE LIKE YOU LOUIS PRIMA** from ***Disney's Greatest Hits 2000***
A singing orangutan gives voice to some species envy in the cartoon movie of *The Jungle Book*.

punches in his condemnation of creationists. In the accompanying video, wheelchair-bound Hawking battles against a "dogma ray" that turns scientists into mindless sheep. Those after more hardcore nerdcore should try MC Hawking's "F*#k the Creationists", with an uncompromising chorus line you could predict from its title. Fellow nerdcore artist **MC Frontalot** includes a more gentle parody of creationists in his *Origin of the Species*, as he adopts the role of creationist minister Reverend Front Aloud. African-American atheist rap artist **Greydon Square** (self-styled as "the black Carl Sagan") manages to weave in isotopic dating, punctuated equilibrium

10 **THE MONKEY'S UNCLE ANNETTE FUNICELLO AND THE BEACH BOYS** from ***Disney's Greatest Hits 2000***
Hilarious song about love across the species, from forgettable Disney movie of the same name.

11 **MONKEY TO MAN ELVIS COSTELLO** from ***The Delivery Man (Special Edition)***
Elvis Costello's expresses a disapproving view of the monkey-to-man transition.

12 **MONKEY MAN AMY WINEHOUSE** from ***Back to Black (Deluxe Edition)***
The most recent of a long line of cover versions of an original by Toots and the Maytals.

13 **LINE IN THE SAND MOTÖRHEAD** from ***Raw Greatest Hits - The Music***
Lemmy gives lessons on the links between macro- and micro-evolution: "Evolution is a mystery, full of change that no one sees".

14 **EVOLUTION KORN** from ***Korn***
Heavy metal meets evolutionary psychology, with an amusing video on the stupid-will-outbreed-us theme from social Darwinism.

15 **L'EVOLUZIONE BANCO DEL MUTUO SOCCORSO** from ***Darwin!***
The ultimate prog-rock celebration of Darwin; a Dante-esque ode to evolution.

16 **NATURAL SELECTION THE GENOMIC DUB COLLECTIVE** from ***Origin of Species in Dub***
Darwin meets dub: vocals by Jamaican scientist Dominic White, music by Garageband (and the author)!

17 **AR HYD Y NOS DJ SG** from ***Clwb Cymru***
An electronic dance version of Charles Darwin's favourite tune that will have the old man twisting in his grave.

18 **EVOLUTION PAUL WELLER** from ***The Best Of... Volume Two***
A mellow instrumental number from one of England's finest songsters.

and the flying spaghetti monster into his track "A Rational Response". Finally, Canadian "lit-hop" artist **Baba Brinkman** sees links between natural selection, medieval poetic jousting and hip hop's freestyling competitions. Brinkman directly engages with evolution, common descent and Dawkinsian memetics in a highly intelligent rap entitled "Natural Selection", while taking a dig at the ID movement.

Ska, reggae and dub

A number of dub tracks from the early days of reggae touch on the ape-to-human transition: "Ape Man" by Augustus Pablo, "Apeman Skank" by the Upsetters and **Lee "Scratch" Perry's** "Super Ape". Perry developed the super ape theme in a number of later tracks: "Return of the Super Ape" and "Super Ape inna Jungle". Jamaican dub poet **Mutabaruka** has recorded a reggae version of Dave Batholomew's "The Monkey", in which monkeys disown their human relatives. More recently, **Dub Syndicate** released a dub called "Natural Selection" in 1996, while reggae's most prolific twosome, **Sly Dunbar** and **Robbie Shakespeare**, included a couple of tracks on their 2005 album *Usuals Suspects* with evolutionary allusions: "Survival of the Dubbiest" and "Pro-Evolution Sucker". In like vein, UK-based reggae-retro band **Prince Fatty** named one of their albums *Survival of the Fattest*. In 2004, working with Jamaican scientist **Dominic White** as The Genomic Dub Collective, the author produced *The Origin of Species in Dub* – our (strictly amateur) attempt to turn *The Origin* into a reggae album. A ska/reggae influence can also be detected on Bruce Springsteen's "Part Man, Part Monkey", which became the unofficial theme song at the Dover trial (see p.287).

Classical, jazz and electronic

Commissioned by the Darwin Symphony Orchestra for the Darwin bicentenary year, British composer **Michael Stimpson** is creating a major three-movement work, *Age of Wonders*, celebrating the life and work of Charles Darwin. Darwin's great-great-granddaughter, **Carola Darwin**, is collaborating with composer **Alejandro Viñao** and librettist Anna Reynolds to create a new opera, *Children of Fire*, about Darwin's *Beagle* voyage and his encounter with the indigenous peoples of Tierra del Fuego. Experimental composer Mark Warhol has created a short chamber music theatre piece, *Voyage of the H.M.S. Beagle* (1999), in which a lone flute represents the vast ocean, while a soprano and baritone recount Darwin's adventure.

Numerous jazz tracks carry the title "*Evolution*", notably those by vibraphonist Roy Ayer on the album *Mystic Voyage* (1993) and guitarist Jeff Golub on his album *Temptation* (2005). The album *Darwin's Dance* from the Gecko Island Jazz Band features eclectic jazz with South American overtones, while woodwind instrumentalist Mark Hollingsworth includes a track entitled "Darwin's Voyage" on his 2007 album *Chasing the Sun*.

Those in search of chilled out music with an evolutionary bent should try the **Future World Orchestra** tracks "The Beagle" and "Origin of Species", originally part of the Bolland brothers' concept album "Darwin (The Evolution)" and now available on the album *The Hidden Files* (2000). In a similar vein is the relaxing instrumental concept album **New Beagle Exploration Journal** (Shin Beagle-Go Tankenki) from Japanese band S.E.N.S.

10 Politics

Darwin's theory of evolution by natural selection was an attempt to describe how the world *is*, not how it *ought* to be. However, in a misguided move from *is* to *ought*, Darwin's metaphorical struggle for existence and Herbert Spencer's slogan "survival of the fittest" were soon transferred to the social and political arena, where they were used to justify neglect and exploitation of the poor and weak (social Darwinism) or to support the application of selective breeding programmes to humans (eugenics).

Social Darwinism

The earliest and most notorious extension of evolutionary thinking to politics was **social Darwinism** – the view that competition between individuals and between nations could, and should, drive social and economic progress in human societies. Inherent to social Darwinism was the suggestion that the richest and most "socially developed" should be allowed to flourish in society, while the poor and the weak should be left to fend for themselves, even if this meant suffering and death. Although the term was used widely only after the 1940s, the heyday of social Darwinism was in the late-nineteenth and early-twentieth centuries. In fact, the term is largely a misnomer, as the underlying philosophy depends chiefly on influences other than Darwin, most of which predate *The Origin of the Species*. One key influence was the *Essay on the principle of population* (1798) by **Thomas Malthus** (see p.37), which argued that human populations inevitably grow to outstrip the supply of resources needed to support them. In subsequent editions of the essay, Malthus advocated late marriage and sexual abstinence as mechanisms to check growth of the population. Extreme interpretations of Malthus's views filtered into British government policy: some have argued that the lack of intervention in the Irish famine of 1845 can be blamed on Malthusianism.

An even stronger influence on social Darwinism was English philosopher **Herbert Spencer** who, as early as 1851, had attacked what he called the "spurious philanthropists", who "[b]lind to the fact that, under the

Herbert Spencer, the English philosopher who coined the phrase "survival of the fittest".

natural order of things, society is constantly excreting its unhealthy, imbecile, slow, vacillating, faithless members ... advocate an interference which not only stops the purifying process, but even increases the vitiation – absolutely encouraging the multiplication of the reckless and incompetent by offering them an unfailing provision and discourages the multiplication of the competent and provident". Spencer subsequently articulated a view of progressive cosmic evolution wedded to the concept of social progress – each new, evolved society better than its predecessor. But although Spencer coined the phrase "survival of the fittest", in Spencer's view of evolutionary progress, natural selection was far less important than Lamarckian inheritance of acquired characters.

Was Darwin himself a social Darwinist? Generally he avoided commenting on such issues, but here and there in his writings, his own social prejudices slip through: "excepting in the case of man himself, hardly any one is so ignorant as to allow his worst animals to breed". But when it comes to the conflict between compassion and cold reason, he is on the side of the angels: "Nor could we check our sympathy, even at the urging of hard reason, without deterioration in the noblest part of our nature… We must therefore bear the undoubtedly bad effects of the weak surviving and propagating their kind; but there appears to be at least one check in steady action, namely that the weaker and inferior members of society do not marry so freely as the sound; and this check might be indefinitely increased by the weak in body or mind refraining from marriage, though this is more to be hoped for than expected."

Social Darwinism is generally linked to *laissez faire* economics and non-interventionist politics. American industrialists readily adopted evolutionary language. **John D. Rockefeller** said, "The growth of a large business is merely a survival of the fittest", while **Andrew Carnegie** claimed that the law of competition "is best for the race, because it insures the survival of the fittest in every department. We accept and welcome, therefore … great inequality of environment, the concentration of wealth, business, industrial and commercial, in the hands of a few, and the law of competition between these, as being not only beneficial, but essential for the future progress of the race". However, rather than always advocating sink-or-swim politics, some followers of social Darwinism engaged in philanthropy, but only when targeted to those best able to benefit from it. Carnegie, for example, used his fortune to set up hundreds of libraries and supported the education of African-Americans. However, there is no doubt that social Darwinism was used to justify many policies that are now judged highly dubious or even morally abhorrent: unregulated capitalism, a brutal criminal justice system, a class system locked in place by social immobility, militarism, colonialism and worst of all, racism and enforced eugenics.

Social Darwinism fails both as hypothesis and policy for a number of reasons. Firstly, it equates evolution solely with competition, when co-operation and altruism are also the products of evolution. As a response to social Darwinism, Russian anarchist **Peter Kropotkin** (1842–1921) wrote *Mutual Aid: A Factor of Evolution* (1902) while in exile in England. Drawing on his experiences on scientific excursions to Siberia, Kropotkin argued that the chief struggle for existence was with a hostile environment

Darwin, evolution, racism and survival

Was Darwin a racist? Darwin certainly used what would now count as non-PC language, calling Fuegians and others "savages". But unlike some of his contemporaries (for example creationist Harvard geologist Louis Agassiz), Darwin was happy to classify all humans together in one species. Darwin accepted that there were differences in temperament and intellect between ethnic groups, but he was far from convinced that these were innate and unchangeable; instead, his emphasis was always on the potential for improvement. He was vehemently opposed to slavery and held people of African origin in high regard.

But there is no escaping that Darwin sometimes slipped into judgemental language that revealed the racist assumptions of his own time. In *The Descent of Man*, Darwin claims that "At some future period, not very distant as measured by centuries, the civilised races of man will almost certainly exterminate, and replace, the savage races throughout the world. At the same time the anthropomorphous apes ... will no doubt be exterminated. The break between man and his nearest allies will then be wider, for it will intervene between man in a more civilised state, as we may hope, even than the Caucasian, and some ape as low as a baboon, instead of as now between the negro or Australian and the gorilla." However, it is important to note that Darwin was not here condoning genocidal destruction of indigenous peoples. In fact, in the *Voyage of the Beagle* he expresses his horror at the gaucho persecution of native Americans: "Who would believe in this age that such atrocities could be committed in a Christian civilized country?"

Whether or not Darwin was a racist has no impact on the veracity of his theory of evolution. In fact, a modern evaluation of evolution provides solid ammunition against racism. **Ernst Mayr** has suggested that evolution's move away from essentialist thinking liberates us from any notion of racial purity: circles and triangles may be defined by fixed essences with no intermediates, but human populations show just as much genetic variation within so-called "races" as between them. The out-of-Africa hypothesis has shown that the blondest Scandinavian and the darkest Australian Aborigine are more closely related to each other than either is to most Africans. In his book *Guns, Germs and Steel* (1997), **Jared Diamond** has documented how European colonial successes resulted from accidents of climate, microbiology, technology and geography, rather than any biological superiority of Europeans.

Curiously, Darwinian notions live on in the African diaspora, particularly in the Rastafarian religion, which places great emphasis on the notion of survival after the horrors of slavery. In his 1980 song "Could you be loved", Rastafarian reggae star Bob Marley even invokes Spencer's reformulation of Darwin's natural selection: "They say only, only, only the fittest of the fittest shall survive; stay alive!"

rather than other individuals. For Kropotkin, "mutual aid" was a more important driving force in evolution than competition: "In the animal world we have seen that the vast majority of species live in societies, and that they find in association the best arms for the struggle for life: understood, of course, in its wide Darwinian sense – not as a struggle for the sheer means of existence, but as a struggle against all natural conditions unfavourable to the species … The mutual protection which is obtained in this case … secure the maintenance of the species, its extension, and its further progressive evolution."

A second problem with social Darwinism is that it confuses economic and social success with biological success and what is natural with what is desirable. In his *Principia Ethica* (1903), the English philosopher **G.E. Moore** (1873-1958) attacked Spencer for committing what he called the **naturalistic fallacy**, conflating a natural property (evolutionary fitness) with a non-natural property (goodness). A similar criticism follows from **Hume's guillotine**, a formulation of the is-ought problem articulated by Scottish philosopher **David Hume** (1711–76). According to Hume, it is a mistake to conclude that one can ever use observations about *what is* (for example, natural selection is a fact) to derive claims about *what ought to be* (for example, natural selection is good and should be encouraged). In fact, a more palatable interpretation of evolution sees natural selection as an enemy of the values of a civilized society. As **Thomas Huxley** stated in his 1893 lecture *Evolution and Ethics*: "Let us understand, once and for all, that the ethical progress of society depends, not on imitating the cosmic process, still less in running away from it, but in combating it."

"…the endeavor to improve the condition under which our industrial population live, to amend the drainage of densely peopled streets, to provide baths, washhouses, and gymnasia … to furnish some provision for instruction and amusement in public libraries … is not only desirable from a philanthropic point of view, but an essential condition of safe industrial development…."

Thomas Huxley, from *The Struggle for Existence*

A third problem with social Darwinism is that it implies a rigid biological determinism in which one's station in life is set by one's innate unimprovable abilities. However, even at the height of social Darwinism, evolutionists like Huxley expressed a more enlightened view of the improvability of humans and human societies. Huxley concluded his 1888 essay, *The Struggle for Existence*: "There is, perhaps, no more hopeful sign of progress among us, in the last half-century, than the steadily increasing devotion which has been and is directed to measures for promoting physical and moral welfare among the poorer classes…" The descendants of those that Spencer counted as the "excrement of society" nowadays enjoy levels of health and wealth, education and intellectual achievement far beyond his imagination. However, these improvements have come about not through the effects of natural selection, but because of deliberate efforts to improve the lot of the poorest and most vulnerable in society through the kind of measures Huxley advocated, twinned with modern innovations such as vaccination and contraception.

Eugenics

Allied to social Darwinism was **eugenics**, a movement and philosophy concerned with the improvement of human populations through interventions that encourage breeding among those considered genetically most desirable (**positive eugenics**) or discourage or prevent reproduction of those judged genetically undesirable (**negative eugenics**). The latter would be achieved though segregation, abortion, contraception, sterilization and, *in extremis*, murder. For its followers – as a banner from a 1921 eugenics conference proclaimed – eugenics was "the self-direction of human evolution".

The roots of eugenics (from the Greek "good in birth") reach back into the ancient world: Plato advocated control of human reproduction by the state, while Sparta and Rome practised infanticide as a way of ridding society of those judged weak or undesirable. The modern formulation of the idea originates in the late nineteenth century with scientific polymath **Francis Galton** (1822–1911). Influenced by the work of his cousin, Charles Darwin, Galton argued that human intellectual talents were inherited and that in protecting the weak and "feeble-minded" and facilitating their reproduction, society was interfering with Darwin's natural selection and placing itself at risk of a "reversion towards mediocrity". He argued that the only way to prevent this decline was through the social equivalent of agricultural selective breeding. In his book *Hereditary Genius* (1869), Galton argued "it would be quite practicable to produce a highly-gifted race of men by judicious marriages during several consecutive generations".

The Aboriginal struggle for existence

In the late eighteenth century, Erasmus Darwin wrote a prescient poem predicting the future success of the Sydney Cove settlement. His grandson Charles Darwin visited Australia with HMS *Beagle* in 1836. Despite the young Darwin's lack of enthusiasm for the country ("I leave your shores without sorrow or regret"), two years later, during the *Beagle*'s next surveying expedition, its then captain, John Clements Wickham, named a coastal settlement in the Northern Territories, **Port Darwin**, after his former shipmate. Darwin was a careful observer of the plight of the indigenous population, observing that "Wherever the European has trod, death seems to pursue the Aboriginal". In Tasmania alone, the **Aboriginal population** had plummeted from 5,000 to 300 in just three decades, and, as Darwin recorded, these were now mostly banished to smaller islands within the Bass Strait. The last full-bloodied Tasmanian, a woman named Truganini, died in 1876, the year that Darwin began his autobiography.

> "Thirty years is a short period, in which to have banished the last aboriginal from his native land ... I do not know a more striking instance of the comparative rate of increase of a civilized over a savage people."
>
> Charles Darwin, from *Voyage of the Beagle*

This ruthless attitude to the native population lasted well into the twentieth century. Beginning in the nineteenth century and continuing up to the 1960s, a eugenics-inspired attempt at social engineering led to many children of Aboriginal descent being removed from their parents. In recent years there have been moves to admit state culpability and achieve reconciliation with what are now known as the **stolen generations** and with Aboriginal peoples in general. In their 1991 hit single "Treaty", the Aboriginal rock band *Yothu Yindi* sang, in Darwinian terms, of common ancestry and future unity: "Now two rivers run their course, separated for so long. I'm dreaming of a brighter day, when the waters will be one." Perhaps fittingly, on Charles Darwin's 199th birthday, in February 2008, an Aboriginal **Welcome to Country** ceremony was performed for the first time at the opening of Australia's parliament. The next day, newly elected prime minister Kevin Rudd delivered a **government apology** to the stolen generations.

During the late-nineteenth and early-twentieth century, Galton's ideas found favour with large sections of society in Britain and elsewhere in the West. Societies, journals and conferences devoted to eugenics flourished. Advocates in the UK included socialists like H.G. Wells and George Bernard Shaw and several of Darwin's descendants including his son Leonard. American supporters included Margaret Sanger, John Harvey

Kellogg, Irving Fisher, Charles Davenport and several US presidents. Theodore Roosevelt proclaimed "the inescapable duty of the good citizen of the right type is to leave his or her blood behind him in the world". However, a bleaker side to the movement soon surfaced as thoughts turned from gentle suggestions about who-should-marry-whom to support for enforced segregation, sterilization and even death, twinned with racist and class-ridden assumptions as to who counted as the unfit. In 1910, Winston Churchill sponsored compulsory sterilization of "the feeble-minded and insane classes". In 1915, Virginia Woolf wrote in her diary: "On the towpath we met and had to pass a long line of imbeciles. It was perfectly horrible. They should certainly be killed." Drawing on analogies with the animal breeder, striving for purebreds, many proponents of eugenics suggested that mixed-race unions should be avoided in the name of racial purity.

Several Western nations adopted eugenic legislation, even in supposedly liberal Sweden where over 60,000 people were sterilized between 1935 and 1975. Over half the states in the USA adopted laws allowing compulsory sterilization, which were applied to tens of thousands or people during the first half of the twentieth century; many states also enacted marriage laws prohibiting the "epileptic, imbecile or feeble-minded" from marrying.

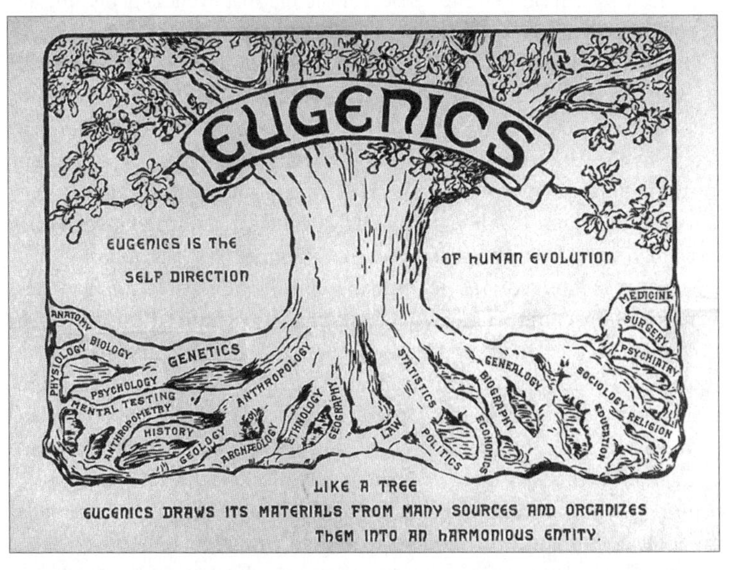

Emblem of the Second International Congress of Eugenics, held in New York in 1921. It was attended by, among others, Leonard Darwin and Herbert Hoover.

However, in Britain, not everyone in public life was in favour of eugenics: G.K. Chesterton, for one, voiced criticism in his book *Eugenics and Other Evils* (1922).

Compulsory sterilization was never sanctioned by the British state. Instead, the government proposed in the 1913 **Mental Deficiency Bill** that the mentally deficient should be segregated away from the rest of society in lunatic asylums, preventing them from breeding. A key opponent of the Bill was Josiah Wedgwood, 1st Baron Wedgwood, radical MP for Newcastle-Under-Lyme and a prominent member of the Darwin-Wedgwood clan. He dismissed the proposed legislation as the work of "eugenic cranks" and made a heroic but futile effort to filibuster the Bill in the British parliament. Much watered down – with no reference to sterilization – the version of the Bill that eventually became an Act of Parliament marks the high point in influence of British eugenicists.

"There will no doubt always remain a class quite outside the pale of all moral influence ... who, if they become parents, are certain to pass on some grievous mental or bodily defect to a considerable proportion of their progeny. Here and here only must the law step in."

Leonard Darwin, letter to the *New York Times* (1912)

In the 1930s and 1940s, the Nazi regime sterilized hundreds of thousands of those they judged "unfit" and killed tens of thousands of the institutionalized disabled, before moving on to the systematic murder of millions of people in the **Holocaust**. Unsurprisingly, when the full horror of Hitler's "final solution" was revealed, eugenics fell from favour, tainted by its association with the Nazis – although this did not prevent Canada, Sweden and several US states from continuing state programmes of sterilization.

Nowadays, interventions in human reproduction based on genetics (genetic counselling and prenatal testing) focus on the effects on individuals rather than implications for society, populations or gene pools (see *The Rough Guide to Genes and Cloning*). Modern medical care means that many of those who, in previous societies, might be prevented from reproducing (by premature death, infertility or other factors) can now do so: everyday examples include use of spectacles to correct poor vision or Caesarian section to ensure safe deliveries. How far these trends are going to influence the course of human evolution remains unclear, but they certainly re-define what it means to be "fit": for example, short-sightedness carries far fewer consequences in Western societies than it might have done in the Pleistocene, when an inability to spot a predator might have been fatal. In addition, action groups and new legislative programmes have pushed back the boundaries of what is socially acceptable (for example, the right of gay or disabled people to reproduce) and of what it means to be mentally or physically disabled. For some, advances in reproductive choice bring ethical concerns, for example as to whether couples with hereditary deafness should be allowed to choose to have a deaf child or whether one should be allowed to choose offspring who have or do not have a "gay gene" (should such a thing exist – the evidence is pretty tenuous). But we are still a long way from Aldous Huxley's *Brave New World* scenario of designer babies as the default option.

From Darwin to Dachau: a non sequitur

In an attempt to discredit evolution through guilt by association, apologists for creationism often link Darwin with Hitler and blame Darwin for the Holocaust. Examples include the 2004 book *From Darwin to Hitler* by the Discovery Institute's **Richard Weikart** and the recent film *Expelled: No Intelligence Allowed.* However, the arguments here are fraught with difficulty. Is it ever fair to hold a historical figure personally responsible for all the future unbidden and unforeseeable consequences of all that he or she has said or done? Was Jesus responsible for the Inquisition or Muhammad for 9/11? Can we blame Newton and his laws of motion for the damage caused by cruise missiles? And even where one can establish a chain of causal links between scientific discoveries and their subsequent abuses, does this mean that we must belittle the discovery or close down future research?

Nazi ideology was derived from a range of ideas and beliefs, which included anti-Semitism, militaristic nationalism, anti-capitalism and anti-communism. The Nazis also blended a distorted German Christian tradition with Nordic mythology and derived their zest for eugenics as much from ancient Sparta as from any modern sources. The influence of evolutionary thinking on Hitler was, if anything, very minor: nowhere in *Mein Kampf* does he mention Darwin, natural selection or biological evolution. In fact, in the first edition of the book, Hitler comes across as a young Earth creationist, claiming at one point that "this planet will, as it did thousands of years ago, move through the ether devoid of men".

The Nazis did cite an evolutionary principle at one crucial point. At the **Wannsee conference** in 1942, they decided on the mass murder of those Jews they hadn't already worked to death on the grounds that "The possible final remnant will, since it will undoubtedly consist of the most resistant portion, have to be treated accordingly, because it is the product of natural selection and would, if released, act as a the seed of a new Jewish revival." But this one fleeting reference is not sufficient to condemn Darwin for Dachau.

Nonetheless, themes associated with eugenics repeatedly surface in modern society. In the 1970s, American millionaire **Robert Klark Graham** established his **Repository for Germinal Choice**, a "genius sperm bank" (reportedly with a smattering of Nobel prize-winning donors), which was responsible for the birth of over two hundred children. In 1994, with their book *The Bell Curve*, American psychologist **Richard Herrnstein** and his co-author Charles Murray unleashed a storm of criticism in suggesting that intelligence is the best predictor of social progress, that there are continuing racial differences in intelligence and that a "cognitive elite" was segregating out from the rest of the population. However, the enduring "devolutionary" idea that the stupid are out-breeding the smart is countered by the **Flynn effect** (named after its discoverer, New Zealander **James Flynn**), the fact that average scores in IQ tests have been steadily improving decade upon decade wherever they have been employed. Advances in birth control now mean that individuals can delay parenthood until they feel that have the intellectual, emotional and financial resources to cope with it. Controversially, American economist **Steven Levitt** (b.1967), in his *Freakonomics* (2005) and elsewhere, has suggested that falling crime rates in the USA during the 1990s were the result of the legalization of abortion in the 1970s, arguing that unwanted children are more likely to become criminals.

Evolution and politics today: an uneasy alliance

In his *Constitutional Government in the United States* (1909) American president **Woodrow Wilson** made a weak analogy between the US constitution and biological evolution, claiming "Living political constitutions must be Darwinian in structure and in practice". It is questionable whether one should go any further in drawing political conclusions from Darwin's biology. Political neo-cons from the Bush era, often closely allied with religious conservative evangelicals, would dismiss evolution as a materialist threat to the accepted social and religious order. However, other conservatives argue that a Darwinian worldview supports political conservatism. A key advocate of this position is **Larry Arnhart**, a professor of political science at Northern Illinois University in the USA and author of *Darwinian Conservatism* (2005). Arnhart believes "Conservatives need Charles Darwin. They need him because a Darwinian science of human nature supports conservatives in their realist view of human imperfectability".

However, Darwin himself was a liberal and many recent writers on evolution, including Stephen Jay Gould, Richard Dawkins, E.O. Wilson, all lean to the left side of the political spectrum. In a short booklet, *A Darwinian Left* (1999), **Peter Singer** argues that the less malleable view of human nature provided by evolutionary psychology is compatible with, and should be incorporated into, the ideological framework of the Left, arguing "To be blind to the facts about human nature is to risk disaster". However, the question of whether it makes sense to talk about a Darwinian outlook in politics, and whether that outlook should be left-wing or right-wing, is dwarfed by another incursion of evolution into public life: the political debate in the USA and elsewhere as to the status of evolution and religion in education, the subject of the next chapter.

11 Religion

Although rightly celebrated as a breakthrough by scientists – and accepted by many people of faith – Darwin's theory of evolution is still reviled by those who view it as irreconcilable with their religious beliefs. In particular, many believers object to the idea that the universe, the Earth and all its inhabitants were not created by a single supreme being. Such "creationists", as they are called, exist in all faiths but are most vocal among fundamentalist US Christians, who have systematically opposed the teaching of evolution in schools. Recently, some creationists have tried to uncouple their opposition to evolution from its religious origins by re-packaging their beliefs as "intelligent design", where natural selection is replaced by some intelligent causal agency, which is not explicitly identified as God. However, time and again, creationists have lost not just the scientific arguments, but also the legal battles.

Resolving a crisis of faith

The Victorian era in Britain was marked by an upsurge in religious belief, followed by a crisis of conviction. By the middle of the nineteenth century, religion occupied centre stage in the public and intellectual life of the country and its expanding empire. But the educated classes were beset by an erosion of faith, stemming as much from internal ethical disputes as from the external influence of science and rationalism. As is evident from the poems of **Alfred Tennyson** and **Matthew Arnold**, this crisis was well under way before Darwin published *The Origin*.

Nonetheless, the publication of *The Origin of Species* in 1859 undoubtedly added fuel to the flames consuming religious faith and certainly provoked much satire. However, its effect on traditional religious belief was overshadowed by the appearance the following year of the book, *Essays and Reviews*. This best-selling volume, written mostly by Anglican clergymen, introduced its British readership to a European rationalist approach to Biblical criticism. In so doing, it stirred up years of argument, particularly over its claim that a belief in miracles was not only irrational,

The sixth day of creation according to Genesis: God makes woman from Adam's rib.

but atheistic (as such miracles would break the lawful edicts issued at Creation). That same year, at the Oxford meeting at which Thomas Huxley debated with Bishop Wilberforce (see box, p.53), up-and-coming Anglican theologian **Frederick Temple** preached a sermon welcoming the insights of evolution.

Between the first publication of *The Origin* and Darwin's 1871 work, *The Descent of Man*, an Anglican bishop, **John William Colenso**, had publicly cast doubt on whether Moses had written the Pentateuch. Around the same time several naturalistic, non-miraculous accounts of the life of Jesus were published, notably **David Strauss's** *Leben Jesu* (translated into English by George Eliot in 1846), **Ernest Renan's** *Vie de Jésus* (1863) and **John Robert Seeley's** *Ecce Homo* (1865). By the 1870s, it was hard for an educated Englishman to retain a traditional Bible-literalist faith; instead the Anglican Church was forced to re-create that faith on a new basis that accommodated both evolution and Biblical criticism. In the 1880s, Temple stated bluntly that "the doctrine of Evolution is in no sense whatever antagonistic to the teachings of Religion". A few years later Temple became Archbishop of Canterbury and evolution largely ceased to excite controversy within the Anglican tradition. Instead, the struggle between evolution and Biblical literalism shifted across the Atlantic to the New World.

Christian creationism in America

Why should the story of modern creationism be so predominately an American story? The reasons stretch back to the first settlers, the Pilgrim Fathers and other dissenters, who fled religious persecution in England, Ulster and Europe. The religions they brought to the New World were decentralized, drawing on the Protestant anti-authoritarian tradition of a-priesthood-of-all-believers. This meant that, when later faced with the challenge of evolution, the decision as to what to believe was down to

The Founding Fathers and religion

US creationists often claim that their country was founded on Christian principles. However, the Founding Fathers were not a conventionally religious group. Instead, they held views on a continuum between **Deism** (the belief that there is a God that created the universe but had nothing more to do with it), **Unitarianism** (the belief in a single divine personality, denying that Jesus is God) and more conformist Protestant Christianity. **George Washington** often skipped communion and dodged church entirely in later life. **Benjamin Franklin** was a self-confessed Deist, as was **Thomas Paine**. In fact, Paine's views on religion, as expressed in *The Age of Reason* (1795), were so offensive that no one would bury him in consecrated ground, instead he was buried on his farm (before a disinterment that, after a curious chain of events, led to his skull residing a few hundred yards from Darwin's Down House in the mid-nineteenth century).

Thomas Jefferson (1743–1826) was sympathetic to the Unitarianism espoused by his friend Joseph Priestley (1733–1804) and by America's second president, John Adams. But although Thomas Jefferson included a reference to the creator in his eloquent preamble to the Declaration of Independence, in his own words, he "belonged to a sect with just one member". In producing what is often called the **Jefferson Bible**, he attempted to strip all supernaturalism from the New Testament, while preserving its moral teachings.

Jefferson's chief legacy in the battle between evolution and creationism was ensuring that the US federal government would never lend support to religion. He drafted the Virginia Statute for Religious Freedom with fellow Founding Father James Madison, before persuading Madison to draft the **First Amendment to the Constitution**, which was ratified by Congress in 1791 and contained what is now called the **Establishment Clause**: "Congress shall make no law respecting an the individual, rather than any pope or archbishop. At the birth of the Great Republic, the **Founding Fathers**, recoiling from a history of religious discrimination in Europe, established freedom of speech and of religion (see box, below). But crucially, the Founding Fathers went one step further – rather than allowing religion into politics, they prohibited state sponsorship of any religion. In so doing, they set Christian creationism on a collision course with the American Constitution.

America's initial responses to evolution

The initial theological response in the US to evolution was mixed. In 1874, **Charles Hodge**, Presbyterian head of the Princeton Theological Seminary, published *What is Darwinism?* in which he concluded that Darwinism, in removing divine design from nature, was "tantamount to atheism".

establishment of religion". In 1802, Jefferson coined a memorable interpretation of the establishment clause in a letter to the Baptists of Danbury, Connecticut, writing that the clause had erected "a wall of separation between church and state". A modern judicial reading of the Establishment Clause interprets it as meaning that "government should not prefer one religion to another, or religion to irreligion" at any level of administration.

US Presidents (from l. to r.) Washington, Jefferson, Theodore Roosevelt and Lincoln. Of the four, probably only Roosevelt was a conventional Christian.

Can a Christian believe in evolution?

The widely read Christian apologist **C.S. Lewis** (1898–1963), although sceptical of evolution, wrote in 1944: "I can't have made my position clear. I am not either attacking or defending Evolution. I believe that Christianity can still be believed, even if Evolution is true." Stephen Jay Gould in a book review once wrote: "Either half my colleagues are enormously stupid, or else the science of Darwinism is fully compatible with conventional religious beliefs." Two of the architects of the modern synthesis (see p.111), Ronald Fisher and Theodosius Dobzhansky, were Christians.

Many of today's Christians are comfortable with an allegorical interpretation of the Genesis creation stories and with a universe that God creates through evolution. And although science relies on a methodological naturalism that excludes the supernatural from its purview, many are happy to accept that science does not entail an all-inclusive naturalism incompatible with religious belief. Anglican Archbishop of Canterbury **Rowan Williams** is of the opinion that "creationism is, in a sense, a kind of category mistake, as if the Bible were a theory like other theories. Whatever the biblical account of creation is, it's not a theory alongside theories... My worry is creationism can end up reducing the doctrine of creation rather than enhancing it." Williams' equivalent in the US Episcopalian Church, **Katharine Jefferts Schori** – an expert on squid evolution – has said that the "vast preponderance of scientific evidence, including geology, paleontology, archaeology, genetics and natural history, indicates that Darwin was in large part correct in his original hypothesis. I simply find it a rejection of the goodness of God's gifts to say that all of this evidence is to

However, his successor at the seminary, **B.B. Warfield**, was happy to accept that God had guided the process of evolution, stating: "I do not think that there is any general statement in the Bible or any part of the account of creation, either as given in Genesis 1 and 2 or elsewhere alluded to, that need be opposed to evolution".

The secularist author Susan Jacoby has claimed that the last quarter of the nineteenth century was a **Golden Age of Freethought** in America, an era when radical orator **Robert Ingersoll** (1833–99) would entertain the public with speeches on Shakespeare, Lincoln and Darwin, and Alfred Russel Wallace took a ten-month tour of the United States, lecturing on Darwinism, biogeography, spiritualism and social reform. However, this was also a period of great religious revival, with the seeds of fundamentalism planted at the first annual **Niagara Bible Conference** in 1878.

Modern creationism was born in the first decade of the twentieth century, when Canadian Seventh-day Adventist **George McCready Price** (1870–1963) attempted to develop a "**flood geology**", in which he claimed be refused because it does not seem to accord with a literal reading of one of the stories in Genesis."

Several evolutionary biologists have written books that attack creationism and/or defend a Christian theistic evolution position. In 1999, Catholic cell biologist **Kenneth Miller** published *Finding Darwin's God: A Scientist's Search for Common Ground Between God and Evolution* in which he patiently sifts through the evidence that the Earth is old and that evolution has happened without any intelligent designer. Miller also shows how science relies on methodological naturalism but does not imply atheism. **Francis Collins** is a physician-scientist who led the publicly funded Human Genome Project and in 2006 published *The Language of God: A Scientist Presents Evidence for Belief*. Here Collins rejects creationism and intelligent design and instead argues for a theistic evolutionary view he calls "BioLogos".

In 2004, American biologist Michael Zimmerman launched the ecumenical **Clergy Letter Project**, aimed at persuading American Christian clergy from various denominations to endorse a letter rejecting creationism and affirming "that the theory of evolution is a foundational scientific truth, one that has stood up to rigorous scrutiny and upon which much of human knowledge and achievement rests". As of May 2008, the project has collected over eleven thousand signatures. In addition, the project has organized two **Evolution Sundays** and one Evolution Weekend, all held as close as possible to Darwin's birthday. Over eight hundred congregations participated in Evolution Weekend 2008.

fossils were evidence of Noah's Flood. In 1910, publication began of *The Fundamentals*, a twelve-volume series of articles, which attempted to defend traditional Protestantism and Biblical literalism against challenges that included liberal theology, Biblical criticism and evolution. In one of the articles, American geologist **George Frederick Wright** (1838–1921) defends the special creation of man, but concludes that the philosopher "Hume is more dangerous than Darwin". In another, the author likens "the teaching of Darwinism, as an approved science, to the children and youth of the schools of the world" to the Roman destruction of the Jewish temple.

The Scopes monkey trial

In the 1920s, the struggle between creationism and evolution came to a head, largely thanks to the influence of Presbyterian lawyer **William Jennings Bryan** (1860-1925), a three-times presidential candidate and former secretary of state, who saw evolution as a threat to morality. In

1921, Bryan delivered several anti-evolution lectures in Virginia, as part of a campaign to prevent the teaching of evolution in American universities and schools. His influence on university education was limited but, unfortunately, Bryan had better luck in persuading state governments to introduce laws banning publicly funded schools from teaching evolution. The most notorious of these laws, the **Butler Act**, was passed in March 1925 in Tennessee and made it illegal for publicly funded teachers "to teach any theory that denies the Story of the Divine Creation of man as taught in the Bible, and to teach instead that man has descended from a lower order of animals". During the debate that preceded the vote on the bill, one senator ironically suggested that they should also "prohibit the teaching that the earth is round". The law was challenged a few months later in the famous **Scopes monkey trial**.

The trial arose from a provocative collusion between principles and profit. The American Civil Liberties Union (ACLU) aimed to establish a test case by advertising that they would support anyone who broke the

Scopes on trial: John Scopes (centre) and Dudley Malone await the jury's verdict.

law. A group of local citizens saw this as a chance to boost the small-town economy of Dayton, Tennessee. They soon recruited local science teacher **John Scopes** (1900–70) to their cause and persuaded Scopes that he had already fallen foul of the law by teaching from the state-approved textbook. Scopes was charged and ordered to appear at the county courthouse on July 10th. William Jennings Bryan volunteered his services to the prosecution, while the ACLU lawyer **Clarence Darrow** (1857–1938) offered to work *pro bono* for the defence, along with **Dudley Malone**, an experienced divorce lawyer.

The trial ran for less than a fortnight, but attracted unprecedented media attention with over two hundred reporters and the first ever radio broadcast of judicial proceedings. Acerbic *Baltimore Sun* reporter **H.L. Mencken** presented the trial to the public in colourful turns of phrase such as "the infidel Scopes" or "the monkey trial". The prosecution called just four witnesses to establish that Scopes had broken the law. The defence tried to field eight expert witnesses to establish that there was no conflict between evolution and the Bible. However, the prosecution argued – and Judge John Raulston accepted – that such expert opinion was irrelevant to the question of whether Scopes had actually taught about evolution. After Bryan snuck in a jibe that humans were descended "not even from American monkeys, but from old world monkeys" defence lawyer Malone delivered a dramatic speech, seen as the highpoint of the trial, passionately arguing that the Bible belonged to the realm of theology and morality rather than to science.

Towards the end, the trial took a bizarre twist as Darrow questioned Bryan as a witness to the authenticity of the Bible. Sparks flew in the resulting exchange, with Darrow using phrases like "your fool religion" and declaring (presciently, given later history): "We have the purpose of preventing bigots and ignoramuses from controlling the education of the United States." To prevent Bryan from summing up, Darrow waived his own right to a closing statement. Instead, Darrow asked the judge to bring the jury in and instruct them to return a guilty verdict, which they did after just nine minutes deliberation. The judge fined Scopes the minimum allowed: just $100.

Bryan died in his sleep a few days after the end of the trial. The following year, the verdict was disallowed on a technicality, so Scopes avoided the $100 fine. Later Scopes even admitted to being unsure whether he had ever taught evolution! Nonetheless, the Butler Act, although never again invoked, stayed on the statute books until 1967. And in the aftermath of the trial, states and school districts,

Creationism: a house divided

Creationism is often presented as a clearly defined alternative to evolution. However, creationism encompasses a wide range of beliefs, represented by rival groups, more at odds with each other rather than united against evolution. Advocates of **young-Earth creationism** (YEC) believe that the Earth is less than ten thousand years old, take the Genesis six-day creation accounts literally and interpret geology in the light of Noah's flood. Spokesmen for this view include Henry Morris (1918–2006), Duane Gish (b.1921), Ken Ham and Kent Hovind (the last currently in prison for tax offences). Organizations supporting YEC include the Dallas-based **Institute for Creation Research**, **Answers in Genesis** (at one time in legal dispute with sister organization Creation Ministries International), and Hovind's **Christian Science Evangelism**.

Supporters of **old-Earth creationism** (OEC) accept the scientific evidence for an ancient Earth, but claim that life was specially created by God. They adopt various strategies to accommodate science with the Bible. **Gap creationists** claim that there was a long gap between the first verse of Genesis ("In the beginning God created the heaven and the earth") and the rest of the creation story, with God recreating the world in six days after the gap. Advocates of this view include disgraced televangelist Jimmy Swaggart.

Day-age creationists interpret the days of creation as long ages of time: "With the Lord, a day is like a thousand years." Prominent day-age creationists include William Jennings Bryan and Canadian-born astronomer Hugh Ross, founder of the organization **Reasons to Believe**. Day-age creationism often overlaps with **progressive creationism**, the notion that God miraculously created new forms of life successively over periods spanning hundreds of millions of years. Progressive creationists reject any descent-with-modification relationship between old and newer kinds of organism. Public arguments between young-Earth creationists and progressive creationists are well documented, particularly between Ham and Ross and between Answers In Genesis and the Ohio-based OEC organization **Answers In Creation**. Turkish creationist Harun Yahya (pseudonym of Adnan Oktar) is also an OEC supporter.

The **intelligent design movement** also spans a wide range of views. Some are simply creationists hiding behind ID as a way of avoiding mention of the "C word". However, ID proponent William Dembski (b.1960) "leaves open as a very live possibility that common descent is the case". More surprisingly, Michael Behe (b.1952), in his most recent book, *The Edge of Evolution*, accepts common descent (including of humans and chimps) and sees no need for ID to explain the emergence of species, genera, etc. Furthermore, Behe dismisses any attempt to treat the Bible as a "science textbook" as "silly" and accepts that "the universe operates by unbroken natural law". Understandably, others in the **Discovery Institute** (ID's premier organization) have been lukewarm in their promotion of Behe's book.

particularly in the southern Bible belt, saw to it that the teaching of evolution was excluded from many American public schools for over a quarter of a century.

From Cold War competition to intelligent design

In 1957 the launch of the Sputnik satellite by the Soviet Union led to Cold War fears that the US was losing out to its communist rival in space and science. One result of this was the founding of the **National Defense Education Act** in 1958, with the aim of funding improvements in the teaching of science and mathematics in the US. That same year, a grant from the National Science Foundation led to the **Biological Sciences Curriculum Study**, which was tasked with modernizing biology textbooks and other teaching material. In the years that followed, evolution re-appeared in school textbooks and science classrooms.

Unfortunately, enthusiasm among American fundamentalists for young-Earth creationism was re-awakened in 1961 by *The Genesis Flood*, a book co-written by Baptist engineer **Henry Morris** and theologian **John Whitcomb**. A few years later, Morris and fellow creationists established the **Institute for Creation Research**, one of several creationist organizations that were to spring up over the coming decades. Many attempts to promote creationism and ban evolution in public schoolrooms followed, but the US courts repeatedly ruled such efforts unconstitutional (see box, p.288). In response, the shape-shifting creationist campaign repeatedly changed tactics. After legal rulings established that teaching biblical creation stories in American public school science classes was unconstitutional, creationists began to promote what they called "**creation science**", a variety of creationism that omitted explicit biblical references. And unable to ban the teaching of evolution, they argued that creation science should be granted **equal time** in the science class. At the end of the 1970s, increasing fundamentalist activism culminated in the election of a president, Ronald Reagan, who supported the equal time stance.

By the 1990s, crushed by constitutional bans on teaching creation science, the religious opponents of evolution re-grouped around a new policy, **intelligent design** (ID). The intellectual godfather of the movement was American law professor **Phillip Johnson** (b.1940), who turned to evangelical Christianity after a failed marriage. Johnson made it his mission in life to fight against the philosophy of naturalism, which for him was epitomized by evolution (although he has never argued *for* creationism or Biblical literalism). Despite a lack of any formal education in biology, in his 1991 book *Darwin on Trial*, Johnson employed a lawyerly approach, aiming to cast "reasonable doubt" on the theory of evolution, while providing little in the way of detailed or testable alternative explanations for biological phenomena.

In early 1992, Johnson met with a number of scientists and philosophers sympathetic to his views at a conference held in Dallas, including several key figures of the later intelligent design movement (IDM): Michael Behe, Stephen Meyer and William Dembski. Four years later, Johnson and his associates established the Center for the Renewal of Science and Culture (simplified in 2002 to the **Center for Science and Culture**) under the auspices of a conservative think tank, the **Discovery Institute**, based in Seattle, Washington. By the late 1990s, Johnson was talking of a **wedge strategy**: "We call our strategy 'the wedge'. A log is a seeming solid object, but a wedge can eventually split it by penetrating a crack and gradually widening the split. In this case the ideology of scientific materialism is the apparently solid log." In 1999, a leaked CRSC document, often referred to as the **wedge document**, revealed that the CRSC's principal goal was not the disinterested advancement of science, but instead "nothing less than the overthrow of materialism and its cultural legacies," with ID replacing "it with a science consonant with Christian and theistic convictions".

Proponents of ID formulated a strategy that they hoped would circumvent constitutional issues, while allowing them to assemble a big tent of supporters: they stripped out most of the tenets of creationism and rather than a "creator", they took care to speak only of an unidentified "intelligent designer". By limiting themselves to vague assertions about intelligent design, they hoped to undermine evolution, but to avoid any association with religion. However, mindful of the risk of legal challenges if they directly advocated teaching of ID, the Discovery Institute and others in the ID movement hit on another subversive tactic: rather than insist that ID be taught, they proposed that schools should merely "**teach the controversy**". Ironically, this phrase originated with a secular left-wing liberal, Chicago English professor **Gerald Graff**, who applied it to questions such as whether *Huckleberry Finn* is a racist or an anti-racist text. Graff was not pleased at its misappropriation: "I felt as if my pocket had been picked when the intelligent design crowd appropriated my slogan." The teach-the-controversy approach provided ammunition for the Ohio education board's stance, adopted in 2002, in which the teaching of intelligent design was allowed but not compulsory. Even George W. Bush was suckered in, stating in 2005 that "both sides ought to be properly taught... so people can understand what the debate is about".

The Dover trial

Dover, Pennsylvania is a small town in Pennsylvania Dutch country, a part of America settled by German speakers in the eighteenth century. During the summer of 2004, creationists on the **Dover Board of Education** decided to target the teaching of evolution in the local high school. They established contact with the **Thomas More Law Center**, a Michigan-based conservative Christian not-for-profit law firm, which agreed to represent the school board should their decisions be subjected to legal challenge. In October 2004, in a six-to-three verdict, the Dover Board of Education voted to add intelligent design to their biology curriculum: "Students will be made aware of the gaps/problems in Darwin's theory and of other theories of evolution including, but not limited to, intelligent design."

The following month, a press release expanded on the new policy, requiring teachers to read a statement in the ninth-grade biology class, which described evolution as "a theory … still being tested … not a fact", while claiming that "Intelligent design is an explanation of the origin of life that differs from Darwin's view" and informing students of the availability of the "reference book, *Of Pandas and People*" (an anti-evolutionary textbook sponsored by the ID movement).

Several dismayed parents from the Dover school district contacted the American Civil Liberties Union (ACLU), an organization that had been a key player in several evolution-related trials, including the Scopes trial. In mid-December, the ACLU filed suit on behalf of eleven parents, headed by **Tammy Kitzmiller**, whose younger daughter Jess was scheduled to be among the first to face the pro-ID statement. Local lawyers Eric Rothschild and Steve Harvey supplied legal support for the plaintiffs *pro bono*, with help from the Pennsylvania ACLU's Vic Walczak. Support with the science was provided by the NCSE's Nick Matzke.

Despite early tentative involvement in the issues, the Discovery Institute kept the trial at arms length, fearful that it would prove to be a test case.

Several of its fellows were pencilled in as defence experts, but several withdrew before making any depositions, leaving just Behe and maverick microbiologist Scott Minnich to feature in the trial.

Courtroom clashes: creationism v. the Constitution

Epperson v. Arkansas, 1968 Little Rock science teacher Susan Epperson asks for a declaratory judgment on the Arkansas equivalent of the Butler Act. A one-day local trial establishes that prohibitions on the teaching of evolution violate the Establishment Clause. On appeal, the Supreme Court rules nine votes to nil in Epperson's favour.

Daniel v. Waters, 1975 Local schoolteachers challenge a Tennessee law requiring equal time for evolution and biblical accounts of creation in the classroom. A Federal District Court rules that the law is unconstitutional in promoting "a clearly defined preferential position for the Biblical version of creation".

McLean v. Arkansas, 1982 A Methodist minister challenges an Arkansas law requiring balanced treatment for "creation science". District court judge **William Overton** in ruling the act unconstitutional, bans creationism from Arkansas schools.

Edwards v. Aguillard, 1987 Parents, teachers and religious leaders challenge "equal time" legislation in Louisiana which, after a series of appeals, leads the Supreme Court to deliver a 7–2 verdict that the Louisiana law violates the Establishment Clause. The Supreme Court decision bans the teaching of creationism in public school science classes across the entire USA.

Peloza v. New Capistrano School District, 1994 Creationists try to argue that "evolutionism" is a religious stance and so should be excluded on constitutional grounds from US schools. A California court not only rejects the argument, but also concludes that the plaintiff's case was frivolous and orders him to pay costs.

Webster v. New Lenox School District, 1990 An Illinois teacher sues a school district for violating his right to free speech by preventing him from teaching "creation science". The court concludes that the school district has a right to restrict teaching to the specified curriculum and that teaching "creation science" in public schools is illegal.

Freiler v. Tangipahoa Parish Board of Education, 1999 Louisiana parents sue the local school board over the requirement that a disclaimer be read out before any discussion of evolutionary biology in the classroom. On appeal, the decision is upheld by a higher federal court and then ratified by the Supreme Court.

Selman v. Cobb County School District, 2005 A Georgia court rules that the creationist strategy of adding "Evolution is a theory, not a fact" stickers to textbooks is unconstitutional and orders "the sticker must be removed from all of the textbooks into which it has been placed". At retrial, the case is settled out of court in favour of the plaintiffs.

Courtroom battles in Harrisburg

The trial was held in Harrisburg from late September to early November 2005. Sitting in judgement was **Judge John E. Jones III** (b.1955), a Republican, born and raised in Pennsylvania and appointed by George W. Bush in 2002. In his opening statement attacking the board's decisions, Rothschild argued that there was no such thing as a minor constitutional violation, set out a definition of creationism based on an early draft of *Of Pandas and People* and stressed that intelligent design was not science. In defence of the board's stance, Patrick Gillen argued that the policy changes were minor and were not motivated by a religious agenda.

Kenneth Miller, a biology professor, textbook author and committed Catholic, was first to testify for the plaintiffs. Miller argued that ID relied on "the argument from ignorance, which is to say that, because we don't understand something, we assume we never will, and therefore we can invoke a cause outside of nature, a supernatural creator or supernatural designer". He pointed out that calling biological systems "designed" equated to a creationist claim, i.e. that they had been created. Miller expressed anger "as a person of faith" with the Dover board's "false duality" ("your Bible-friendly theory" versus "your atheist theory"), which meant that children were "forced to choose between God and science".

In their testimony, the plaintiffs laid out their views as ordinary men and women, stressing their own religious convictions and providing eyewitness accounts of the board's creationist agenda. Philosophy professor **Barbara Forrest** gave an account of the history of the intelligent design movement, stressing its origins in creationism and detailing the religious subtext to its wedge strategy. Drawing on behind-the-scenes work by Nick Matzke, Forrest read from a 1986 draft of *Of Pandas and People* (my italics): "*Creation* means that the various forms of life began abruptly through the agency of an intelligent *creator*..." A crude find-and-replace job, after the Edwards v. Aguillard verdict, had turned this into "*Intelligent design* means that the various forms of life began abruptly through an intelligent *agency*..." Although not raised in the trial, Matzke also spotted a "missing link" in the evolution of creationism into intelligent design – a passage where an imperfect edit had turned "creationists" into "cdesign proponentsists".

Later in the trial, palaeontologist Kevin Padian provided the courtroom with an account of cladistic classification and a detailed critique of common creationist/ID fallacies about the fossil record. Padian hammered home his message by reference to the established literature on the evolution of the middle ear and a flurry of recent reports of fossilized feathered

The Catholic response to evolution

The Catholic Church has a long tradition of ecclesiastics who have rejected a literal reading of the Bible. Early church father **Origen** (c.185–c.254) wrote: "What intelligent person will suppose that there was a first, a second and a third day, that there was evening and morning without the existence of the sun and moon and stars? ... these things, by means of a story which did not in fact materially occur, are intended to express certain mysteries in a metaphorical way." In the fifth century **Augustine** of Hippo (354–430) argued that the Bible should not be interpreted literally if it contradicts what everyone knows from common sense and experience.

In the 1890s the Inquisition silenced American priest John Zahm, when he tried to harmonize evolution and Catholic teachings in his *Evolution and Dogma*. However, within fifty years of the publication of *The Origin of Species*, Pope Pius X had ratified a decree that "special creation" applied only to man, not to the other species. In 1950, Pius XII accepted the legitimacy of scientific investigation into the evolution of the human body, but affirmed a divine origin for the human soul. In 1996, **Pope John Paul II** (1920–2005) spoke up for evolution: "new findings lead us toward the recognition of evolution as more than an hypothesis ... convergence in the results of these independent studies ... constitutes in itself a significant argument in favour of the theory", but went on to reject any materialistic explanation for the human soul.

Before becoming **Pope Benedict XVI** (b.1927), Joseph Ratzinger endorsed Catholic acceptance of scientific views of the age of the universe, common descent of all living organisms and "support for some theory of evolution to account for the development and diversification of life on earth". He also wrote of "the inner unity of creation and evolution and of faith and reason", claiming "We cannot say: creation or evolution, inasmuch as these two things respond to two different realities."

Shortly after Ratzinger became pope in 2005, **Christoph Schönborn**, Archbishop of Vienna, unleashed a storm of controversy when he wrote in *The New York Times*: "The Catholic Church, while leaving to science many details about the history of life on earth, proclaims that by the light of reason the human intellect can readily and clearly discern purpose and design in the natural world, including the world of living things. Evolution in the sense of common ancestry might be

dinosaurs. Additional expert witnesses for the plaintiffs included theologian John Hought and philosopher Robert Pennock.

Monty Python's flying creationism

First up for the defence in the Dover trial was **Michael Behe** (b.1952), biochemistry professor and ID supporter, whose favourite arguments true, but evolution in the neo-Darwinian sense – an unguided, unplanned process of random variation and natural selection – is not." Schönborn's critics included the Vatican's chief astronomer, George Coyne, who weighed in heavily against ID: "Intelligent design isn't science even though it pretends to be. If you want to teach it in schools, intelligent design should be taught when religion or cultural history is taught, not science." Schönborn later clarified that he has no problem with theistic evolution and admitted that "Darwin undoubtedly scored a brilliant coup, and it remains a great oeuvre in the history of ideas. With an astounding gift for observation, enormous diligence, and mental prowess, he succeeded in producing one of that history's most influential works."

The teaching of evolution has proven uncontroversial in Catholic schools (or perhaps because it is too controversial, it is simply ignored). Several Catholics have adopted frontline positions in defence of evolution (most notably Kenneth Miller). However, the Catholic Church still allows enough wiggle room for creationists to stay in the church and at least one American Catholic lay organization, The Kolbe Center for the Study of Creation, promotes young-Earth creationism.

Cardinal Christoph Schönborn.

on the bacterial flagellum had already been demolished by Kenneth Miller (see box, p.304). While trying to defend ID as science, Behe conceded "there are no peer-reviewed articles by anyone advocating for intelligent design". He admitted that his definition of "theory" was so loose as to include astrology and that the plausibility of the argument for ID depended upon the extent to which one believes in the existence of God.

From the Wizard of Oz to the flying spaghetti monster

In the film version of *The Wizard of Oz*, the state of Kansas is immortalized in Dorothy's line, "Toto, I've a feeling we're not in Kansas anymore." When in 1999, the Kansas State Board of Education voted to drop evolution from the state's science teaching standards, Maryland physics professor Bob Park responded on his news website *What's New*, with the gibe "Uh, sorry Dorothy, it's Kansas all right – Oz is not this strange." But when, six years later, the Kansas State Board of Education approved the teaching of intelligent design, unemployed physics graduate Bobby Henderson came up with a more unusual response. Henderson wrote an open letter to the education board, professing his belief in a supernatural creator called the **flying spaghetti monster** (or FSM). Henderson claimed that his belief system was just as valid as the notion of intelligent design, and expressed his hopes for equivalent treatment: "I think we can all look forward to the time when these three theories are given equal time in our science classrooms across the country, and eventually the world; one-third time for intelligent design, one-third time for flying spaghetti monsterism, and one-third time for logical conjecture based on overwhelming observable evidence."

Henderson's letter didn't win over any hearts and minds at the education board, but it did spawn a complex parody religion, dedicated to the worship of a **spaghadeity** made of pasta and meatballs. Followers became known as **Pastafarians**, while those that waver in their faith are termed **spagnostics**. The religion's founder established a canonical website (venganza.org) and in 2006 published *The Gospel of the Flying Spaghetti Monster*. Pastafarian iconography has flourished, with a noodly recreation of Michelangelo's *Creation of Adam* and a proliferation of FSM logos (parodies of the Jesus fish). The central teachings of the religion include the assertion that there is an invisible and undetectable flying spaghetti monster, who created the entire universe after drinking heavily (accounting for the world's numerous examples of unintelligent design), and who has subsequently tested the faith of Pastafarians by continually seeding the universe with erroneous evidence for evolution. Pastafarian heaven contains beer volcanoes and a stripper factory – Pastafarian hell is similar, except that the beer is stale and the strippers all have a sexually transmitted disease. Pirates are venerated, particularly the mythical **Captain Mosey**. Pastafarians celebrate **International Talk Like a Pirate Day** on 19 September. In a parody of anti-evolutionist claims that Darwinism leads to immorality, Pastafarians are united in the belief that "global warming, earthquakes, hurricanes, and other natural disasters are a direct effect of the shrinking numbers of pirates since the 1800s". The FSM has been publicised in contexts as diverse as a New York Dolls video, British sitcom *The IT Crowd*, *Playboy* magazine, Greydon Square's atheist rap music and an episode of *South Park*. The movement even has its first martyr in Bryan Killian, an American student who was suspended from high school for wearing pirate regalia to mark his Pastafarian faith.

A Pastafarian take on the creation of Adam.

During cross-examination, Behe reiterated a claim that "the scientific literature has no detailed testable answers on how the immune system could have arisen by random mutation and natural selection". In response, as summarized by the trial judge: "Behe... was presented with fifty-eight peer-reviewed publications, nine books, and several immunology textbook chapters about the evolution of the immune system; however, he simply insisted that this was still not sufficient evidence of evolution..." Rothschild's cross-examination then laid into Behe's claim that "intelligent design theory focuses on the proposed mechanism of how complex biological structures arose". The resulting exchange, often parodied as **Monty Python's flying creationism**, included a trail of evasions from Behe in response to Rothschild's repeated requests for a detailed description of any such mechanism.

The case for the defence also relied on the testimony of sociology professor **Steve Fuller** who argued that the ID concept had a long pedigree among scientists and that the context in which a scientific discovery originates was irrelevant to its validity (i.e. ID's religious origins didn't matter). Bizarrely, Fuller also advocated an "affirmative action" programme for intelligent design. Several board members also testified for the defence, revealing their ignorance of what was meant by intelligent design. Bill Buckingham told of his addiction to the painkiller OxyContin, while Alan Bonsell was caught **lying under oath** when he claimed not to know who was involved in the donation of copies of *Of Pandas and People* to the Dover High School.

A few days after the end of the trial, in fresh elections to the Dover School Board, those who had voted for the ID policy were thrown out of office. The election results drew a barbed response from American TV evangelist Pat Robertson: "I'd like to say to the good citizens of Dover: if there is a disaster in your area, don't turn to God – you just rejected Him from your city." Although the new school board was opposed to the ID policy, it was still keen to have a verdict on the case and so allowed the legal proceedings to continue. However, the change in the board's make-up means that there will never be any appeal against the judge's decision.

Wise words from Judge John E. Jones III

Anyone expecting a Republican Bush appointee to engage in a whitewash was in for a shock. Just over six weeks after the end of the Dover trial, Judge Jones delivered his 139-page decision, in which he concluded, "**the Board's ID Policy violates the Establishment Clause**". In dismissing ID as religion, not science, Jones pulled no punches: "For the reasons that follow, we conclude **that the religious nature of ID** would be readily apparent to an objective observer, adult or child … writings of leading ID proponents reveal that the designer postulated by their argument is the God of Christianity … The overwhelming evidence at trial established that ID is a religious view, a mere re-labelling of creationism, and not a scientific theory … ID cannot uncouple itself from its creationist, and thus religious, antecedents." And as if one knockout blow were not enough, Jones found that "**ID fails on three different levels**, any one of which is sufficient to preclude a determination that ID is science. They are: (1) ID violates the centuries-old ground rules of science by invoking and permitting supernatural causation; (2) the argument of irreducible complexity, central to ID, employs the same flawed and illogical contrived dualism that doomed creation science in the 1980s; and (3) ID's negative attacks on evolution have been refuted by the scientific community." He also saw through the teach-the-controversy tactic: "This tactic is at best disingenuous, and at worst a canard. The goal of the IDM is not to encourage critical thought, but to foment a revolution which would supplant evolutionary theory with ID." However, Jones saved his bluntest condemnation for the members of the school board, whom he called liars: "the Dover School Board members' testimony … was marked by selective memories and outright lies under oath … It is ironic that several of these individuals, who so staunchly and proudly touted their religious convictions in public, would time and again lie to cover their tracks and disguise the real purpose behind the ID policy."

Jones criticized the "breathtaking inanity of the Board's decision ... The students, parents, and teachers of the Dover Area School District deserved better than to be dragged into this legal maelstrom, with its resulting utter waste of monetary and personal resources..."

Creationist Kudzu

It would be wrong to conclude that the Kitzmiller v. Dover verdict has delivered the knockout blow to creationism. In April 2008, proponents of ID released an outlandish documentary, *Expelled: No Intelligence Allowed*, a mishmash of anti-science propaganda, misinformation and misquotes. Lauri Lebo, a journalist who covered the Dover trial, points out the film fails even by its own logic: "The first half of the movie is devoted to explaining how intelligent design is not religion" and then "the filmmakers seem to completely forget their earlier message. The rest of the movie is devoted to proving that atheistic scientists hate God and are trying to suppress intelligent design because, well, it's all about belief in God". The film also drew a lawsuit from Yoko Ono over unauthorized use of Lennon's *Imagine* and a sharp rebuke from the Jewish Anti-Defamation League: "The film ... misappropriates the Holocaust and its imagery ... Using the Holocaust in order to tarnish those who promote the theory of evolution is outrageous and trivializes the complex factors that led to the mass extermination of European Jewry." At the time of writing (mid-2008), the sixty-day news feed on the NSCE website documents ongoing court cases or new legislative attempts to open the door to creationism in Florida, Michigan, Alabama, Louisiana and Missouri. Like kudzu or bindweed cut back in the garden, the contagion of creationism simply springs up somewhere else. The great American orator Ingersoll once wrote that Darwin "broke the chains of superstition and filled the world with intellectual light". In the fight against those who would re-apply the chains and extinguish the light, as Jefferson put it, "the price of freedom is eternal vigilance".

Creationism in Europe

Although creationism has gained most ground in the USA, it has also made inroads into Europe. Opinion polls put support for young-Earth creationism as high as 20 percent in some European countries. In Germany, the organization **Wort und Wissen** promotes creationism with the pseudoscientific "textbook" *Evolution: Ein kritisches Lehrbuch*, which dismisses macroevolution, but thanks to its low price and attractive design outsells conventional textbooks. One of its authors, Siegfried Scherer, was a fellow of the Discovery Institute, until he fell out with the organization over its policy of promoting ID in schools.

Elsewhere in Europe, creationism has surfaced in the news repeatedly in recent years. In 2002, controversy erupted in England over government backing for schools funded by Christian millionaire car dealer Peter Vardy, in the face of claims that the schools included creationism in their curriculum. In 2004, Italian education minister **Letizia Moratti** removed evolution from middle-school curricula, fearing that it might encourage a materialist view of life. That same year, Serbian education minister **Ljiljana Čolić** ordered Serbian schools to suspend the teaching of evolution unless they also taught creationism. The ban prompted an outcry from Serbian scientists and local bishops. A few days later, the Serbian prime minister reversed the decision and Čolić was forced to resign. The following year, the Netherlands was ridiculed as the "Kansas

Is Darwin kosher? Jewish responses to evolution

In the Middle Ages, several Jewish scholars, including **Maimonides** (1135–1204), **Nahmanides** (1194–c.1270) and **Gersonides** (1288–1344), suggested that the account of creation given in the Torah should not be read literally, but interpreted symbolically. One of Nahmanides' disciples, **Isaac ben Samuel** of Acre, calculated the age of the universe as 15 billion years, reckoning that each day in the eyes of God equates to a thousand solar years. In the early nineteenth century, several rabbis attempted to reconcile new findings from palaeontology with the Jewish mystical tradition of **Kabbalah**.

Around the same time, the **Jewish Enlightenment** (Haskalah) brought European Jews into contact with the wider intellectual life of the continent. On reading *The Origin*, Polish-born **Naphtali Halevi** (1840–94) sought to reconcile evolution with the Torah. In 1876, Halevi sent Darwin a long essay in Hebrew, *Toldot Adam*, which Darwin mentioned in his autobiography: "an essay in Hebrew ... showing that the theory is contained in the Old Testament". In a covering letter, Halevi addresses Darwin: "To the Lord, the Prince, who 'stands for an ensign of the people' (Isa. xi. 10), the Investigator of the generation, the 'bright son of the morning' (Isa. xiv. 12), Charles Darwin, may he live long!" In the essay, Halevi makes an argument, drawn from a rather idiosyncratic analysis of word use in the Torah, that there were no irreconcilable contradictions between Darwin's evolution and the Genesis account of creation.

Modern Jewish responses to evolution span a range of opinions. Evolution has elicited little controversy in the Conservative and Reform communities. However, many Orthodox rabbis remain young-Earth creationists. **Menachem Mendel** of Europe", after education minister **Maria van der Hoeven** suggested that discussion of intelligent design in schools might promote dialogue between religious groups.

In 2006, the **League of Polish Families**, a far-right coalition partner in the government of Jarosław Kaczyński, launched an aggressively anti-evolution campaign. In response, the Polish Academy of Sciences published an open letter in Polish newspapers, defending evolution. Fortunately, the league lost all its seats in the subsequent Polish election. Also in 2006 the creationist group Truth in Science sent information packs to every secondary school in the UK. In response, the **British Centre for Science Education** was set up. In late 2006, schoolgirl Maria Shraiber and her father launched a lawsuit against Russia's education authorities over the compulsory teaching of evolution in schools; a court in St Petersburg rejected the suit a few weeks later.

In late 2007, a committee reporting to the Council of Europe issued a report on the threat of creationism in European schools, which concluded

Schneerson (1902–94), the last world leader of the Lubavticher Hassidim, was strongly opposed to evolution and most Hassidim share his view, including ex-Lubavitcher TV celebrity rabbi, Shmuley Boteach. **Moshe Feinstein** (1895–1986), a key arbiter in Jewish law for the ultra-Orthodox (Haredi) community, ruled that even reading about evolution was forbidden. In the early 1990s, an Israeli dairy, hoping to capitalize on the *Jurassic Park* craze was threatened with loss of kosher status for selling milk with free silver dinosaur stickers – the rabbi who awarded the kosher status protested, "This is like seeping sacrilege ... Dinosaurs symbolize a heresy of the creation of the world because they reflect Darwinistic theories."

Natan Slifkin is an Orthodox rabbi whose books on natural history have earned him the nickname "**the Zoo Rabbi**". Ten years after his move to Israel in 1995, a group of Haredi Rabbis banned his books in their communities. Opinion among Orthodox Jews rapidly polarized, with intense opposition to the ban matched by a decision by Slifkin to publish an updated version of his views as *The Challenge of Creation*. **Gerald Schroeder**, an Israeli physicist from the Modern Orthodox tradition, has attempted to use relativistic time to reconcile the Biblical accounts with a universe that is billions of years old. Schroeder also squares Genesis with human evolution by equating the creation of Adam to the appearance of writing. However, Schroeder remains wedded to the idea of intelligent design. Fortunately, the Jewish community in the US, keen to maintain separation between church and state, has leant little or no support to the teaching of ID or the teach-the-controversy stance in American public schools.

Muslim responses to evolution

Early in the twentieth century, Muslim poet, philosopher and politician **Sir Muhammad Iqbal** (1877–1938), venerated as *Muffakir-e-Pakistan* (Thinker of Pakistan), argued for reconciliation between Islam and evolution, claiming Muslim priority for the very idea of evolution. Although most Muslims still adopt a literalist interpretation of the Qur'an, Islam views Genesis, with its young-Earth chronologies, as a corrupted version of the divine message. As Qur'anic creation accounts are less precise in their timescales than those in Genesis, liberal Muslims argue that theistic interpretations of evolution are compatible with Islam, so long as they allow for the special creation of man. Conservative Muslims, particularly Wahabis, reject evolution completely. For them, the idea of random variation (the seed corn of natural selection) conflicts with the view that everything happens according to the will of Allah.

In 2006, the national science academies of several Muslim countries, including Turkey, Indonesia and Pakistan, signed up to an international statement of support for the teaching of evolution. Nonetheless, in recent years, there has been a resurgence of creationist activism in Indonesia, Malaysia and among Muslim minorities in the West. In the UK, Muslim university students are now more likely to challenge evolution in the classroom than Christians. However, the epicentre of Islamic creationism is Turkey, where the creation-evolution debate is embedded in a nationwide struggle between secular modernism and Islamic traditionalism. The most notorious Turkish creationist is Adnan Oktar (b.1956), who defends old-Earth creationism under the pseudonym **Harun Yahya.** In the late 1980s, Oktar spent over a year in a psychiatric hospital, after writing his first book *Judaism and Freemasonry*. Since then Oktar has gone on record as a Holocaust denier and has dismissed intelligent design as "another of Satan's distractions". In 2006, Oktar published his *Atlas of Creation*, a lavishly illustrated, poorly argued tome, which was sent unsolicited to scientists and schoolteachers throughout Europe and the United States. Within his native Turkey, Oktar has attempted to stifle criticism through several lawsuits. However, at the time of writing (May 2008), Oktar has just been sentenced to three years in jail for a variety of crimes including engaging in illegal threats and creating an illegal organization for personal gain.

Turkish creationist Adnan Oktar (also known as Harun Yahya).

"creationism could become a threat to human rights which are a key concern of the Council of Europe ... The war on the theory of evolution and on its proponents most often originates in forms of religious extremism which are closely allied to extreme right-wing political movements ... some advocates of creationism are out to replace democracy by theocracy."

What is wrong with creationism?

If the reader has got this far in the book and, having worked through all the positive evidence for evolution already presented, still wants to ask this question, then nothing I can say is likely to have any effect. Plus, in a *Rough Guide to Evolution*, there is precious little space for a detailed rebuttal of every point ever raised by creationists. The reader is instead directed to the vast existing literature on this topic (see p.329).

Nonetheless, it is worth outlining a few key objections. Firstly, creationism does not just undermine biology – biological evolution is not sealed off from the rest of science. Instead, it is supported by what Darwin's contemporary, the Cambridge philosopher William Whewell, called a consilience of inductions, in other words, independent evidence from the rest of the natural sciences reinforces our grounds for accepting evolution. As Pope John Paul II put it: "In fact it is remarkable that this theory has had progressively greater influence on the spirit of researchers, following a series of discoveries in different scholarly disciplines. The convergence in the results of these independent studies – which was neither planned nor sought – constitutes in itself a significant argument in favor of the theory." This means if you wish to throw away evolution in favour of young-Earth creationism, you also have to throw away archaeology, comparative linguistics, Earth and planetary sciences, astronomy and cosmology. In effect, you are signing up to the mother of conspiracy theories, in which all professional scientists are liars or fools. It is no coincidence that American creationism comes from a land where large segments of the population also entertain the idea of alien abduction, faked Moon landings, and their own government's involvement in the assassination of JFK, the Oklahoma bombings or 9/11. Or that Turkish creationist Harun Yahya's *Atlas of Creation* advertises a catalogue of publications advocating Jews-and-freemasons-control-the-world theories.

Although there is no space here to discuss every claim made by creationists, as we have seen in Chapter 1, discrediting the idea that the Earth is less than ten thousand years old is easy, using evidence from sources as diverse as radioisotopes, tree rings, lake sediments and ice cores. However, as an alternative to defending the indefensible, some young-Earth creationists accept that the evidence points to an ancient Earth, but argue that God created the Earth with an "appearance of age". This is often called the **Omphalos hypothesis**, after an 1857 book of that title by English naturalist Philip Gosse. Gosse argued that even if creation occurred from nothing, the creator would necessarily leave traces of previous existence that had never actually occurred. Although Adam was never hooked up to a placenta, he required a navel (*omphalos* in Greek) because it made him a complete human being. Similarly, God must have created trees with rings that they never grew and rocks with a fossil record of life that never actually existed. In *The Genesis Flood*, Whitcomb and Morris extend the argument to include the creation of light that appears to be coming from stars more than ten thousand light years away. This kind of thinking has drawn adverse responses from Catholic scientist Kenneth Miller and Orthodox Rabbi Natan Slifkin, who both reject it as depicting God as a **dishonest charlatan**. A secular response, last Thursdayism, proposes that by this logic, the world might just as easily have been created last Thursday, but with the appearance of age such as false memories and fictitious history books. There is even a parody religion, **The Church of Last Thursday**.

Old-Earth creationists accept that the Earth and the universe are much older than ten thousand years, but still reject common descent. Instead, they envisage divine intervention at various stages in geological time resulting in the sudden appearance of new types of plants and animals. The numerous transitional forms in the fossil record and abundant evidence for common descent allow us to dismiss this argument on scientific grounds (see Chapter 4). However, for Kenneth Miller, this view of life also fails theologically: Miller parodies the implied deity as "**God the magician**" and points out how many miraculous interventions would be required just to create all the known members of the elephant family. And given that so many species become extinct so often, it implies that the creator is not very good at his or her job.

...and with intelligent design?

Although its proponents would have us believe that intelligent design is something new, in fact it is merely a re-packaged version of the watch-maker argument visited in Chapter 1. In Darwin's day, this argument was applied to everyday entities like the eye; ID proponents like Behe now attempt to blind us with science by applying it to objects less accessible to the non-expert, such as molecular complexes or biochemical pathways. But the argument still fails for the same old reasons and more.

Firstly, the suggestions that this or that microscopic or biochemical entity appears to be "intelligently designed" haven't take us any further forward scientifically. Austrian-born philosopher Karl Popper laid down one of the main criterion for scientific enquiry in the mid-twentieth century, namely that, for a hypothesis to be scientific, it has to be falsifiable. In other words, the hypothesis has to generate predictions and there has to be some observation you can make, or experiment you can perform, that could show that these predictions are wrong. A fertile hypothesis will generate many testable predictions. Such a hypothesis becomes an established theory once repeated attempts to falsify its predictions have failed. And the best theory will explain the most facts with the fewest assumptions (a principle philosophers call **inference to the best explanation**).

At heart, ID is a lawyer's argument, trying to cast "reasonable doubt" on evolution by bringing up some supposed gap in the theory. But unlike evolution by natural selection, intelligent design is not a fertile *scientific* hypothesis and it generates hardly any testable predictions, as evidenced by the lack of any productive experimental research programme on the part of the ID movement. In fact, when attempts have been made to test the few predictions that flow from the ID claims, the predictions have failed (see box, p.304). In addition, the ID/watchmaker argument has little or no explanatory power; instead of providing mechanistic explanations of how systems work, it retreats into agnosticism, a shrug of the shoulders, we don't know why the designer did things one way rather than another. Darwin dismissed this approach as unscientific: "On the ordinary view of the independent creation of each being, we can only say that so it is; that it has pleased the Creator to construct all the animals and plants in each great class on a uniformly regulated plan: but this is not a scientific explanation."

Another shortcoming of the ID proposal is a lack of mechanism. As noted above, during cross-examination in the Dover trial, Behe refused even to speculate what mechanisms might be at work in intelligent design. Did the designer miraculously create the genes encoding the designed system from nothing? Or did he stack the odds in their favour by designing just the right variants for natural selection to work on? Darwin dismissed this latter option in *Variation of Plants and Animals under Domestication* (1868), by making an analogy with an architect's use of rocks that had fallen off a cliff: "Can it be reasonably maintained that the Creator intentionally ordered ... that certain fragments should assume certain shapes so that the builder might erect his edifice?"

Another problem with intelligent design is that in effect it boils down to an argument from incredulity: because I cannot see how this system could have evolved, it must be therefore have been designed. Unlike those using the watchmaker argument, ID proponents pretend that they are not arguing on religious grounds, as far as they are concerned, the intelligent designer could be anyone, even an alien intelligence. But to look for external interference in a process that can be explained without it breaks a key rule-of-thumb known as **Occam's razor**, named after English

Don't want to believe in evolution?

Is it possible to be a rationalist (a believer in the laws of logic) but not believe in evolution? Just about! There are several philosophical showstoppers that bring rational argument to a halt. But they are all pretty mind-bending! The first showstopper is **metaphysical solipsism**: the belief that you, the reader, is all there is and that this book and this author, this world and the evolution of life in it, are all just figments of your imagination. However, it is scarcely possible to hold this belief in your mind for even a minute and, as English philosopher Bertrand Russell once pointed out, solipsism "is rejected in fact even by those who mean to accept it. I once received a letter from an eminent logician... saying that she was a solipsist, and was surprised that there were no others. Coming from a logician and a solipsist, her surprise surprised me."

One modern variant on solipsism is the **brain-in-a-vat** idea, taken seriously by Berkeley philosopher Barry Stroud. In this scenario, your brain has been removed from your body, placed in a vat of life-sustaining liquid and your neurons hooked up to a supercomputer that provides you with a virtual reality indistinguishable from any "real" reality. So, the argument goes, if you are in a vat, all your conclusions about evolution in the real world are false. And, as you have no way of knowing whether you are in a vat or not, this leaves you free to doubt the reality of evolution.

But why suppose you ever had a body in the first place, why not suppose you are a disembodied brain created yesterday with false memories of a biological world built by evolution? Some cosmologists are seriously discussing the idea of **Boltzmann brains**, self-conscious entities that arise from random fluctuations in vacuum energy (named after Austrian physicist Ludwig Boltzmann, who suggested that the whole universe resulted from such a fluctuation). If the universe lasts long enough, such entities are inevitable, say the cosmologists. But why stop at a brain – viewing yourself as a **Boltzmann-brain-in-a-vat** breaks none of the laws of physics and also gets you off the hook of having to believe in evolution.

scholastic philosopher William of Ockham (c.1288–c.1347). Occam's razor states that the preferred explanation of any phenomenon should make as few assumptions as possible.

Let's illustrate this with an example from outside biology: **Stonehenge**, a prehistoric circle of standing stones in southern England. That prehistoric Britons could even erect the stones is remarkable enough, but more surprising still is the geological evidence that the stones originated 120 miles away in Wales. However, Occam's razor means that we do not rush to a supernatural explanation (that the wizard Merlin summoned up the stones by magic) or even an extraterrestrial explanation (aliens helped make the monument). Instead, there are plenty of mundane explana-

Another showstopper, popularized by the *Matrix* films, is **the simulation hypothesis**. According to this viewpoint, we are all living in a simulated reality, run on a computer powerful enough to create a internally consistent simulation, so detailed that it could not be distinguished from "real" reality. Swedish philosopher Nick Bostrom argues that it is more likely than not that we are living in such a simulation. His argument rests on the assumption that any sufficiently advanced civilization capable of creating simulations that contained intelligent individuals would be unlikely to restrict itself to a single simulation, instead, it would run billions of them. Thus, he asks, why suppose that we are the one civilization that develops the simulations rather than one of the billions run in simulation? Richard Dawkins points out that this merely pushes the need for evolution back stage as the only plausible source of the intelligences running the simulation.

Mathematical physicist **Frank Tipler** has controversially attempted to interweave cosmology, simulation and religion. He posits that as the universe comes to an end in a singularity, the computational capacity of the universe will outrun time, so that an intelligent civilization could run an infinite simulation within a finite time. Tipler terms this final state of infinite information **the omega point**. Recently, Tipler has come to identify his omega point with God and to equate the associated infinite simulation with the resurrection of the dead. But why not assume we are already dead in Tipler's sense, i.e. already living in his omega point simulation and thus free to dispense with any direct evolutionary explanation for our own origins?

How is an evolutionary biologist to respond to all this? The obvious response is to adapt a line from George Orwell and say that you have to be a real philosopher to believe all that, no scientist could be so foolish! In fact, insofar as none of these scenarios is verifiable, they fall outside the realm of science and bring no additional explanatory power. The evolutionary show is not over yet!

From the bacterial flagellum to the face on Mars

When confronted with William Paley's watch-watchmaker analogy and its modern variants, how is a scientist to proceed? One obvious way forward is to look more closely at the watch or its equivalent and see how many independent lines of evidence support the hypothesis of design. Thus, a watch doesn't look artificial just because of the intricacy of its mechanism; there are other discontinuities from nature. For example, it contains materials (glass, metals) of a purity and surfaces of a smoothness never seen in nature.

For years, creationists have highlighted the bacterial flagellum, a microscopic structure that allows bacteria to swim, as evidence of the creator's handiwork. At the start of the ID movement, biochemist Michael Behe re-badged it as an "irreducibly complex" system, a structure so complex that it could only function when all its components are present. For Behe, this means that the flagellum cannot have arisen by gradualistic Darwinian evolution from something

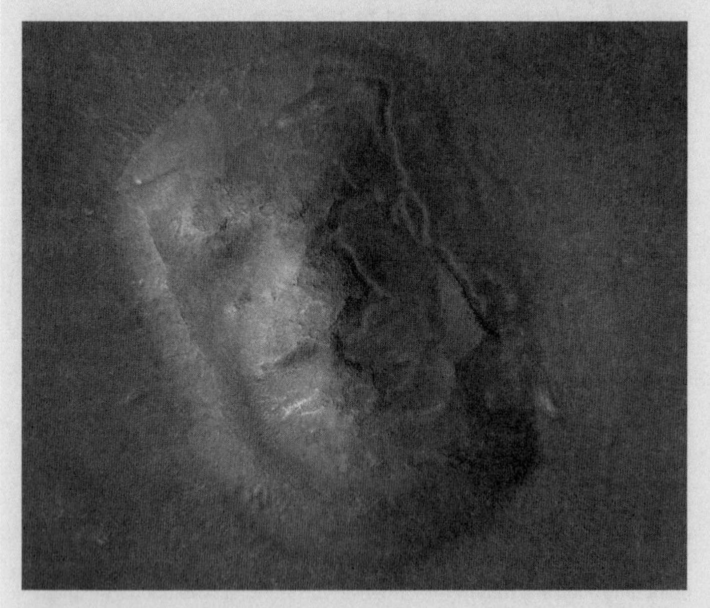

Is this the face of a humanoid or an ape? Neither, it's a Martian mesa, as this 2006 photograph reveals.

tions that rely on processes we already know about. For example, we can hypothesize that the stones were quarried in Wales then hauled to England, or that they were transported to England by glaciers during the Ice Age. Maybe even a single person could have erected the stones simpler. Fellow ID enthusiast William Dembski compares evidence of design in nature to the faces on Mount Rushmore. However, a more accurate view of the supposedly designed flagellum reveals it is more like the face on Mars. This Martian equivalent of Mount Rushmore, seen by some as a legacy of an alien civilization, was first revealed by a Viking 1 photographic survey of the surface of Mars in 1976. However, after analysis of the higher resolution Mars Global Surveyor data from 2006, NASA stated that "a detailed analysis of multiple images of this feature reveals a natural looking Martian hill whose illusory face-like appearance depends on the viewing angle and angle of illumination". No face, no artifact, just a completely natural mesa.

During the Dover trial, Kenneth Miller hacked away at the notion that the flagellum is irreducibly complex, pointing out that it houses a subsystem capable of performing a useful function (protein secretion) in the absence of the rest of the structure. After the Dover trial, Nick Matzke and I published a paper reviewing additional evidence in favour of viewing bacterial flagella as evolved, rather than designed, entities. We concluded that unlike a watch, there are no obvious discontinuities between bacterial flagella and the rest of the natural world. Flagella are made from variants of proteins that do jobs elsewhere – no equivalent of smooth surfaces or pure metals here! Even Behe accepts that all flagella are descended from a common ancestor, but there is no reason to stop there. Instead, an excellent case can be made for plausible precursors of the flagellum or simpler modules that performed other functions (even flagella that don't move are pretty good at helping bacteria stick to things).

What we didn't address was the theological problem of why would a beneficent deity design bacterial flagella, when they help bacteria cause some pretty unpleasant diseases, from cystitis to syphilis. In any case, since Nick and I wrote our paper, things have gotten even worse for Behe and company. In 2007, two research groups provided a dramatic example of how un-"irreducibly complex" the flagellum was by showing that it could still function even without what everyone had assumed was the engine that drove its assembly, and a third team of scientists showed how a precursor of the flagellum might have look, when they discovered a cut-down version on the surface of a bacterium that lived inside aphids. The bacterial flagellum is no more designed than the face on Mars; both have been sculpted by natural forces – one by the wind, the other by natural selection.

From *The Origin of Species* to the origin of bacterial flagella *Nature Reviews in Microbiology* 4:784-90

(retired construction worker Wally Wallington has been exploring this possibility using only pre-modern materials and approaches). Similar arguments apply whenever an evolutionary problem is subjected to similar scrutiny.

Richard Dawkins: biology's bard or Darwin's Rottweiler?

Born in Kenya but schooled in England, Richard Dawkins completed an undergraduate degree in zoology at the University of Oxford, then stayed on to gain a D. Phil (Oxford's equivalent of a PhD) under Danish ethologist Nikolaas Tinbergen. After a brief spell in California, Dawkins returned to join the faculty at Oxford in 1970, where he has remained ever since. He rose to fame in 1976 with his best-selling *The Selfish Gene*, notable not just for what it said, but because of the clarity of its prose, from the opening chapter: "The argument of this book is that we, and all other animals, are machines created by our genes" to the stunning conclusion: "We, alone on earth, can rebel against the tyranny of the selfish replicators." The book that followed, *The Extended Phenotype* (1982), marked Dawkins' last effort as a professional scientist, before a chrysalis-into-butterfly transformation into arguably the world's finest popular science writer. In the *Blind Watchmaker* (1986), Dawkins demolished Paley's watchmaker argument, placing DNA in the driving seat of evolution: "It is raining DNA outside. On the bank of the Oxford canal ... is a large willow tree, and it is pumping downy seeds into the air ... The whole performance ... is in aid of one thing and one thing only, the spreading of DNA around the countryside." In *River out of Eden* (1995), he hammers home the reality of common descent: "The river of DNA has been flowing through our ancestors in an unbroken line that spans not less than three thousand million years." In *Unweaving the Rainbow* (1998), his arresting opening blows away what he calls the anaesthetic of familiarity: "We are going to die, and that makes us the lucky ones. Most people are never going to die because they are never going to be born."

Over time, alongside the first Dawkins, keen to convey the wonders of nature, grew a second Dawkins. Irritated by creationists and drawing on the analogy of computer viruses, by 1991 Dawkins had come to see religion as a "virus of the mind". This new strident anti-religious stance led Oxford theologian Alister McGrath to dub Dawkins "Darwin's Rottweiler". But after 9/11, Dawkins' contempt for religion reached new heights: "Many of us saw religion as harmless nonsense ... September 11th changed all that. Revealed faith is not harmless nonsense, it can be lethally dangerous nonsense." In 2006, Dawkins established his not-for-profit organisation, The Richard Dawkins Foundation for Reason and Science and published *The God Delusion*, which went on to sell over a million copies. Unlike Stephen Jay Gould, who, though a non-believer, advocated that science and religion were "non-overlapping magisteria", Dawkins has no time for any compromise with religion. This has brought him a huge fandom among atheists, but has alienated not just creationists, but more moderate religious believers.

In his private life, Oxford don Dawkins is curiously entangled with that quintessentially English extra-terrestrial scientist Doctor Who. Twice-divorced Dawkins met his third wife, Lalla Ward (b.1951) through his friend Douglas Adams, who wrote for the show. Ward starred as the Doctor's companion Romana from 1979 to 1981 and in 2008 Dawkins made a guest appearance in the series as himself.

Richard Dawkins: eloquent exponent of evolution, fierce critic of religion.

There are many religious grounds for rejecting the intelligent design/watchmaker-in-disguise proposition. One is that it equates to a **God-of-the-gaps** argument. In implicating God as an explanation for those aspects of the natural world that science cannot yet explain, it makes belief in God as a hostage to fortune – as science progresses to explain more of the world, God's role tends to shrink as a result. Another religious objection comes from Kenneth Miller, who dismisses the IDM view of the designer as "**God the mechanic**", a cack-handed creator periodically called into fix his own his creation because he didn't get it right first time. Another way to look at the God of ID is as someone who invents a rich and complex game like chess (i.e. a rule-governed universe) and then decides to cheat at his own game.

What is so wrong with teaching the controversy?

As I hope I have shown in this book, there is no scientific controversy about evolution. Creationists attempt to dismiss it as "just a theory", rather than a fact. But this is based on a misunderstanding of what is meant by a "theory" in science. Scientists use the term "theory" to describe any coherent well-validated model that explains some aspect of the natural world. Examples include the Newton's theory of gravity, Dalton's atomic theory, Pasteur and Koch's germ theory of infection, as well as Darwin's theory of evolution. In many ways, theories are better than facts, because theories add explanation to what we observe. Although Popper's view of science portrays theories as always provisional and always open to refutation, in reality good theories soon harden into facts. None of today's chemists or bacteriologists really keeps an open mind as to the existence of atoms or germs, just as no biologist ever doubts that humans and chimpanzees share a common ancestor.

Why not "teach the controversy" anyway, to show students how to distinguish accepted science from pseudoscience or discarded science? But, in that case, why not also discuss astrology in the astronomy class, homeopathy or phlogiston theory in the chemistry lab or discarded theories about the flat or hollow earth in geography? Why stop at the Bible? Why not see how today's science measures up to Chocktaw or Chinese creation stories? The simple answer is that today's school and university science curricula are already too full to the brim with genuine science for us to waste time, year in year out, "slaying the slain", a favourite phrase of Huxley's, which prompts me to close with this parody from an 1861 edition of the magazine *Punch*:

> "To twice slay the slain,
> By dint of the Brain,
> (Thus Huxley concludes his review)
> Is but labour in vain,
> Unproductive of gain,
> And so I shall bid you 'Adieu'!"

Part 4
Resources

12

Darwin's places

Although his voyage on HMS *Beagle* made Charles Darwin a well-travelled man, after his return to Britain he did not leave his homeland again. The following guide concentrates on those places within the UK that are most closely associated with Darwin and his family.

Shrewsbury

Charles Darwin was born and brought up in this county town of Shropshire. Pronounced Shroozeberry or Shoozeberry by the locals and Shrozeberry by nearly everyone else, it is a handsome medieval city well worth a visit. In the centre of town, on the High Street you will find the plain and simple **Unitarian Church**, where the young Darwin worshipped with his mother. At the north end of the High Street, at Mardol Head, the **Darwin Gate** is a dramatic seven-metre public sculpture installed in 2004. Heading down Mardol Head and Shoplatch brings you to **Bellstone**, named after a granite boulder that you can reach through the iron gates to the Morris Hall. Here, each year at midday on Darwin's birthday, a toast is drunk to the great man.

Further up Bellstone on the left is Claremont Hill where at **no.13** the young Darwin studied at a school run by the Reverend George Case (now a block of flats). At the end of Claremont Hill sits the magnificent neoclassical **St Chad's Church**, completed just seventeen years before the baby Charles was baptized there. Across the road from the church is Quarry Park and the **Dingle**, a quarry-turned-garden, where the young Darwin fished for newts. Returning back along Claremont Hill and turning left into Barker Street brings you to **Shrewsbury Museum and Art Gallery** (formerly a seventeenth-century merchant's home and warehouse), which hosts displays on the Darwin family.

Back in the town centre, Pride Hill leads to Castle Street where, opposite the Castle, is the **old Shrewsbury School** building (now the town's **library**), which Charles attended as a boarder from 1818 to 1825.

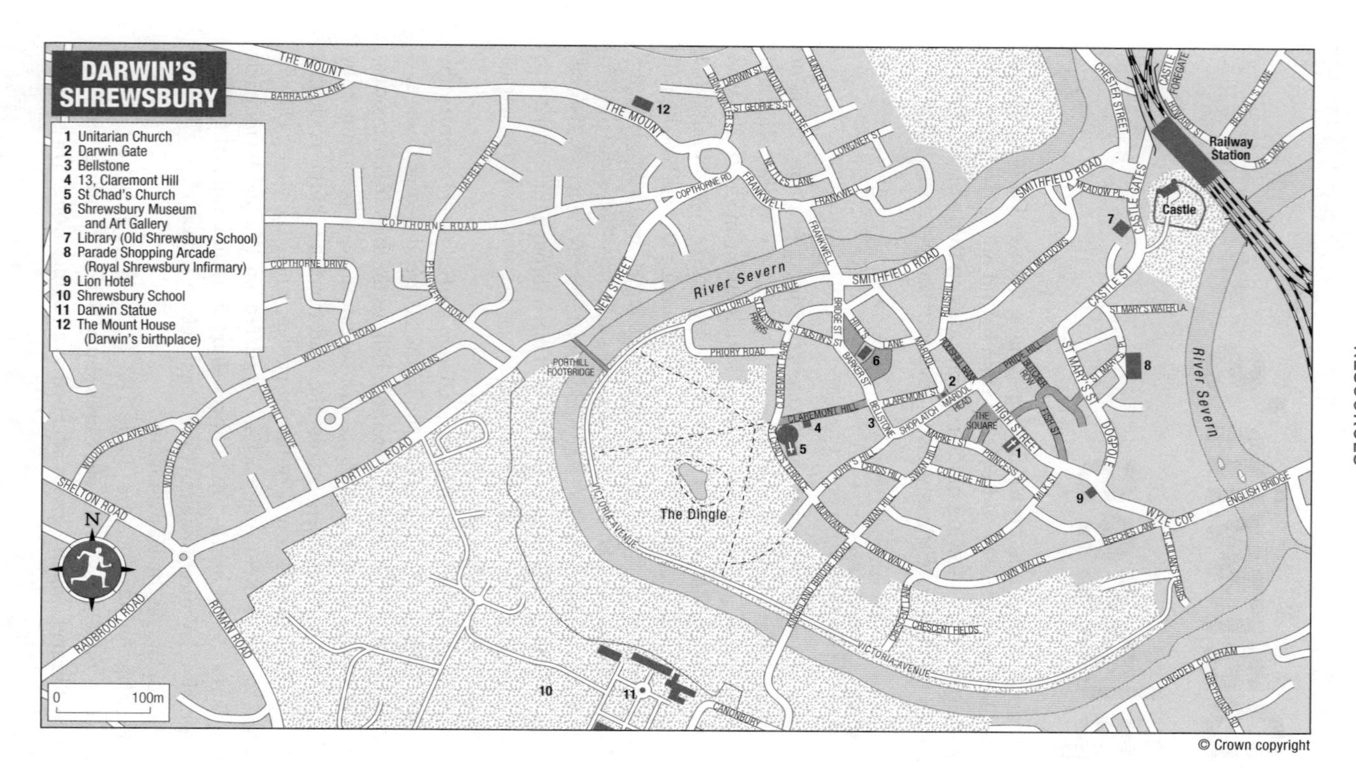
DARWIN'S SHREWSBURY
1 Unitarian Church
2 Darwin Gate
3 Bellstone
4 13, Claremont Hill
5 St Chad's Church
6 Shrewsbury Museum and Art Gallery
7 Library (Old Shrewsbury School)
8 Parade Shopping Arcade (Royal Shrewsbury Infirmary)
9 Lion Hotel
10 Shrewsbury School
11 Darwin Statue
12 The Mount House (Darwin's birthplace)
Railway Station
Castle
River Severn
The Dingle
English Bridge
Porthill Footbridge
0 100m
N
© Crown copyright

Just to the left of the entry gate is a plaque commemorating the school and its pupils, while upstairs in the music library you can see Darwin's old classroom. The imposing statue of an elderly seated Darwin by Horace Montford was erected in 1897. Other local Darwin sites include **The Royal Shrewsbury Infirmary** (now the Parade Shopping Arcade) where Robert Darwin worked as a doctor and **The Lion Hotel**, a seventeenth-century coaching inn close to the river on Wyle Cop, where Darwin would have caught the stagecoach on his way to join HMS *Beagle*.

Across the river from the town centre, in the Kingsland district, is the **new Shrewsbury School**, which has recently erected a statue of the young Darwin, hat in hand, exploring the Galápagos (viewable only with the school's permission).

On the same side of the river, but further north is **The Mount House**, an impressive three-storey redbrick house, built for Charles's father Robert in 1800. Darwin was born in an upstairs room above and directly to the left of the front entrance. Sadly, opportunities to view the interior of the house are limited, as it accommodates the offices of the district valuers.

More in the Midlands

About seven miles northwest of Shrewsbury is the village of **Montford**, where Darwin's parents, Robert and Susannah, are buried in St Chad's churchyard. In the neighbouring county of Staffordshire is the village of Maer and **Maer Hall**, the Wedgwood family home from 1802 to 1843 and the site of Darwin's courtship of his cousin Emma. The two were married in the nearby church of **St Peter**.

Further south in the county is the city of **Lichfield**, where Erasmus Darwin, Charles's distinguished grandfather, lived in a fine Georgian house (now a museum) close to the cathedral. Erasmus regularly travelled the sixteen miles to Birmingham in order to meet up with fellow members of the Lunar Society at **Soho House**, home of his friend the industrialist Matthew Boulton and now also a museum. Another thirty miles south in Worcestershire is the spa town of **Great Malvern**, visited by Charles Darwin for health reasons in 1849, when he stayed at **The Lodge** (now Hill House), and again in 1851 with his desperately sick daughter Annie. She died while staying at **Montreal House** and is buried in the graveyard of the medieval Priory Church under the cedars of Lebanon. Both The Lodge and Montreal House are on Worcester Road.

Cambridge

Darwin spent three years as a student at Cambridge University, reading theology while avidly pursuing his interest in natural history. Two plaques mark the site of Darwin's first lodgings above a tobacconist's shop in **Sidney Street** (the site now occupied by a branch of Boots). Across the street at the start of **St Andrew's Road** is the entrance to **Christ's College**, where Charles studied, along with his cousin William Darwin Fox. Darwin moved into rooms in college in October 1829, occupying them until he graduated in 1831. The rooms (first floor, middle staircase) now house a cameo relief of Darwin placed there by his son George in 1885. The college grounds also contain a bronze bust of Darwin given by a group of Americans in his centenary year, while the dining hall houses a portrait by W.W. Ouless and a stained glass window.

On the left-hand side of St John's Street (coming from the direction of Bridge Street) is the old **School of Divinity**, where a plaque commemorates Darwin's period of study there. Further along in the same direction, you arrive at King's Parade and **King's College** with its spectacular, late Gothic chapel, where Darwin enjoyed listening to the college choir. On his return from the *Beagle* voyage, Darwin lodged for three months at **22 Fitzwilliam Street** (near the Fitzwilliam Museum), his stay commemorated by a stone plaque.

After Charles's death in 1882, his widow Emma spent her winters at **The Grove**, now part of **Fitzwilliam College** (the author's alma mater), which is situated a mile or so north of the city centre on the Huntingdon Road. Emma's son Horace lived next door on the site now occupied by **New Hall**. Horace and his brother George are both buried in the nearby **Ascension Parish Burial Ground**.

George, who was Plumian Professor of Astronomy at the University, bought Newnham Grange as a family home in 1885. Situated at the river's edge at one corner of the city centre, it now forms the heart of **Darwin College**, home to an assortment of Darwinian artefacts on loan from the Darwin family. A blue plaque on the building commemorates George's daughter **Gwen Raverat**, who left a beautiful memoir of her childhood here (*Period Piece: A Cambridge Childhood*). The **Cambridge University Botanic Garden**, founded by Darwin's mentor John Stevens Henslow in 1831, is also well worth a visit. For details of what to see in Cambridge during the bicentenary, visit www.darwin2009.cam.ac.uk.

London

Darwin's momentous interview with Robert Fitzroy took place in September 1831 at the **Old Admiralty** (also called the Ripley Building), a fine Georgian building at the top of Whitehall. Nearby in Carlton House Terrace you will find the current home of the **Royal Society of London**, which Darwin joined as a fellow in 1839. Heading north into Pall Mall brings you to the classical splendour of the **Athenaeum Club**; Darwin frequently dined and wrote letters from here after being accepted as a member in 1838.

Halfway along the north side of Piccadilly is **Burlington House**, home to many learned institutions, including the **Linnaean Society**, where the Darwin-Wallace paper was read in 1858 (although not quite in the society's current location). Today, Darwin's portrait hangs there alongside that of Wallace. Heading up nearby **Albemarle Street** brings you almost immediately to no.50, the head office of **John Murray** (now part of Hodder Headline), first publisher of *The Origin of Species*. Further up, on the same side, is **Brown's Hotel** (formerly Brown's and St George's Hotel) where Huxley founded the X Club, a group of like-minded Darwin supporters, in November 1864.

The Athenaeum Club opened in 1830; Darwin became a member in 1838.

After the *Beagle* voyage, Darwin moved to London in March 1837 and lived in Soho on the south side of **Great Marlborough Street**, first at his brother Erasmus's house at no.43 (now no.48, the home of Schott the music publishers) and later a few doors along at no.36 (on the site of the current no.41). On the last day of 1838, and a month before his wedding, Darwin moved again, this time about a mile further north to **no.2 Upper Gower Street**, which he and Emma referred to as "Macaw Cottage" on account of its garish furnishings. A blue plaque marks the site, now

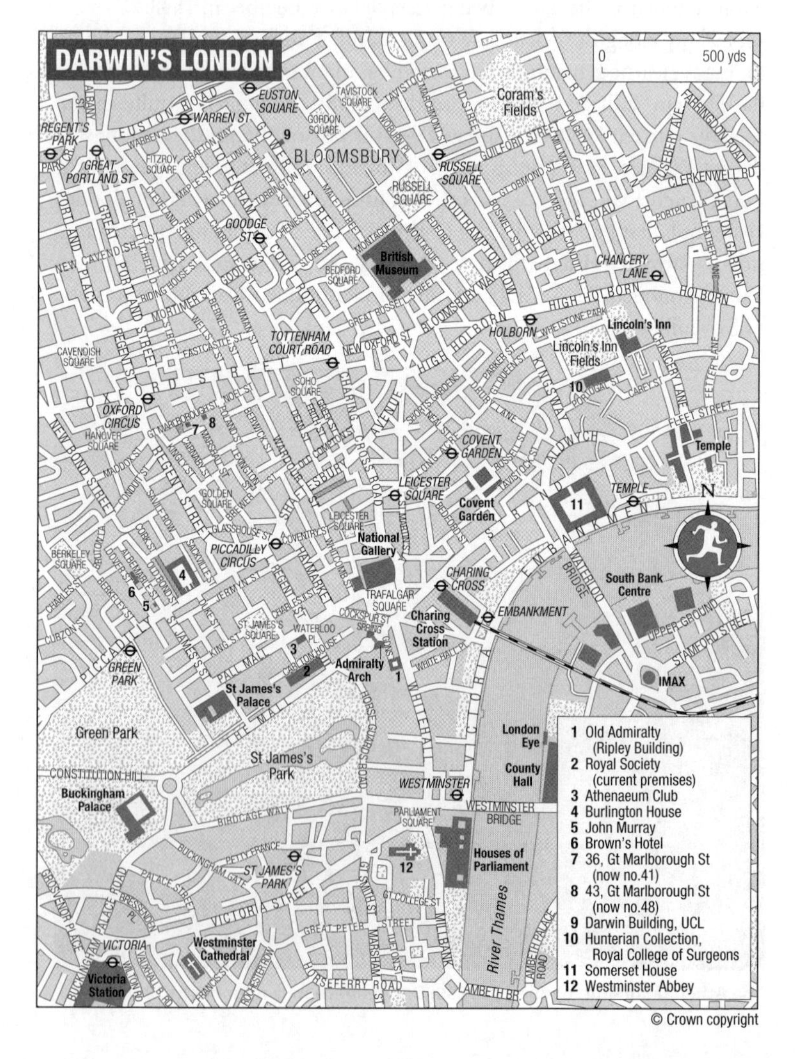

occupied by the **Darwin Building** of University College London. Inside the building, to the right in a lobby by the stairs, is a wooden sign erected by Darwin's great-grandson, Richard Keynes. University College also contains the **Grant Museum**, a zoological collection named after Darwin's early mentor Robert Grant (see p.18). Also associated with the young Darwin is **Somerset House** in the Strand, which in the 1830s was home to the **Geological Society of London** (where Darwin served as Secretary and Vice-President) and the **Royal Society**. Both societies relocated, in the latter half of the nineteenth century, to Burlington House, with the Royal Society moving to its current location only in 1967.

Five minutes further north, on the south side of Lincoln's Inn Fields, is the **Hunterian Museum** of the Royal College of Surgeons. This once housed Darwin's collection of fossil vertebrates, which Richard Owen helped him to identify. The specimens were subsequently moved to the **Natural History Museum** in South Kensington, an institution founded by Owen. An imposing white-marble statue of a seated Darwin, finished by Sir Joseph Boehm in 1885, has recently re-assumed its original pride of place at the top of the main staircase in the museum's Central Hall – displacing a statue of Owen, who in later life assumed the role of Darwin's rival.

And finally, Darwin is buried in **Westminster Abbey**, where you should prepare for the grandeur of the majestic Gothic architecture and disappointment at the plain marble slab and simple bronze relief bust that mark his last resting place in the north choir aisle.

Downe, Kent

Down House, Luxted Road, in the village of Downe, was Darwin's home from 1842 until his death (the nearest stations are Chelsfield and Orpington). The house is maintained by English Heritage as a fascinating museum of Darwin's life, with a greenhouse and gardens that have been restored to their appearance in Victorian times. The garden includes the famous **Sandwalk**, where Darwin liked to walk and think. Plans are currently afoot to make Down House a World Heritage site (www.darwinatdowne.co.uk). In the nearby village churchyard of **St Mary the Virgin** resides the joint grave of Darwin's wife Emma and his brother Erasmus, plus two of Darwin's children. In the village high street is the **George and Dragon** pub, where Darwin regularly attended meetings of the Downe Friendly Society (a benefit society for the local community).

13

Events, activities and retail

Darwin200

2009 marks the two-hundredth anniversary of Charles Darwin's birth (on 12 February) and 150 years since the publication of *The Origin of Species* (24 November). Thousands of celebratory events are planned across the globe. What follows is just a brief selection of events in Darwin's home country. For the most up-to-date information visit darwin200.org and darwin-online.org.uk.

The Natural History Museum, London

Situated in London's South Kensington, the Natural History Museum (www.nhm.ac.uk) is acting as the umbrella organization for all Darwin200 events. From late 2008, the museum is hosting the **Darwin Exhibition**. A remarkable celebration of Charles Darwin's ideas and their impact, the exhibition has already been seen at museums in New York, Boston, Chicago and Toronto. In February, the museum unveils the new Darwin-inspired roof canopy, an artwork by Tania Kovats in the form of a cross-section of an ancient oak tree, cut lengthways and inserted into the ceiling.

Cambridge

Numerous events are planned for Darwin's university town of **Cambridge**. These include the **Darwin Lecture Series** (www.dar.cam.ac.uk/lectures) at Darwin College in March, the **Darwin 2009 Festival** in July and a Darwin-themed Cambridge Music Festival in November. **Christ's College** (www.christs.cam.ac.uk), where Darwin studied, are planning various events throughout the year including a Galápagos Conservation Trust dinner, a colloquium on the reception of Charles Darwin in Europe, the refurbishment of Darwin's rooms and the unveiling of a new bronze statue of Darwin by Anthony Smith. Darwin's zoological specimens will be on display at the **Cambridge Museum of Zoology** (www.zoo.cam.ac.uk/museum), there's a Darwin the Geologist exhibition at the **Sedgwick Museum** (darwinthegeologist.org) and a Teaching Charles Darwin display at the **University Botanic Garden** (www.botanic.cam.ac.uk). The **Fitzwilliam Museum** (www.fitzmuseum.cam.ac.uk) will host an exhibition, Endless Forms: Charles Darwin, Natural Science and the Visual Arts, from June to October, and the University Library is displaying HMS *Beagle*-related manuscripts and letters (www.darwin-project.ac.uk).

Shrewsbury and other UK sites

The **Shrewsbury Festival** is an annual event that has been held in Darwin's home town every February since 2003. Activities include academic lectures, talks in churches, nature walks, museum exhibitions and a public toast to Darwin on his birthday. Extra events are scheduled for 2009, including a Darwin Song Cycle project as part of the local folk festival and the unveiling of a new Darwin memorial and "geo garden" (www.darwinshrewsbury.org).

Many other institutions are planning Darwin events or exhibitions. In London these include Kew Gardens, London Zoo and the Hunterian Museum at the Royal College of Surgeons. In the west of England there are events at Plymouth Museum and Art Gallery (Darwin's Voyage of Discovery) and a lecture series at the Bath Literary and Scientific Institution (Darwin and Beyond), while the Lyme Regis Fossil Festival is staging a one day event in May entitled Evolution Rocks! In Scotland both the National Museum and the National Library (both in Edinburgh) are planning events, and Darwin-related discussions will be included in the Cheltenham Science Festival in June. The UK's Research Councils (www.darwin.rcuk.ac.uk), the Royal Institution (www.rigb.org) and the Wellcome Trust (www.wellcome.ac.uk) are all running Darwin200 projects.

Finally, during 2009, The Charles Darwin Trust and English Heritage plan to open a new education room and exhibition at Darwin's home, **Down House**, while the **Beagle Project** (www.thebeagleproject.com) aims to launch a working replica of HMS *Beagle*.

Ten more things to look out for in 2009

- Magazine special issues on Darwin: *The Lancet, Nature* and *The Journal of Victorian Studies*.
- Ten new Darwin stamps from the Royal Mail.
- A new TV documentary from David Attenborough for the BBC.
- Darwin and his daughter hit the big screen in *Origin*, a film scripted by John Collee from Randal Keynes's book *Annie's Box*.
- A possible film version of *Evolution's Captain*, the account of Fitzroy and Darwin's difficult relationship.
- An online tree of life with which you create a musical experience (www.sonarboria.org).
- "Darwin's leftovers": a knitted art installation in the form of odds and ends from Darwin's study, created by the Stroud knitting group.
- A four-part series on Charles Darwin for BBC Radio 4.
- A new Darwin-related dance work from the Rambert Dance Company.
- A graphic novel by Simon Gurr and Eugene Byrne, which will be distributed to over 100,000 people free of charge.

Darwin in the US

A celebration of Charles Darwin's life and legacy in science and society is held on or close to his birthday on 12 February each year. Initiated spontaneously at several US universities over recent decades, Darwin Day received a new impetus in 2004 with the establishment of a not-for-profit organization, **Darwin Day Celebration**, and an associated Darwin Day website (darwinday.org), which features educational material and a register of Darwin Day events. The Darwin Day movement has expanded in recent years to take in over eight hundred events across the globe and has gathered a lot of media interest. The format of Darwin Day events is as broad as the organizers' imaginations: everything from formal lectures and debates to film festivals and fancy-dress parties. Some combine Darwin Day with a celebration of Abraham Lincoln, who was born on the same day (see p.19), and some even advocate a public holiday on Darwin–Lincoln's birthday. **Evolution Sunday** (now renamed **Evolution Weekend** to welcome members of all religions) aims to get clergy and congregations to discuss evolution and celebrate Darwin's legacy during the day of worship closest to Darwin's birthday. At Salem in Massachusetts, **Salem State College** has held a week-long Darwin Festival every February since 2000.

Darwinian retail

Many see evolution as a lifestyle choice as much as a theory, a position largely driven by the creation–evolution debate in the US. So, if you decide you do want to join Darwin's posse, what are you to do? First, you will need to wear one variety or another of the **Darwin fish** (see box, below). Then, for the seriously committed Darwinian, there are a whole host of purchasing possibilities. First stop should be **Cafépress**, an online store which stocks tens of thousands of evolution-related designs (www.cafepress.com/buy/evolution) on clothes, bags, mugs, coasters, ornaments, greetings cards, bumper stickers etc – you can even have Darwin on your underwear! Many of the ideas and slogans are extremely witty, such as a chimp as Ché Guevara (Viva La Evolución); "My ancestors survived the end-Cretaceous and all I got was this lousy T-shirt!"; "If God didn't want us to teach evolution, then why did he create smart people?"

The evolution of a fish

Early Christians employed a simple image of a fish as one of their symbols, using the Greek word for fish, "icthys", as an acrostic for "Jesus Christ, Son of God, Saviour".

Some time in the early 1980s, atheists and anti-creationists began to spoof the fish symbol by introducing a version with legs, the Darwin fish. Since then, the fishy iconography of evolution has undergone an adaptive radiation, spawning a procreating fish (an evolution fish mounting a Jesus fish); a fish that gets up on two legs; an evolution fish devouring a creation fish; a Lamarck fish (complete with giraffe neck) and a pair of reconciliation fish (a Jesus fish kissing a Darwin fish). Plus, for Pastafarians, there is a fish-like emblem of the Flying Spaghetti Monster (see p.293) and even a pirate fish.

If you want to festoon your walls with Darwin posters, then **AllPosters.com** is the place to go with a selection of mainly nineteenth-century images of the great man (and also HMS *Beagle*) framed or unframed. For videogamers there is a wide range of evolutionary options including **Darwinia** (www.darwinia.co.uk), in which you have to destroy a viral infection which is threatening Darwinians with extinction, and the ultra-cool **Spore** (www.spore.com), in which the player takes control of the evolution of a species as it journeys from primordial goo to become a sentient civilization-builder.

Drinkers have three beverages from which to choose. There's a dry white wine from Oregon called **Evolution**, "evolved", as the publicity suggests, from nine different grape varieties, and a similarly named Merlot from the Western Cape made by New Zealander Rhyan Wardman for Origin Wines. Finally, beer drinkers should get hold of some **Evolution Amber Ale** from Utah's Wasatch Beers (www.wasatchbeers.com), which, so the company claims, is "intelligently designed for intelligent beer drinkers".

Books, films and websites

What some have called the "Darwin Industry" has spawned a huge literature, on Darwin, his precursors and his legacy. What follows is just a brief selection of what is available in print, on screen and online about Darwin and evolution.

Darwin

Precursors

Evolution: The History of an Idea Peter J. Bowler (2003) A scholarly but approachable survey of the evolution of evolution.

The Structure of Evolutionary Theory Stephen Jay Gould (2002) A 1433-page opus that interweaves Gould's controversial views on how evolution works with heavy history and light-hearted anecdotes.

www.gutenberg.org Download the works that influenced Charles Darwin: by Bacon, Malthus, Erasmus Darwin, Paley, Hutton and Lyell, all available free-of-charge.

Life

Autobiography Charles Darwin, edited by Neve and Messenger (2002) Darwin's own account of his life that mixes the personal with the professional.

Darwin: A Very Short Introduction Jonathan Howard (2001) A handy analysis of Darwin and his work that you can read in an evening.

The Reluctant Mr Darwin David Quammen (2006) Entitled *The Kiwi's Egg* in the UK, an engaging account of Darwin's life and work after leaving HMS *Beagle*.

Introducing Darwin Jonathan Miller and Borin van Loon (2006) Darwin's life and work presented in the style of a graphic novel.

Charles Darwin John van Wyhe (2009) An illustrated description of how Darwin's own experiences fed into the theories for which he became known, complete with images of diaries and handwritten drafts.

Charles Darwin: Voyaging (1995); Charles Darwin: The Power of Place (2003) Janet Browne This two-volume biography provides an authoritative and readable guide to the man and his milieu.

The History of My Shoes and the Evolution of Darwin's Theory Kenny Fries (2007) A gay man with disabled legs provides a sympathetic narrative of Darwin's and Wallace's stories interwoven with his own.

The Sandwalk Adventures Jay Hosler (2003) A graphic novel, with Darwin seen from the viewpoint of a mite living in his eyelashes!

Young readers

Charles Darwin David King (2006) An informative and well-illustrated short biography aimed at young readers.

The Tree of Life Peter Sis (2003) Another introduction for kids with riotous imagery likely to bedazzle.

Inside the "Beagle" with Charles Darwin Fiona MacDonald and Mark Bergin (2005) A well-illustrated guide to Darwin's *Beagle* voyage that got the thumbs up from my 11-year-old.

Who Was Charles Darwin? Deborah Hopkinson and Nancy Harrison (2005) Another informative guide for 9–13-year-olds.

The Voyage of the Beetle Anne H. Weaver and George Lawrence (2007) A beetle accompanies Darwin around the world and tells his story.

Works

The Works of Charles Darwin in 29 Volumes edited by Paul H. Barrett and R. B. Freeman (1990) A near-complete set of Darwin's published works available from NYU Press.

Penguin Classics have published several of Darwin's works including *The Voyage of the Beagle*, edited by Browne and Neve (1989), *The Origin of Species*, edited by Bynum and Browne (2008), *Expression of Emotions* edited by Messenger and Neve (2008), and *The Descent of Man*, edited by Desmond and Moore (2004). Penguin has also published a 128-page abridged version of the Origin: *On Natural Selection* (2005).

The Portable Darwin edited by Duncan Porter and Peter Graham (1993) A handy selection of Darwin's work in a single volume.

On the Origin of Species audiobook A five-hour extract edited and read by Richard Dawkins, available via iTunes.

www.darwin-online.org.uk Created by John van Wyhe, **Darwin Online** provides a one-stop searchable archive of all Darwin's published writings and many of his unpublished papers. With over 50,000 pages of text and 150,000 images, it allows users to check for themselves what Darwin said on any given issue without relying on experts.

www.darwinproject.ac.uk Founded in 1974 by **Frederick Burkhardt**, the Darwin Correspondence Project collates and documents thousands of Darwin's letters, producing a new printed volume every one to two years (about half way through an estimated thirty volumes).

Context

How to Read Darwin Mark Ridley (2005) Advice for the novice on how to approach *The Origin* and *The Descent of Man*.

Darwin's Origin of the Species: A Biography Janet Browne (2007) Darwin's biographer tells the story of the book that changed the world.

Darwin Philip Appleman (2001) Selections from Darwin's work with the writings of his forebears, critics and intellectual descendants.

The Tangled Bank Stanley Edgar Hyman (1974) A wonderful examination of the imaginative and rhetorical power of Darwin's prose.

Top ten films with an evolutionary twist

1. Fantasia (1940) In the fourth section of this animation classic, Disney presents a stunning history of the Earth from the planet's formation to the end of the dinosaurs, with Stravinsky's *Rite of Spring* as soundtrack.

2. Inherit the Wind (1960) Spencer Tracy and Fredric March play the warring attorneys, and Dick York the defendant in this fictionalized account of the Scopes monkey trial, directed by Stanley Kramer.

3. Planet of the Apes (1968) Charlton Heston's lost astronaut Taylor faces the simian equivalent of the Scopes trial.

4. 2001: A Space Odyssey (1968) Alien intervention and Ardey's killer-ape hypothesis collide in Arthur C. Clarke's retelling of human evolution.

5. Evolution (2001) In this kooky sci-fi comedy flick, a meteor strike brings fast-evolving aliens that threaten an American town.

6. Ice Age (2002) A computer-animated children's film, in which a sabre-toothed tiger, a sloth and a mammoth struggle to return a lost human child to his tribe.

7. Darwin's Nightmare (2004) A documentary detailing the catastrophic evolutionary and social consequences of the introduction of the Nile perch into Lake Victoria.

8. Idiocracy (2006) A dark comedy set five hundred years from now in a dysgenic future where the dumb have outbred the smart.

9. The Darwin Awards (2006) The Darwin Awards (www.darwinawards.com) honour those who improve the gene pool by inadvertently and idiotically removing themselves from it. The film weaves some of the best stories into an improbable scheme to cut insurance-company costs.

10. Evolution: The Musical (2008) "Charles H. Darwin" meets Jesus in this Rocky Horror re-mix of the creation-versus-evolution debate.

Evidence for evolution

Introductory books

Evolution: A Very Short Introduction Brian Charlesworth (2003) A whistle-stop tour of the subject that you can complete in just a few hours.

The Blind Watchmaker Richard Dawkins (1986) Evolutionary biologist turned science writer eloquently explains how natural selection provides a compelling alternative to the watch-maker argument.

Climbing Mount Improbable Richard Dawkins (1996) Exploits fitness landscapes (see p.110) to explain how remarkable complexity can arise from the cumulative effects of numerous small, often unseen, improvements.

Evolution: What the Fossils Say and Why It Matters Donald Prothero

and Carl Buell (2007) A powerful defence of evolution that emphasizes the importance and richness of the fossil record.

The Making of the Fittest Sean Carroll (2006) The perfect complement to Prothero and Buell, here is a readable run-through of evolution at the molecular level.

Darwin's Dangerous Idea Daniel C. Dennett (1996) A philosopher of science describes Darwin's natural selection as a mindless algorithm surveying design space and bringing meaning to the universe.

Evidence and Evolution Elliott Sober (2008) The ultimate intellectual workout as to what counts as evidence for evolution, complete with a description of Bayesian logic.

Glorified Dinosaurs Luis M. Chiappe (2007) A well-written, beautifully illustrated text on the evolution of birds.

Your Inner Fish Neil Shubin (2008) A palaeontogist provides insights into the art of his science while describing how our evolutionary history has shaped our bodies.

At the Water's Edge Carl Zimmer (1999) An outstanding science writer describes how life came ashore and then went back into the sea.

Evolution (in Action): Natural History through Spectacular Skeletons Jean-Baptiste de Panafieu, Patrick Gries (2007) A breathtaking collection of animal skeleton photographs linked to descriptions of evolutionary concepts.

Textbooks

Evolution Mark Ridley (3rd edition, 2003) A superb undergraduate textbook, with an attractive layout. www.blackwellpublishing.com/ridley

Evolution Nicholas Barton, Derek Briggs, Jonathan Eisen, David Goldstein, Nipam Patel (2007) A weighty student tome from an inter-displinary team of authors. www.evolution-textbook.org

Documentaries

Evolution US Public Broadcast Service A boxed set of four DVDs providing a superb overview of the evidence for evolution.

David Attenborough's Life series BBC Television Evolution permeates this series of series from *Life on Earth* (1979) to *Life in Cold Blood* (2008). Attenborough's new Darwin documentary, *Tree of Life*, will be broadcast by the BBC in 2009.

Websites

www.pbs.org/evolution An authoritative and accessible guide to evolution that accompanies a series of Public Broadcast Service documentaries.

evolution.berkeley.edu A similarly authoritative, well-crafted one-stop source for information on evolution.

www.talkorigins.org A superb archive of key facts and incisive comments on all aspects of evolution and creationism.

wiki.cotch.net The EvoWiki, an online encyclopedia in the style of the Wikipedia, with over 3,000 pages.

nhm.ac.uk/nature-online/evolution A colourful overview of evolution from London's Natural History Museum, complete with many short movies.

nationalacademies.org/evolution A series of reports and podcasts on evolution aimed at the public from the US National Academies.

www.nescent.org Stacks of educational resources and links on evolution.

www.newscientist.com/channel/life/evolution *New Scientist* magazine's extensive back catalogue of articles on all aspects of evolution.

www.hhmi.org/biointeractive/evolution Includes great lectures on evolution from Ken Miller and others.

Blogs and podcasts

The Loom (blogs.discovermagazine.com/loom) Probably the best blog focused primarily on evolution, by science writer Carl Zimmer.

The Panda's Thumb (pandasthumb.org) The definitive blog on the evolution versus creationism debate, complete with informative and often heated discussions.

Pharyngula (scienceblogs.com/pharyngula) Another highly rated blog on evolution and creationism.

The Rough Guide to Evolution (roughguides.blogspot.com) Companion blog to this volume, contains a long blog-roll detailing numerous other evolution blogs.

Evolution 101 (www.drzach.net) A podcast that covers topics as diverse as junk DNA and the female orgasm.

Evolution since Darwin

The Eclipse of Darwinism Peter Bowler (1992) Documents the doldrums of Darwin's influence.

A Reason for Everything Marek Kohn (2005) A review of the efforts and eccentricities of the English evolutionists inspired by Darwin.

Frogs Flies and Dandelions Menno Schilthuizen (2001) A popular account of speciation.

Dawkins vs. Gould: Survival of the Fittest Kim Sterelny (2007) A history of the Dawkins–Gould clashes.

Evolution and the Theory of Games John Maynard Smith (1982) A summary of evolutionary game theory from the master of the field.

Deep Time: Cladistics, the Revolution in Evolution Henry Gee (2001) An accessible introduction to the impact of cladistics in biology.

Palaeobiology and human evolution

Books

Life on a Young Planet Andrew H. Knoll (2004) A popular introduction to palaeobiology.

The Evolution and Extinction of the Dinosaurs David Fastovsky and David Weishampel (2005) Popular dinosaur books abound, particularly for younger readers. This provides a weightier adult alternative.

Palaeobiology II Derek Briggs and Peter Crowther (2001) A textbook introduction to the subject.

Life of the Past This impressive series of books from Indiana University Press includes *Dawn of the Dinosaurs: Life in the Triassic* by Nicholas Fraser (2006) and *After the Dinosaurs: The Age of Mammals* by Donald Prothero (2006).

The Last Human Gary Sawyer, Viktor Deak, Esteban Sarmiento, Richard Milner (2007) A lavishly illustrated guide to twenty-two species of extinct human/hominin.

Seven Million Years Douglas Palmer (2006) A broad introduction to the evolution of humans and our relatives since our divergence from chimps.

Homo Britannicus Chris Stringer (2006) The story of human life in Britain, stretching back 700,000 years.

Smithsonian Intimate Guide to Human Origins Carl Zimmer (2005) A well-illustrated guide to human evolution, by one of America's finest science writers.

The Complete World of Human Evolution Chris Stringer and Peter Andrews (2005) A comprehensive guide to human evolution that includes guides to archaeological sites and anatomical comparisons between fossils and living human relatives.

Human Evolution: Trails from the Past Camilo Cela-Conde and Francisco Ayala (2007) A multi-disciplinary review of human evolution, encompassing physical anthropology and molecular biology, psychology and philosophy.

Reconstructing Human Origins Glenn Conroy (2004) A scholarly introduction to human evolution, drawing on physical and genetic evidence and palaeoclimatology.

Principles of Human Evolution Robert Foley and Roger Lewin (2004) A superb textbook aimed at university students.

From Lucy to Language Donald Johanson, Blake Edgar, David Brill (2001) The stunning full-page photographs of hominin skulls more than make up for the poorly laid out text.

Websites

www.palaeos.com A massive amateur effort full of information about extinct animals.

www.nhm.ac.uk nature-online/life The UK's Natural History Museum website on palaeobiology.

dinobase.gly.bris.ac.uk The University of Bristol's online dinosaur database.

www.talkorigins.org/faqs/homs Documents the evidence for human evolution and the latest discoveries.

www.bbc.co.uk/sn/prehistoric_life The BBC's introduction to palaeobiology, including human origins.

www.becominghuman.org An informative site that includes an interactive documentary, educational exhibits and research tools.

anthropology.si.edu/humanorigins The Smithsonian's guide to human origins.

www.amnh.org/exhibitions/atapuerca Covers finds in Atapuerca and much more.

Blogs

palaeoblog.blogspot.com A blog on palaeobiology.

johnhawks.net/weblog The premier blog on palaeoanthropology, with archives packed full of expert commentary.

scienceblogs.com/afarensis Offers a broad-ranging view of anthropology, politics and society.

Documentaries

Walking with Dinosaurs produced by Tim Haines and Jasper James (1999) Ground-breaking BBC TV documentary using CGI and anamatronics. Haines and James went on to make *Walking with Beasts* (2001) for the BBC, and *Walking with Cavemen* (2003), *Prehistoric Park* (2006) and *Primeval* (2007) for Impossible Pictures. All provide a lively introduction to palaeobiology, although the Wikipedia entries for each show point out various inaccuracies.

Politics, psychology and philosophy

Books

The Cambridge Companion to Darwin Jonathan Hodge, Gregory Radick (2003) An excellent guide to the wider influence of Darwin's thought, particularly on politics.

Evolutionary Psychology: A Beginner's Guide Robin Dunbar, John Lycett, Louise Barrett (2005) A brief introduction to evolutionary psychology.

The Blank Slate Steven Pinker (2000) A lively introduction to evolutionary psychology.

The Adapted Mind Jerome Barkow, Leda Cosmides, and John Tooby (1996) A formal pioneering position statement on evolutionary psychology.

Adapting Minds David Buller (2006) Provides an alternative view, highly sceptical of evolutionary psychology.

In the Name of Eugenics Daniel Kevles (1995) The definitive work on the history of eugenics.

Darwin Tim Lewens, 2007 A well-crafted introduction to Darwin's influence on philosophy.

Religion and creationism

Books

God's Funeral A.N. Wilson (2000) An analysis of the Victorian crisis of faith that places Darwin's impact in the context of numerous other secularizing predecessors.

Evolution vs. Creationism Eugenie C. Scott (2008) A clear-headed comparison between the science of evolution and creationist pseudo-science.

The Creationists Ronald L. Numbers (2006) The definitive encyclopaedic treatment of the phenomenon of creationism.

Science, Evolution, and Creationism (2007) Published as a book from the US National Academy of Sciences this is also available online from nationalacademies.org/evolution.

The Counter-Creationism Handbook Mark Isaak (2006) A catalogue of counter-arguments against common creationist claims, also available online (www.talkorigins.org/indexcc).

Intelligent Thought John Brockman (2006) A superb multi-disciplinary set of short essays by scholars from the sciences and humanities challenging Intelligent design.

Scientists Confront Creationism Andrew Petto and Laurie Godfrey (2008) Scientists dissect and dismiss the arguments of the creationist and ID movements.

Why Darwin Matters Michael Shermer (2007) Founder-editor of *The Skeptic* magazine explains why Darwin, evolution and science matter.

Summer of the Gods Edward J. Larson (1999) The definitive recent work on the Scopes Trial and its relevance to contemporary America.

Trials of the Monkey Matthew Chapman (2000) A Darwin descendant explores the American response to his ancestor's ideas from the Scopes Trial up to today.

The Devil in Dover Lauri Lebo (2008) A poignant account of a local journalist's struggle with the events and beliefs surrounding the Dover trial.

The Evolution Dialogues Catherine Baker and James B. Miller (2006) Aims at helping educators reconcile Christianity and evolution in the classroom.

Evolution on cult TV

1. The Simpsons *Lisa the Skeptic* (Season 9, Episode 8)
Lisa asks Stephen Jay Gould (who provides his own voice) to test a sample of a fossil angel, which later turns out to be just a marketing ploy for a new mall. The show paid tribute to Gould after his death with a dedication in the Season 13 finale.

2. The Simpsons *The Monkey Suit* (Season 17, Episode 21)
With creationism on the march in Springfield, Lisa holds secret evolution classes, only to face trial in "Lisa Simpson v. God". Marge, rallying to her daughter's side, becomes so engrossed in *The Origin of Species* that she is even caught reading it in the shower!

3. The Simpsons *Homerazzi* (Season 18, Episode 16)
A not-to-be-missed opening sequence detailing the evolution of Homer Simpson with a parody of the Disney film *Fantasia*.

4. South Park *Go God Go* (Season 10, Episode 12)
Forced to teach it against her wishes, schoolteacher Garrison derides evolution as "five monkeys having butt sex with a fish-squirrel". Richard Dawkins, hired to provide balance, wins Garrison over intellectually and sexually.

5. Family Guy *Untitled Griffin Family History* (Season 4, Episode 27)
Peter Griffin presents an evolutionary view of life's and his family's history. But the state of Kansas obliges him to provide balance with a creationist version, in which TV witch Jeannie summons up animals, cars, petrol pumps, Jesus and Santa Claus.

6. Lil' Bush *Evolution / Press Corps Dinner* (Season 1, Episode 5)
After flunking evolution in class, a born-again Lil' Bush adopts the Bible as "a

Documentaries

Flock of Dodos Randy Olsen (2006) An amusing look at the evolution–ID debate from biologist-turned-filmmaker Randy Olson.

Judgment Day: Intelligent Design On Trial (2007) This PBS film introduces the issues behind the Dover trial. Online: www.pbs.org/wgbh/nova/id.

Websites

www.ncseweb.org US National Centre for Science Education.

www.bcseweb.org.uk British Centre for Science Education, the NCSE's sister organization.

www.csicop.org/intelligentdesign watch A view of ID from the Committee for the Scientific Investigation of Claims of the Paranormal.

www.noanswersingenesis.org.au An anti-creationist website.

www.aibs.org/mailing-lists/the_aibs-ncse_evolution_list_server.html A site providing information on the struggle against creationism.

richarddawkins.net Dawkins' own website, packed full of information, and guaranteed to upset religious believers.

handbook for not having to know about anything", before proclaiming the "no-evolution revolution" in a rap parody of Eminem and 50 Cent.

7. Guinness commercial *noitulovE*

A minute-long retrospective on tetrapod evolution advertising Ireland's best-known stout, in which three Guinness drinkers end up as mudskippers with a "good things come to those who wait" perspective.

8. Friends *The One Where Heckles Dies* (Season 2, Episode 3)

When flaky Phoebe announces she doesn't believe in evolution, Ross weighs in with ridicule: "Evolution is not for you to buy, Phoebe. Evolution is scientific fact… like gravity," to which Phoebe responds: "Oh, okay, don't get me started on gravity… lately I get the feeling that I'm not so much being pulled down as I am being pushed!"

9. Sopranos *The Fleshy Part of the Thigh* (Season 6, Episode 4)

In hospital recovering from a gunshot wound, Tony gets into discussion with a creationist who tells him that humans and dinosaurs co-existed and that "evolution is just Satan's way of coming between man and God." One of Tony's associates is doubtful "No way. *T. rex* in the Garden of Eden? Adam and Eve would be runnin' all the time, scared shitless!"

10. Heroes

A whole TV series steeped in references to evolution, albeit a near-magical distortion of anything Darwinian. Mohinder contrasts human evolution with that of cockroaches, while Darwin permeates his relationship with his father and with flatmate Eden. In a dream sequence, Charles Deveaux says that Peter Petrelli is a Darwinian, while Claude Rains later compares Peter to Darwin's pigeons. The haunting theme tune is entitled "Natural Selection".

Glossary

Adaptation A structural or behavioural feature of an organism that makes it well suited to its environment.

Allele One of the alternative forms of the same gene.

Allopatric speciation Speciation that occurs when populations within a species cannot interbreed because they have become geographically isolated.

Amino acid A fundamental molecular building block of proteins.

Analogous features Features or functions that look similar, but which have not evolved from similar features in a common ancestor.

Atavism The reappearance in an organism of characteristics that were present in its remote ancestors.

Balancing selection A form of natural selection that works to maintain variation within a population.

Base A component of a DNA molecule that defines its sequence.

Biogeography The study of the geographical distribution of plants and animals.

Biological species concept The definition of a species as a set of organisms that can interbreed.

Bonobo A species of highly sexually active ape that lives south of the Congo River.

Cambrian Explosion The apparently sudden appearance of many large and diverse animals in the fossil record during the Cambrian Period.

Catastrophism The hypothesis that Earth has been affected by sudden, short-lived, global catastrophic upheavals, which led to the extinction of some species.

Cell The fundamental structural and functional unit of living organisms.

Chloroplast A structure found in the cells of green plants and some algae where photosynthesis occurs.

Chromosome A structure in the cell that carries genetic information; usually consists of a thread of DNA.

Clade A set of organisms containing an ancestor and all its descendants; a monophyletic group.

Cladism/cladistics A system of taxonomy in which organisms are classified purely on evolutionary grounds, rather than on grounds of similarities.

Cladogram An evolutionary tree that focuses on the order in which different groups have branched off from a lineage irrespective of the rate of evolutionary change.

Codon A triplet of bases in DNA that code for one amino acid via the genetic code.

Co-evolution A pattern of evolution in which evolutionary changes in one species influence the evolution of the other species.

Common descent The theory that each group of organisms is descended from a common ancestor; more generally, the idea that all life on Earth is derived from a single common origin.

Convergence/convergent evolution The process by which analogous features evolve independently in distinct lineages.

Creationism The belief that all living things on Earth were created separately, more or less as they are now, by a supernatural creator.

Deep time A geological notion of the Earth's age that stretches back far beyond Biblical estimates.

Directional selection Selection causing a change towards one extreme of a distribution of character states.

Disruptive selection Selection favouring extremes in either direction from the population average.

DNA/deoxyribonucleic acid The molecule from which genes are built.

Dominant The property of an allele (A) where the phenotype of the heterozygote (Aa) is the same as that of the homozygote (AA).

Eugenics The theory and policy of improving human populations by controlled breeding.

Eukaryote An organism that wraps its DNA in a membrane-bound sac, the nucleus.

Eutherian A kind of mammal that possesses a placenta.

Evo-devo A branch of evolutionary biology that aims to use insights from genetics, developmental biology and the fossil record to improve our understanding of the evolution of new species and new body-plans.

Evolution The process of genetic change within a lineage over time, beginning with changes in the frequency of alleles in a population from generation to generation and culminating in the birth of new species.

Evolutionary arms race A situation in which predator and prey, evolve continuously and progressively, each attempting to outdo the other.

Fitness The success of an individual, allele or genotype in surviving and reproducing, as assessed by its genetic contribution to subsequent generations.

Frequency-dependent selection A variety of natural selection in which the fitness of a phenotype depends on its frequency relative to other phenotypes in a given population.

Gene The fundamental unit of inheritance, a DNA sequence that provides a discrete function, typically by encoding a protein.

Genetic code The code that relates triplets of bases in DNA to amino acids in the proteins.

Genetic drift The variation in the frequencies of alleles in a population that occurs by random chance, rather than through the action of natural selection.

Genome The full complement of genetic material in a cell or organism.

Genotype The genetic profile of an individual organism, consisting of all its alleles.

Germ cells The cells in a multi-cellular organism that produce the egg and sperm/pollen and so form an undying lineage perpetuated by reproduction.

Heredity The process by which characteristics are passed from one generation to the next.

Heritable Capable of being transmitted from one generation to the next (applied to phenotypic characters).

Heritability The degree to which variation in a phenotypic character in a population is due to differences in genotypes that can be inherited by offspring.

Heterozygote advantage/heterosis A situation in which being heterozygous brings increased fitness over either homozygous state.

Homeotic mutation A mutation that causes a tissue or structure to grow in a place more appropriate to another, e.g. a foot in place of an antenna in a fly. Homeotic genes, particularly hox genes, control the patterns of development within an animal's body.

Hominin Any human relative whose last common ancestor with modern humans post-dates the split between the human and chimpanzee lineages.

Homology Similarity between two features that arises because of their descent from a common ancestor and often occurs despite differences in function.

Homoplasy The presence of similar characters in two different groups of organisms where the similarity is not due to common descent but has arisen as the result of convergent evolution.

Intelligent design The argument that some complex biological structures have been designed by an unidentified non-human intelligence.

Intersexual selection A form of sexual selection in which males compete for the attention of the females.

Intrasexual selection A form of sexual selection in which males compete aggressively among themselves for access to females.

Lamarckian inheritance/Lamarckism The now discredited idea that an organism can pass on characteristics acquired during its lifetime to its offspring.

Linnaean classification A hierarchical system for classifying and naming groups of organisms, devised by Swedish naturalist Linnaeus, in which species are designated by two Latin words indicating genus and species; for example, *Homo sapiens*.

LUCA Acronym for the last universal common ancestor. It could be a loose community rather than a single organism.

Macroevolution Any evolutionary change at or above the level of species, often used vaguely to mean any kind of evolution on a grand scale or over a long period of time.

Meme A word coined by Richard Dawkins for a unit of culture (an idea, a song etc) passed from one person to another; the cultural equivalent of a gene.

Mendelian inheritance A mode of inheritance common in multi-cellular organisms in which an individual has two copies of any given gene, one from each parent, which are passed down in equal proportions to the sperm or egg cells.

Messenger RNA/mRNA A kind of RNA that acts as a go-between between DNA and the synthesis of new proteins.

Metabolism The chemical processes that underlie the energy-releasing breakdown of molecules and the synthesis of new molecules that are necesary to maintain life.

Microevolution Evolutionary change below the level of species, particularly changes in gene frequencies within a population.

Mitochondrion A structure in eukaryotic cells that produces energy and is passed to offspring only from mothers.

Modern synthesis The merger of Darwin's theory of evolution by natural selection with Mendel's theory of inheritance.

Monophyletic group A group of organisms that consists of a common ancestor and all its descendants and no other organisms.

Monotreme An egg-laying mammal; for example, the duck-billed platypus.

Mutation A change in the genetic material.

Natural selection The differential survival and reproductive success of individuals in a variable population that depends on heritable differences in adaptation to the environment, such that favourable variations are preserved and injurious variations eliminated.

Neo-Darwinism Darwin's theory of evolution by natural selection shorn of any association with the inheritance of acquired characteristics; often used to describe the modern synthesis of Darwinism and Mendelism.

Neutral theory The theory that most molecular evolution involves selectively neutral changes, ie changes in DNA or protein sequence that have no effect on fitness.

Nucleotide One of the building blocks of DNA and RNA.

Nucleus A membrane-bound region of the eukaryotic cell that houses the DNA.

Outgroup A group assumed to be less closely related to all of the taxonomic groups under consideration than any of them is to each other; for example, a bird when classifying mammals.

Palaeoanthropology The scientific study of human evolution using evidence from fossils.

Palaeobiology The scientific study of ancient organisms drawing on fossil evidence, modern biology and Earth sciences.

Palaeontology The scientific study of fossils.

Parapatric speciation Speciation in which a new species arises from a population that has entered a new habitat within the same geographical area as the parent species.

Paraphyletic group A group of organisms that contains a single ancestral organism together with some, but not all, of its descendants; for example, the class reptiles, which traditionally excludes birds and mammals.

Parsimony A scientific principle that states that the simplest explanation of a phenomenon is the best one; when applied to evolutionary trees, a methodological preference for the branching pattern that requires the smallest number of evolutionary changes.

Peripatric speciation Speciation that occurs when a small population becomes isolated, typically on the edge of the range of the ancestral.

Phenotype The observable anatomical, physiological, behavioural and biochemical characteristics of an organism, produced by the interaction of the organism's genotype and environment during growth and development.

Phylogenetics The study of evolutionary relationships between groups of organisms, particularly of the patterns of lineage branching.

Phylogeny An organism's evolutionary genealogy.

Phylum A high-level taxonomic rank used in classifying organisms, coming below kingdom.

Plate tectonics The theory that the Earth's surface is made up of rigid plates, whose movements explain continental drift, mountain building, earthquakes, volcanoes, and aspects of biogeography.

Polyphyletic group A group of organisms that includes descendants from more than one common ancestor that have been grouped together on account of homoplasy.

Polyploid An organism that posssesses more than two sets of genes and chromosomes.

Prokaryote An organism in which the cell lacks a nucleus; the term includes bacteria and archaea.

Protein A molecule made up of a sequence of amino acids; life's building and action molecule.

Punctuated equilibrium A pattern of evolution in which species, once in existence, do not change much over long periods of time and then are suddenly replaced by quite different species.

Radiometric dating A technique for determining the absolute age of a fossil or rock by comparing the amount of an original radioactive material to the amount of its decay product.

Recessive The property of an allele (a) where the phenotype of the heterozygote (Aa) is different from that of the homozygote (aa), but instead resembles the phenotype of the homozygote of the dominant allele (AA).

Ribosome The site of protein synthesis in the cell, in which mRNA sequences are translated into protein sequences.

RNA/ribonucleic acid A nucleic acid similar to DNA that acts as an intermediary in the process of translating DNA sequences into protein.

Sexual selection A type of natural selection which favours traits that give a competitive edge over rivals in attracting or keeping a mate.

Social Darwinism The doctrine that competition between individuals or groups drives progress in human societies.

Speciation The formation of one or more new species.

Species A fundamental unit of biological classification often defined as a population of organisms capable of interbreeding and producing fertile offspring.

Stabilizing selection A variety of natural selection that tends to hold the form of a population constant, discriminating against those with extreme values for a given phenotype.

Sympatric speciation The kind of speciation which occurs in populations that live together with an overlapping geographical distribution.

Symplesiomorphy A shared ancestral feature, which is discounted as unhelpful in cladistics; for example, gills in fish.

Synapomorphy A shared new feature, present in the last common ancestor of a group but not in any earlier ancestors nor in any outgroup; used in cladistic classifications: for example, feathers in birds.

Taxonomy The theory and practice of classifying and naming groups of living organisms.

Tetrapod A group of vertebrates possessing four legs; includes amphibians and their legless variants.

Transitional fossil A fossil that illustrates the transition between two groups of organisms, combining traits from an older, ancestral group with traits of a more recent group.

Uniformitarianism The hypothesis that geological features can be best explained by the cumulative action of natural processes operating today.

Vertebrates A group of animals that possess a backbone and an internal skeleton made of bone or cartilage.

Vestigial Pertaining to a feature of an organism that represents a simpler evolutionary remnant of a formerly well developed, fully functional feature.

Index

A

B

C

D

E

F

P

Q

R

Picture Credits

4 The London Art Archive/Alamy; 7 Christopher and Sally Gable/DK Images; 8 The London Art Archive/Alamy; 10 © Dave Souza; 18 Corbis; 22 Cambridge University Library/Corbis; 25 Hulton-Deutsch Collection/Corbis; 27 Bettmann/Corbis; 28 The Print Collector/Alamy; 31 Rob Reichenfeld/DK Images; 34 William Perlman/Star Ledger/Corbis; 37 Bettmann/Corbis; 40 David Ball/Corbis; 43 © Mark Pallen; 44 Mary Evans Picture Library/Alamy; 51 Bettmann/Corbis; 53 Hulton-Deutsch Collection/Corbis; 55 Bettmann/Corbis; 57 © Lewis Bush; 61 Bettmann/Corbis; 64 Geoff Brightling/DK Images; 69 Kim Taylor and Jane Burton/DK Images;70 Bettmann/Corbis; 75 Kennan Ward/Corbis; 79 Andrew Darrington/Alamy; 82 © Ted Daeschler/Academy of Natural Sciences/VIREO; 86 Brandon Cole Marine Photography/Alamy; 88 Udo Richter/epa/Corbis; 90 Classic Image/Alamy; 92 Michael Nicholson/Corbis; 94 Arne Hodalic/Corbis; 103 Harry Taylor/DK Images/Natural History Museum, London; 106 Pictorial Press Ltd/Alamy; 117 Wally McNamee/Corbis; 119 Jason Spencer/Alamy; 124 University of Sussex; 129 Moviestore Collection; 143 © Evan Hurd; 145 Worldwide Picture Library/Alamy; 150 NASA; 157 Chris Boydell/Australian Picture Library/Corbis 162 Colin Keates/DK Images; 167 Kevin Schafer/zefa/Corbis; 169 Lynton Gardner/DK Images/American Museum of Natural History; 177 Brandon D. Cole/Corbis; 183 Patrick Robert/Corbis; 185 Fred Spoor 189 Time & Life Pictures/Getty Images; 190 Jonathan Blair/Corbis; 195 & 197 Bettmann/Corbis; 198 Colin Keates/DK Images/Natural History Museum, London; 202 M. Balan/DK Images; 208 Bettmann/Corbis; 212 & 218 NASA; 223 © Joe Staines; 227 Time & Life Pictures/Getty Images; 231 DK Images; 239 Getty Images; 244 Private collection; 250 The Print Collector/Alamy; 255 Moviestore Collection; 265 Elliott and Fry/public domain; 271 American Philosophical Society; 277 DK Images; 279 Jon Spaull/DK Images; 282 Bettmann/Corbis; 291 Christian Bier; 293 Niklas Jansson; 298 Patrick Robert/Corbis; 304 NASA; 307 © Lalla Ward; 315 © Joe Staines; 322 © Wasatch Beers

Map credits

Reproduced by permission of Ordnance Survey on behalf of HMSO.

Ordnance Survey Licence number 100020918.